FUNCTIONAL ANALYSIS, HOLOMORPHY AND APPROXIMATION THEORY II

NORTH-HOLLAND
MATHEMATICS STUDIES 86

Notas de Matemática (92)
Editor: Leopoldo Nachbin
Centro Brasileiro de Pesquisas Físicas
and University of Rochester

Functional Analysis, Holomorphy and Approximation Theory II

Proceedings of the Seminário de Análise Funcional, Holomorfia e Teoria da Aproximação, Universidade Federal do Rio de Janeiro, August 3-7, 1981

Edited by

Guido I. ZAPATA
Instituto de Matemática
Universidade Federal do Rio de Janeiro

1984

NORTH-HOLLAND – AMSTERDAM • NEW YORK • OXFORD

ISBN: 0 444 86845 3

Publishers:

ELSEVIER SCIENCE PUBLISHERS B.V.
P.O. BOX 1991
1000 BZ AMSTERDAM
THE NETHERLANDS

Sole distributors fot the U.S.A. and Canada:

ELSEVIER SCIENCE PUBLISHING COMPANY, INC.
52 VANDERBILT AVENUE
NEW YORK, N.Y. 10017

Library of Congress Cataloging in Publication Data

Seminário de Análise Funcional, Holomorfia e Teoria da Aproximação (1981 : Universidade Federal do Rio de Janeiro)
Functional analysis, holomorphy, and approximation theory II.

(North-Holland mathematics studies ; 86) (Notas de matemática ; 92)
1. Functional analysis--Congresses. 2. Holomorphic functions--Congresses. 3. Approximation theory--Congresses. I. Zapata, Guido I. (Guido Ivan), 1940- . II. Series. III. Series: Notas de matemática (Amsterdam, Holland) ; 92.
QA320.S456 1981 515.7 83-25454
ISBN 0-444-86845-3

PRINTED IN THE NETHERLANDS

In Memory of

SILVIO MACHADO

Born on September 2, 1932 in Porto Alegre, RS, Brazil
Died on July 28, 1981 in Rio de Janeiro

FOREWORD

This volume is the Proceedings of the Seminário de Análise Funcional, Holomorfia e Teoria da Aproximação, held at the Instituto de Matemática, Universidade Federal do Rio de Janeiro (UFRJ) in August 3-7, 1981. The participant mathematicians and contributors are from Argentina, Brazil, Canadá, Chile, France, Hungary, Italy, Mexico, Spain, Rumania, United States and West Germany.

"Functional Analysis, Holomorphy and Approximation Theory" includes papers either of a research, or of an advanced expository, nature and is addressed to mathematicians and advanced graduate students in mathematics. Some of the papers could not actually be presented at the seminar, and are included here by invitation.

The members of the Organizing Committee - J.A. Barroso, M.C. Matos, L.A. Moraes, J. Mujica, L. Nachbin, D. Pisanelli, J.B. Prolla and G.I. Zapata (Coordinator) - would like to thank the Conselho de Ensino para Graduados e Pesquisa (CEPG) of UFRJ, Conselho Nacional de Desenvolvimento Científico e Tecnológico (CNPq), and I.B.M. do Brasil for direct financial contribution. We would also like to acknowledge the indirect financial contributions from Universidade Federal do Rio de Janeiro, Coordenação de Aperfeiçoamento de Pessoal de Nível Superior (CAPES), Financiadora de Estudos e Projetos (FINEP), as well as other universities and agencies.

We are happy to thank Professor Sergio Neves Monteiro, president of CEPG of UFRJ for his personal support and understanding; Professor Paulo Emidio de Freitas Barbosa, Dean of the Centro de Ciências Matemáticas e da Natureza (CCMN) of UFRJ, in whose facilities the seminar was comfortably held; and Professor Leopoldo Nachbin for his constant friendly support and understanding. We also thank Wilson Luiz de Góes for a competent typing job.

Finally, let us tell with emotion that one mathematician was sorely missed at the seminar, Silvio Machado, a member of its Organizing Committee, who passed away after a heart attack in July 28 of 1981, just a few days before the opening of the meeting. The loss caused by his death will surely be long felt by our community, in particular by his friends in which many participants of the seminar are included. As a posthumous homage, all of us wish to dedicate these Proceedings to the memory of Silvio Machado.

Guido I. Zapata

TABLE OF CONTENTS

Functional Analysis, Holomorphy and Approximation Theory II, G.I. Zapata (ed.)

ON GENERALIZED TOEPLITZ KERNELS AND THEIR RELATION WITH A PAPER OF ADAMJAN, AROV AND KREIN

Rodrigo Arocena

SUMMARY

We consider the relation of the so called generalized Toeplitz kernels with some theorems of Adamjan, Arov and Krein, concerning the unicity of the best uniform approximation of a bounded function, canonical approximating functions and the parametrization of the approximations.

I. INTRODUCTION

Let K be a kernel in Z, the set of integers, that is, a function in $Z\times Z$. K is said a Toeplitz kernel if $K(j,n) = K(j+1,n+1)$, $\forall\ (j,n) \in Z\times Z$, or, equivalently, if there exists a sequence $\{c(n): n \in Z\}$ such that $K(j,n) = c(j-n)$. K is positive definite, p.d., if $\sum_{j,n} K(j,n)s(j)\overline{s(n)} \geq 0$ whenever $\{s(j): j \in Z\}$ is a sequence of finite support. Let T be the unit circle, $e_n(t) = e^{int}$ and $\hat{m}(n) = \int_T e_{-n}dm$, where m is a complex Radon measure in T and integrals are over T unless otherwise specified. The classical Herglotz theorem says that a Toeplitz kernel K is p.d. iff $K(j,n) = \hat{m}(n-j)$, $\forall\ (j,n) \in Z\times Z$, where m is a positive Radon measure in T, and, moreover, m is unique.

In this paper we deal with the following extension of the notion of Toeplitz kernel.

(I.1) DEFINITION. K is a generalized Toeplitz kernel, GTK, if $K(j,n) = K(j+1,n+1)$, for every $j,n \in Z-\{-1\}$.

This definition is equivalent to saying that there exist four sequences K_{rs}, $r,s = 1,2$, such that $K(j,n) = K_{rs}(j-n)$, $\forall\ (j,n) \in Z_{rs} \overset{\text{def}}{=} Z_r \times Z_s$, where $Z_1 = \{n \in Z: n \geq 0\}$, $Z_2 = Z - Z_1$.

For these kernels there is a natural extension of the Herglotz theorem. Let $M = (m_{rs})$, $r,s = 1,2$, be a matrix of complex Radon measures in T; we write $K \sim M^{\wedge}$ if $K_{rs}(j,k) = \hat{m}_{rs}(k-j)$, $\forall\ (j,k) \in Z_{rs}$. As usual, M is said to be positive, $M \geq 0$, if for every $A \subset T$, $(m_{rs}(A))$ is a positive definite numerical matrix. Then the following theorem holds, a proof of which will be given in next section.

(I.2) THEOREM. Let K be a GTK. Then K is p.d. iff $K \sim (m_{rs})^{\wedge}$ for some positive matrix measure (m_{rs}).

If K and $M = (m_{rs})$ are as in (I.2) we say that M belongs to the class $\mathfrak{m}(K)$.

Let us fix the following notations. dt is Lebesgue normalized measure in T; $L^p \equiv L^p(T,dt)$, $1 \leq p \leq \infty$; $\hat{f}$ is the Fourier transform of $f \in L^1$. $H^p = \{f \in L^p: \hat{f}(n) = 0, \forall\ n < 0\}$, $H^p_- = \{f \in L^p: \hat{f}(n) = 0, \forall\ n \geq 0\}$; P is the set of trigonometric polynomials, $P_+ = P \cap H^1$, $P_- = P \cap H^1_-$. $M = (m_{rs})$ will always indicate a matrix of four $(r,s = 1,2)$ Radon measures in T such that $m_{12} = \bar{m}_{21}$. Set

$$\text{(I.3)} \qquad M(f_1,f_2) = \sum_{r,s=1}^{2} \int_T f_r \bar{f}_s \, dm_{rs} .$$

As is well known $M \geq 0$ iff $M(f_1,f_2) \geq 0$, $\forall (f_1,f_2) \in P \times P$; now, if $M^{\wedge} \sim K$ and K is p.d. it is easy to see that $M(f_1,f_2) \geq 0$ holds for every $(f_1,f_2) \in P_+ \times P_-$; in this case we say that M is weakly positive and we write $M \succ 0$. Conversely, if $M \succ 0$ and $M \sim K^{\wedge}$, then K is p.d.; if $M' = (m'_{rs})$ is such that $M' \sim K^{\wedge}$, we write $M \sim M'$. Then (I.2) implies the following.

(I.4) THEOREM. If $M \gg 0$, there exists a positive matrix measure M' such that $M' \sim M$.

If M and M' are as in (I.4) we say that M' is a positive lifting of M.

Theorems (I.2) and (I.4) were established in [9]. In [4] and [5] they were applied to uniform approximation by analytic functions and related moment problems, of the type considered by Adamjan, Arov and Krein [1]. These questions led to the problem of the unicity of $M \geq 0$ such that $\hat{M} \sim K$, that is, of finding conditions that ensure that $\mathfrak{m}(K)$ contains only one element, which is not always true, contrary to what happens in the classical case. In [6], a different and simpler proof of the generalized Herglotz theorem (I.2) was given, which can be extended to the vectorial case [7] and leads to a better unicity condition. That approach happens to be much closer to the ideas of Adamjan, Arov and Krein. In particular, it enables the consideration of the second paper [2] that those authors dedicated to Hankel operators, generalizing their principal theorems and extending their concepts and methods to obtain new results concerning generalized Toeplitz kernels. That is the subject of this paper.

In sections II and III we review the proofs given in [6] of the existence of elements in $\mathfrak{m}(K)$ and of the unicity condition, adding some details that will be needed in the sequel. In sections IV to XI some new results are presented, concerning the following subjects and their applications.

a) Conditions that ensure that the class $\mathfrak{m}(K)$ contains only one element, including an extension of a theorem of [2] concerning the unicity of the best uniform approximation of a bounded function by analytic functions.

b) Definition, characterization and properties of the canonical elements of $\mathfrak{m}(K)$, and, in particular, generalization of some theorems on canonical functions given in [2].

c) A remarkable parametrization of $\{h \in H^\infty: \|f-h\|_\infty \leq r\}$, with given $f \in L^\infty$ and $r > 0$, is established in [2]. By means of it the class $\{M' \geq 0: M' \sim M\}$, with given $M > 0$, is parametrized.

II. CONSTRUCTION OF ALL THE POSITIVE MATRIXES ASSOCIATED TO A POSITIVE DEFINITE GENERALIZED TOEPLITZ KERNEL

From now on, K will always be a p.d. GTK. Let H_K be the Hilbert space defined by the linear space P and the metric given by $\langle e_n, e_m \rangle_K \stackrel{\text{def}}{=} K(n,m)$. Set H_j = the closed linear hull in H_K of $\{e_n: n \neq j\}$, $j = 0,-1$, and $V: H_{-1} \to H_o$ the linear isometry given by $Ve_n = e_{n+1}$. We say that U belongs to the class $\mathfrak{u}(K)$ if U is a unitary extension of V to a Hilbert space $H_U \supset H_K$.

We shall now see that every $U \in \mathfrak{u}(K)$ generates a matrix $M(U) \in \mathfrak{m}(K)$. For every $v \in H_U$, $J(n,m) \stackrel{\text{def}}{=} \langle U^n v, U^m v \rangle_{H_U}$ is a p.d. ordinary Toeplitz kernel; choosing successively $v = e_o, e_{-1}, e_o+e_{-1}, e_o+ie_{-1}$, we get four such kernels, K_{11}, K_{22}, F, G, which, by the classical Herglotz theorem, are given respectively by the Fourier transforms of four positive measures, m_{11}, m_{22}, u, v. Set $m = \frac{1}{2}(u+iv-(1+i)(m_{11}+m_{22}))$, $dm_{12} = e_{-1}dm$, $m_{21} = \bar{m}_{12}$. It is easy to see that the matrix $M(U) \stackrel{\text{def}}{=} (m_{rs})$, $r,s = 1,2$ verifies:

$$\text{(II.1)} \qquad \hat{m}_{11}(-n) = \langle U^n e_o, e_o \rangle_{H_U}, \quad \hat{m}_{12}(-n) = \langle U^{n-1} e_o, e_{-1} \rangle_{H_U}$$

$$\hat{m}_{21}(-n) = \langle U^{n-1} e_{-1}, e_o \rangle_{H_U}, \quad \hat{m}_{22}(-n) = \langle U^n e_{-1}, e_{-1} \rangle_{H_U}, \quad \forall\, n \in Z.$$

If $f_1 = \Sigma\, a_j e_j$, $f_2 = \Sigma\, b_j e_j$ belong to P, the previous equalities imply that $\sum_{r,s=1,2} \int_T f_r \bar{f}_s dm_{rs} = \| \Sigma a_j U^j e_o + \Sigma b_j U^{j+1} e_{-1} \|^2_{H_U}$, so $M(U) \geq 0$. Analogously, a straight forward verification shows that $K(j,k) = \hat{m}_{rs}(k-j)$, $\forall (j,k) \in Z_{rs}$. So we have proved the following.

(II.2) PROPOSITION. Each $U \in \mathfrak{U}(K)$ gives rise, by means of formulae (II.1), to a matrix $M(U) \in \mathfrak{M}(K)$.

Obviously, V always admits a unitary extension to $H_K \oplus l^2$, so $\mathfrak{U}(K)$ is not empty, neither is, by the above proposition, $\mathfrak{M}(K)$. So we have proved the generalized Herglotz theorem (I.2).

Now we shall consider the reciprocal of (II.2), that is, to each $M = (m_{rs}) \in \mathfrak{M}(K)$ we shall associate $U(M) \in \mathfrak{U}(K)$ in such a way that $M[U(M)] = M$. Since $M \geq 0$, the linear space $P \times P$ and the metric

(II.3) $$\langle (f_1,f_2),(g_1,g_2) \rangle_M = \sum_{r,s=1,2} \int_T f_r \bar{f}_s dm_{rs}$$

define a Hilbert space H_M. Since $M^{\wedge} \sim K$ the correspondence $e_n \to (e_n,0)$ if $n \geq 0$ and $e_n \to (0,e_n)$ if $n < 0$ defines an isometry from H_K to H_M, so we can identify the former with a closed subspace of the latter. Let $U(M)$ be the unitary operator in H_M given by:

(II.4) $$U(M)(f_1,f_2) = (e_1 f_1, e_1 f_2), \quad \forall (f_1,f_2) \in P \times P.$$

Then $U(M)$ extends V, so $U(M) \in \mathfrak{U}(K)$. Moreover, $\langle U(M)^n e_o, e_o \rangle_{H_{U(M)}} = \langle (e_n,0),(e_o,0) \rangle_M = \hat{m}_{11}(-n)$; from analogous verifications in the other cases, the following result is apparent.

(II.5) PROPOSITION. If $M \in \mathfrak{M}(K)$, formulae (II.3) and (II.4) define a unitary operator $U(M) \in \mathfrak{U}(K)$ such that $M[U(M)] = M$.

Summing up, the construction presented in this section gives all the matrices $M \geq 0$ such that $M^{\wedge} \sim K$.

III. THE UNICITY CONDITION

The very definition of $\mathfrak{m}(K)$ and the theorem of F. and M. Riesz imply the following.

(III.1) LEMMA. Let $(m_{rs}) \in \mathfrak{m}(K)$. Then $(u_{rs}) \in \mathfrak{m}(K)$ iff $m_{rr} = u_{rr}$, $r = 1,2$, $u_{12} = m_{12} + h\,dt$, $u_{21} = m_{21} + \bar{h}\,dt$, where $h \in H^1$.

Consequently the problem of unicity - that is, of knowing when $\#[\mathfrak{m}(K)] = 1$ - is the problem of knowing when K detrmines $\hat{m}_{12}(n)$ for every $n \geq 0$. Now, a straight forward calculation shows that:

(III.2) If $e_{-1} \in H_{-1}$, then $\langle e_o, V^{n+1}e_{-1}\rangle_K = \hat{m}_{12}(n)$, $\forall\, n \geq 0$;

if $e_o \in H_o$, then $\langle V^{-n-1}e_o, e_{-1}\rangle_K = \hat{m}_{12}(n)$, $\forall\, n \geq 0$.

Obviously, any of the two conditions in (III.2) implies that K determines m_{12} and, consequently, there is only one $M \in \mathfrak{m}(K)$. We shall see, reciprocally, that if neither of those two conditions holds, $\mathfrak{m}(K)$ contains more than one element.

(III.3) NOTATION. Let $j = -1,0$ and $e_j \notin H_j$; let v_j be the unit vector in H_K, perpendicular to H_j and well determined by the condition $c_j \overset{\text{def}}{=} \langle e_j, v_j\rangle_K > 0$; let u_j be the orthogonal projection of e_j on H_j. Define the linear operator $V_t\colon H_K \to H_K$, by

(III.4) $$V_{t/H_{-1}} \equiv V, \quad V_t(v_{-1}) = e^{it}v_o .$$

It is clear that $V_t \in \mathfrak{U}(K)$ and that every unitary extension of V to H_K has this type. Set $(m^t_{rs}) = M(V_t)$; from (II.2), $H_{V_t} = H_K$ and the above notation it follows that

(III.5) $$\hat{m}^t_{12}(0) = \langle u_o, Vu_{-1}\rangle_K + c_o c_{-1} e^{-it} .$$

Since $c_o c_{-1} \neq 0$, if $e^{it} \neq e^{it'}$, $M(V_t) \neq M(V_{t'})$, so:

(III.6) THEOREM. For every K, GTK, the following conditions are equivalent: a) $\mathfrak{m}(K)$ contains more than one element; b) $e_j \notin H_j$, $j = -1, 0$.

Let us recall (see (I.1)) that every GTK is given by four sequences K_{rs}, $r,s = 1,2$; if $K_{11} = K_{22}$ (as is usual in the applications) it is easy to see that $\langle e_j, e_k \rangle = \overline{\langle e_{-j-1}, e_{-k-1} \rangle}_K$, $\forall\ (j,k) \in Z \times Z$; then $B(ae_j) = \bar{a} e_{-j-1}$ defines an antilinear isometry in H_K, such that $B(e_o) = e_{-1}$, $B(H_o) = H_{-1}$ so $\mathrm{dist}(e_{-1}, H_{-1}) = \mathrm{dist}(e_o, H_o)$. Consequently:

(III.7) COROLLARY. Let K be a p.d. GTK such that $K_{11} = K_{22}$. Then the following conditions are equivalent: a) $\mathfrak{m}(K)$ contains more than one element; b) $e_{-1} \notin H_{-1}$; c) $e_o \notin H_o$.

IV. EXTENSION OF ADAMJAN, AROV AND KREIN UNICITY THEOREM

We shall now see how the method employed in section 2 of [2], for proving a theorem concerning the unicity of the best approximation in L^∞ by function of H^∞, can be used to extend that theorem to some GTK, by considering our previous results.

To each K, GTK, we associate two Toeplitz forms, T_1' and T_2', and one Hankel form, H', in the following way:

(IV.1) $T_1'(e_j, e_n) = K_{11}(j-n)$ if $j,n \geq 0$; $H'(e_j, e_{-n}) = K_{12}(j+n)$ if $j \geq 0$, $n < 0$; $T_2'(e_j, e_n) = K_{22}(j-n)$ if $j,n < 0$.

The method under consideration applies to GTK such that the above forms are bounded, that is, given by bounded operators $T_1\colon H^2 \to H^2$, $H\colon H^2 \to H^2_-$, $T_2\colon H^2 \to H^2_-$. Now, we have the following well known (see for example [14]) property of Toeplitz operators.

(IV.2) PROPOSITION. Let $\{b_n : n \in Z\} \subset C$ and T be the bilinear form in $P_+ \times P_+$ given by $T(e_j, e_n) = b_{j-n}$. T can be extended to a bounded form in $H^2 \times H^2$ iff $\exists\, f \in L^\infty$ such that $\hat{f}(-n) = b_n$, $\forall\, n \in Z$, and in that case $\|T\| = \|f\|_\infty$.

The corresponding result for Hankel operators is the following.

(IV.3) NEHARI'S THEOREM [13]. Let $\{b_n\}_{n=1}^{\infty} \subset C$ be given. Then there exists a bounded operator $H: H^2 \to H^2_-$ such that $(He_j, e_{-k}) = b_{j+k}$, $\forall\, j \geq 0$, $k > 0$, iff there exists $f \in L^\infty$ such that $\hat{f}(-n) = b_n$, $\forall\, n > 0$. In that case $\|H\| = \operatorname{dist}(f, H^\infty)$.

This theorem is a simple consequence of the generalized Herglotz theorem applied to the GTK given by $K_{11}(n) = K_{22}(n) = \|H\|\delta(n)$, $\forall\, n \in Z$, and $K_{12}(n) = b_n$, $\forall\, n > 0$. (See [5]).

From (IV.2) and (IV.3) we have the following.

(IV.4) PROPOSITION. Let K be a p.d. GTK. The bilinear forms defined by (IV.1) are given by bounded operators iff $K \sim M\hat{}$, with $M = (w_{rs}dt) \succ 0$ and $w_{rs} \in L^\infty$.

When K is as in (IV.4) the unicity condition (III.6) can be given in terms of the operators T_1, T_2, H. As in [2], the proof rests on the following.

(IV.5) KREIN'S LEMMA. Let A be a bounded non-negative operator in a Hilbert space E; let E_A be the Hilbert space obtained by completing the linear manifold E with respect to the metric $(g,g')_A \overset{\text{def}}{=} (Ag,g')$. Then in order that, for any h ($\in E$), the linear functional $F_h(g) = (g,h)$, $h \in E$, be continuous in E_A it is necessary and sufficient that $\lim_{\varepsilon \downarrow 0} ((A + \varepsilon I)^{-1}h,h) < \infty$, or, equivalently, that $h \in A^{1/2}E$.

(IV.6) LEMMA. Let K be a p.d. GTK. Set $F_j(f) = (f,e_j) \equiv \hat{f}(j)$,

$\forall$ $f \in P$. Then $e_j \notin H_j$ iff F_j is continuous in H_K, $j = -1,0$.

PROOF. F_j continuous in H_K implies that there exists $v_j \in H_j$ such that $F_j(f) = \langle f, v_j \rangle_K$, so $\langle e_n, v_j \rangle_K = 0$, $\forall$ $n \neq j$, and $\langle e_j, v_j \rangle_K = 1$; consequently $0 \neq v_j \in H_K \ominus H_j$, so $e_j \notin H_j$. Reciprocally, if there exists $v_j \in H_K \ominus H_j$ such that $\langle e_j, v_j \rangle_K = 1 = (e_j, e_j)$, then $F_j(f) - \langle f, v_j \rangle_K = 0$, for every $f \in P$ such that $\hat{f}(j) = 0$, and therefore $\exists$ c, constant, such that $F_j(f) = c\langle f, v_j \rangle_K$, $\forall$ $f \in P$; since $1 = (e_j, e_j) = c$, $|F_j(f)| \leq \|v_j\|_K \|f\|_K$, $\forall$ $f \in P$.

(IV.7) THEOREM. Let K be a p.d. GTK such that its associated Toeplitz and Hankel forms (IV.1) are given by bounded operators T_1, T_2, H. Then $\mathfrak{m}(K)$ contains more than one element iff the following hold:

(1) $\lim\limits_{\varepsilon \to 0^+} ([(T_1+\varepsilon I) - H^*(T_2+\varepsilon I)^{-1}H]^{-1} e_o, e_o) < \infty$;

(2) $\lim\limits_{\varepsilon \to 0^+} ([(T_2+\varepsilon I) - H(T_1+\varepsilon I)^{-1}H^*]^{-1} e_{-1}, e_{-1}) < \infty$.

If K verifies also $K_{11} = K_{22}$, then (1) holds iff (2) does.

PROOF. Let p_+, p_- be the orthogonal projectionoperators of L^2 onto H^2, H^2_-, respectively. Set $A = (T_1+H)p_+ + (T_2+H^*)p_-$; then A is a bounded operator in L^2 and it is easy to see that

(i) $(Af, g) = \langle f, g \rangle_K$, $\forall$ $f, g \in P$.

Now, we know (III.6) that $\#[\mathfrak{m}(K)] > 1$ iff $e_j \notin H_j$, $j = -1,0$, and this is equivalent (IV.6) to F_j being continuous in H_K, $j = -1,0$. So from (i) and Krein's lemma (IV.5) we get that $\#[\mathfrak{m}(K)] > 1$ iff:

(ii) $\lim\limits_{\varepsilon \to 0^+} ((A+\varepsilon I)^{-1} e_j, e_j) < \infty$, $j = -1,0$.

Let $(f_1, f_2) \in H^2 \times H^2_-$ be such that $(A+\varepsilon I)(f_1+f_2) = e_{-1}$; then $[(T_2+\varepsilon I) - H(T_1+\varepsilon I)^{-1}H^*] f_2 = e_{-1}$, and this last equality implies that condition (2) is equivalent to (ii), with $j = -1$. The result follows.

EXAMPLE. Let $f \in L^\infty$ and $s = \mathrm{dist}(f,H^\infty)$; for every $r > 0$ $\exists$ $h_r \in H^\infty$ such that $|f-h_r| \le s+r$, a.e. so $\begin{pmatrix} s+r & f-h_r \\ \overline{f-h_r} & s+r \end{pmatrix} \ge 0$, which implies $\begin{pmatrix} s+r & f \\ \bar{f} & s+r \end{pmatrix} > 0$, $\forall\ r > 0$; therefore $M \overset{\text{def}}{=} \begin{pmatrix} s & f \\ \bar{f} & s \end{pmatrix} > 0$.

Let $K \sim \hat{M}$; every matrix in $\mathfrak{m}(K)$ has the form $\begin{pmatrix} s & f-h \\ \overline{f-h} & s \end{pmatrix} \ge 0$, with $h \in H^1$, which is equivalent to $h \in H^\infty$ and $\|f-h\|_\infty = s$ ([4]). Consequently, $\#[\mathfrak{m}(K)] = 1$ iff f has a unique best approximation in H^∞. For this kernel K we have $T_1 = sI$ in H^2, $T_2 = sI$ in H^2_-, so from (IV.7) see get the following.

(IV.8) COROLLARY. Let $f \in L^\infty$, $s = \mathrm{dist}(f,H^\infty)$ and H be the Hankel operator given by $(He_j,e_{-k}) = \hat{f}(-j-k)$, $\forall\ j \ge 0$, $k > 0$. There is only one $h \in H^\infty$ such that $\mathrm{dist}(f,H^\infty) = \|f-h\|_\infty$ iff the following holds:

$$\lim_{r \downarrow s} ([r^2I-H^*H]^{-1}e_o,e_o) = \infty.$$

This last result constitutes theorem (2.1) of [2], which is proved there by the method the extension of which to GTK has been presented in this section.

V. CHARACTERIZATION OF CANONICAL MATRICES

For K, GTK such that $\#[\mathfrak{m}(K)] > 1$, we shall use notation (III.3) and also the following.

(V.1) $\quad g = g(K) = \langle u_o,Vu_{-1}\rangle_K$, $\quad \rho = \rho(K) = c_oc_{-1}$.

Then, from (II.1), it follows that:

(V.2) $\quad \hat{m}_{12}(0) = g + \rho\langle v_o,Uv_{-1}\rangle_{H_U}$ for every $U \in \mathfrak{U}(K)$ and $(m_{rs}) = M(U)$.

Clearly, $|\langle v_o,Uv_{-1}\rangle_{H_U}| \le 1$, so $U \in \mathfrak{U}(K)$ can be chosen in such a way that $\langle v_o, Uv_{-1}\rangle_{H_U}$ takes any complex value of modulus not bigger than one. That implies the following.

(V.3) $$\{\hat{m}_{12}(0):\ (m_{rs}) \in \mathfrak{m}(K)\} = \{z \in C:\ |z-g| \leq \rho\}.$$

(V.4) DEFINITIONS. Let K be a p.d. GTK such that $\#[\mathfrak{m}(K)] > 1$. $U \in \mathfrak{u}(K)$ is a canonical element of $\mathfrak{u}(K)$ if $H_U = H_K$. $M \in \mathfrak{m}(K)$ is a canonical matrix if there exists a canonical $U \in \mathfrak{u}(K)$ such that $M = M(U)$.

From (III.4) it follows that the set of canonical elements of $\mathfrak{u}(K)$ is the same as the set $\{V_t:\ t \in [0,2\pi)\}$.

We have the following characterization of canonical matrices.

(V.5) THEOREM. Let K be a p.d. GTK such that $\#[\mathfrak{m}(K)] > 1$ and $M = (m_{rs}) \in \mathfrak{m}(K)$. The following condtions are equivalent:
(a) M is a canonical matrix; (b) $|\hat{m}_{12}(0)-g| = \rho$; (c) if $U \in \mathfrak{u}(K)$ and $M = M(U)$, then $U(H_K) = H_K$; (d) $H_K \approx H_M$.

PROOF.

(a) implies (b): Since $(m_{rs}) = M(V_t)$ for some t, (b) follows from (III.5).

(b) implies (c): In this case (V.2) says that $|\langle v_o, Uv_{-1}\rangle_{H_U}| = 1$, so Uv_{-1} is parallel to v_o; then $U(H_K) = V(H_{-1}) + \{cUv_{-1}:\ c\in C\} = H_K$.

(c) implies (d): Since $U(M)(H_K) = H_K$. $(0,e_n) = U(M)^{n+1}(0,e_{-1})$ and $(e_j,0) = U(M)^j(e_o,0)$ belong to H_K for every $n \geq 0$ and every $j < 0$, so $P\times P \subset H_K$.

(d) implies (a): Follows immediately from the definitions.

Note that, since $U[M(V_t)]$ is canonical, it must be $U[M(V_t)] = V_{t'}$, so $M(V_t) = M(V_{t'})$ and, by (III.5) and (III.4), $t = t'$. Then:

(V.6) $$U[M(V_t)] = V_t\ , \quad \forall\ t \in [0,2\pi).$$

VI. SOME AUXILIARY MATRICES

Let $M = (m_{rs})$ $r,s = 1,2$, be an hermitean matrix of complex Radon measures in T; set $dm_{rs} = w_{rs}dt + dv_{rs}$, where v_{rs} is singular with respect to Lebesgue measure. From now on $W = W(M)$ and $L = L(M)$ will denote the matrices $W = (w_{rs})$ and $L = (v_{rs})$. The following equality is evident.

(VI.1) $$\sum_{r,s=1,2} \int_T f_r \bar{f}_s dm_{rs} = \sum_{r,s=1,2} \int_T f_r \bar{f}_s w_{rs} dt + \sum_{r,s=1,2} \int_T f_r \bar{f}_s dv_{rs}$$
$$\forall\ f_1, f_2 \in P.$$

So $W \geq 0$ and $L \geq 0$ imply $M \geq 0$. Conversely, if $M \geq 0$, let $A \subset T$ be a Lebesgue null set such that support $v_{rs} \subset A$, $r,s=1,2$; then for any $B \subset T$ the numerical matrices $(v_{rs}(B)) = (m_{rs}(B \cap A))$ and $(\int_B w_{rs} dt) = (m_{rs}[B \cap (T-A)])$ are positive, so $L \geq 0$ and $W \geq 0$. (VI.1) also shows that $W > 0$ and $L \geq 0$ imply $M > 0$. Conversely, if the latter holds let $M' \sim M$ be such that $M' \geq 0$; then $W' \overset{\text{def}}{=} W(M') \sim W$ and $L' \overset{\text{def}}{=} L(M') = L$, so $W > 0$ and $L \geq 0$. Then:

(VI.2) PROPOSITION. The following relations hold: a) $M > 0$ iff $W > 0$ and $L \geq 0$; b) $M \geq 0$ iff $W \geq 0$ and $L \geq 0$; c) $M \sim M'$ iff $W(M) \sim W(M')$ and $L(M) = L(M')$; d) if $M \geq 0$ then $\|(f,g)\|_M^2 = \|(f,g)\|_W^2 + \|(f,g)\|_L^2$, $\forall\ (f,g) \in P \times P$.

The last assertion follows from (VI.1), too. Let K be a p.d. GTK, $M \in \mathfrak{m}(K)$ and $W = W(M)$; denote by K' the GTK associated to $W^\wedge$. The above considerations show that K' is p.d., does not depend of the matrix $M \in \mathfrak{m}(K)$ used in its definition, and, moreover:

(VI.3) $\mathfrak{m}(K) = \mathfrak{m}(K') + L$, with $K \sim M^\wedge$, $K' \sim W^\wedge$, $W = W(M)$ and $L = L(M)$.

Also, since the v_{rs} are singular measures, $\text{dist}_{H_L}(e_j,P_j) = 0$, where P_j is the linear space generated by the e_k with $k \neq j$; then, by (VI.2d), $\text{dist}_{H_K}(e_j,P_j) = \text{dist}_{H_{K'}}(e_j,P_j)$.

So the problems of unicity and of describing all the elements of $\mathfrak{m}(K)$ can be restricted to the GTK generated by function matrices (w_{rs}). We can even suppose that $w_{11} = w_{22}$. In fact, if $M = (m_{rs})$ is hermitean and $m_{11}, m_{22} \geq 0$, set $w = (w_{11} \cdot w_{22})^{1/2}$ and $W_o = \begin{pmatrix} w & w_{12} \\ w_{21} & w \end{pmatrix}$; for any matrix $N = (u_{rs})$ and any function h set $N(h) = \begin{pmatrix} u_{11} & u_{12}+h\,dt \\ u_{21}+\bar{h}\,dt & u_{22} \end{pmatrix}$; set $H^1(M) = \{h \in H^1 : M(h) \geq 0\}$.

Then:

(VI.4) PROPOSITION. The following relations hold: a) $M \succ 0$ iff $W_o \succ 0$ and $L \geq 0$; b) $M \geq 0$ iff $W_o \geq 0$ and $L \geq 0$; c) $h \in H^1(M)$ iff $h \in H^1(W_o)$.

PROOF. Clearly $W_o \geq 0$ iff $W \geq 0$ so (b) follows from (VI.2). Obviously $M \succ 0$ iff $\exists\, h \in H^1(M)$; also, $W_o \succ 0$ and $L \geq 0$ iff $\exists\, h \in H^1(W_o)$ and $L \geq 0$. Now $h \in H^1(M) \Leftrightarrow W(h) \geq 0$ and $L \geq 0$ $\Leftrightarrow W_o(h) \geq 0$ and $L \geq 0 \Leftrightarrow h \in H^1(W_o)$ and $L \geq 0$; so (a) and (c) have been proved.

The consideration of canonical matrices can also be restricted to function matrices. In fact:

(VI.5) PROPOSITION. Let K and K' be as in (VI.3) and $\#[\mathfrak{m}(K)] > 1$. Then M is canonical in $\mathfrak{m}(K)$ iff W is canonical in $\mathfrak{m}(K')$.

PROOF. Since, for every $(f_1,f_2) \in P{\times}P$, $\text{dist}^2_{H_M}((f_1,f_2), P_+{\times}P_-) = \text{dist}^2_{H_W}((f_1,f_2), P_+{\times}P_-)$, $H_K \approx H_M$ iff $H_{K'} \approx H_W$. From (V.5) the result follows.

VII. A REPRESENTATION OF THE HILBERT SPACE ASSOCIATED TO A POSITIVE FUNCTION MATRIX

We shall now see how a construction employed in [2] can be extended to give a representation of H_M, when $M \geq 0$. Set $W = (w_{rs}) = W(M)$ and $p = w_{11} - |w_{12}|^2/w_{22}$, assuming from now on that $w_{12}(t)/w_{22}^{\frac{1}{2}}(t) = 0$ whenever $w_{22}(t) = 0$.

Set $E = L^2 \oplus L^2(p\,dt)$. It is easy to see that a linear isometry B from H_W to E is defined by

$$\text{(VII.1)} \qquad B(f_1, f_2) = (f_1 w_{12}/w_{22}^{\frac{1}{2}} + f_2 w_{22}^{\frac{1}{2}}, f_1), \quad \forall\, (f_1, f_2) \in P \times P.$$

We shall now determine the range of B; it is not hard to see that $(F,G) \in E \ominus B(H_W)$ is equivalent to $\bar{F} w_{12}/w_{22}^{\frac{1}{2}} + \bar{G} p = 0$ and $w_{22}^{\frac{1}{2}} \bar{F} = 0$, dt-a.e., that is, to $F = 0$, w_{22} dt-a.e. and $G = 0$, $p\,dt$-a.e.; consequently:

$$\text{(VII.2)} \qquad B(H_W) = \chi_{\{w_{22}>0\}} L^2 \oplus L^2(p\,dt).$$

Moreover:

(VII.3) THEOREM. Let $M \geq 0$, $W = (w_{rs}) = W(M)$, $L = L(M)$, $p = w_{11} - |w_{12}|^2 \chi_{\{w_{22}>0\}}/w_{22}$. Then the following hold.

a) $H_W \approx \chi_{\{w_{22}>0\}} L^2 \oplus L^2(p\,dt)$, where the isometric isomorphism is given by (VII.1).

b) $H_M \approx \chi_{\{w_{22}>0\}} L^2 \oplus L^2(p\,dt) \oplus H_L$.

c) If, also, $K \sim \hat{M}$ is such that $\#[\mathfrak{m}(K)] > 1$, then $H_W \approx L^2 \oplus L^2(p\,dt)$.

(a) has been proved already; (b) follows from (a) and (VI.2); as to (c), it stems from:

(VII.4) LEMMA. Let $M > 0$, $(w_{rs}) = W(M)$, $w = (w_{11} \cdot w_{22})^{\frac{1}{2}}$. If M

has more than one positive lifting, then $\log w_{11}$ and $\log w_{22} \in L^1$ or, equivalently, $\log w \in L^1$.

This lemma is an obvious consequence of (VI.5) and the following.

(VII.5) PROPOSITION. Let $w \geq 0$ be an integrable and not trivial function. Then the following conditions are equivalent: (i) There exists a function f such that $\begin{pmatrix} w & f \\ \bar{f} & w \end{pmatrix}$ is weakly positive and has more than one positive lifting; (ii) $\log w \in L^1$.

PROOF.

(i) implies (ii): If h_1 **and** h_2 are distinct elements of H^1 such that $w \geq |f+h_1|$ and $w \geq |f+h_2|$, $\log w \in L^1$ because $2w \geq |h_1-h_2|$.

(ii) implies (i): $\exists\, h \in H^1(T)$ such that $|h| = w$ a.e., so h is not trivial and $\begin{pmatrix} w & h \\ \bar{h} & w \end{pmatrix} \geq 0$; take $f \equiv 0$.

VIII. ANOTHER FORMULATION OF THE UNICITY CONDITION

From the above representation of H_M we can obtain a condition equivalent to $M \geq 0$ having only one positive lifting.

(VIII.1) THEOREM. Let $M \geq 0$, $(w_{rs}) = W(M)$, $w = (w_{11} \cdot w_{22})^{\frac{1}{2}}$ and $p = w - |w_{12}|^2/w$. The following properties are equivalent.

a) There exists a positive matrix $M' \neq M$ such that $M' \sim M$.

b) $\log w \in L^1$ and at least one of the following conditions holds:

i) $\log p \in L^1$; ii) there exist $h_1 \in H^1$ and a not trivial h_2 belonging to $H^1 \cap L^2(\frac{1}{w}\, dt)$ such that $\dfrac{|wh_1 - w_{12}\bar{h}_2|^2}{w(|w|^2-|w_{12}|^2)} \chi_{\{p>0\}}$ belongs to L^1 and that $w(t)h_1(t) - w_{12}(t)\bar{h}_2(t) = 0$ whenever $p(t) = 0$.

PROOF. Because of (VI.4) we may assume that $M = \begin{pmatrix} w & w_{12} \\ w_{21} & w \end{pmatrix}$. Suppose (a) holds. Then $\log w \in L^1$ (VII.4) and, consequently, $H_M \approx L^2 \oplus L^2(p\,dt)$. Since unicity does not hold, there exists a not trivial $(F,G) \in [L^2 \oplus L^2(p\,dt)] \ominus B(H_o)$, which is equivalent to

$$(*) \qquad F\bar{w}_{12}/w^{\frac{1}{2}} + Gp = \bar{h}_1 \quad \text{and} \quad Fw^{\frac{1}{2}} = h_2, \quad \text{with} \quad h_1, h_2 \in H^1.$$

If $F = 0$ a.e., $\|h_1\|_1 > 0$, so $\log w \geq \log p = \log|h_1|^2 - \log[|G|^2 p] \in L^1$. If $F \in L^2$ is not trivial, (*) shows that $h_2 \in L^2(\frac{1}{w}\,dt)$ and is not trivial; moreover, from (*) it also follows that $Gpw = h_1 w - \bar{h}_2 w_{12}$; since $G \in L^2(p\,dt)$, the last equality ensures the validity of (ii). So (a) implies (b).

Now assume that (b) holds. If $\log p \in L^1$, let h_1 be an outer function in H^1 such that $p = |h_1|$ a.e., and set $G = \bar{h}_1/p$; then $(0,G)$ is a not trivial element of E, orthogonal to $B(H_o)$ because of (*), but not to $B(e_o,0)$, so $e_o \notin H_o$ and unicity does not hold. If (ii) holds, let $h_1 = e_n h_1'$ with $h_1' \in H^1$ and $\hat{h}_1'(0) \neq 0$; set $h_2' = e_n h_2$, $F = h_2'/w^{1/2}$, G such that (*) holds with h_1, h_2 repalced by h_1', h_2', so (F,G) is not trivial and belongs to $E \ominus B(H_o)$; on the other hand $\langle (F,G), B(e_o,0)\rangle_E = \hat{h}_1'(0) \neq 0$ and so $e_o \notin H_o$. The result follows.

(VIII.2) COROLLARY. Let $f \in L^\infty$, $r = \text{dist}(f,H^\infty) > 0$ and $h \in H^\infty$ be such that $r = \|f-h\|_\infty$. Then such h is not unique iff at least one of the following conditions holds: (i) $\log[r-|f-h|] \in L^1$; (ii) there exist $h_1 \in H^1$ and a not trivial $h_2 \in H^2$ such that $\frac{|h_1 - (f-h)\bar{h}_2|^2}{r^2 - |f-h|^2}\chi_{\{r^2-|f-h|^2>0\}} \in L^1$ and that $\frac{h_1}{\bar{h}_2} = \frac{f-h}{r}$ whenever $r = |f-h|$.

(VIII.3) COROLLARY. Let f be unimodular and dist $(f,H^\infty) = 1$. There exists a not trivial $h \in H^\infty$ such that $\|f-h\|_\infty = 1$ iff $f = ug/\bar{g}$, with u interior and $g \in H^2$ and outer.

REMARK. Clearly, in (VIII.1) h_2 may be assumed to be outer.

EXAMPLE. If $w_{rs} = w$, $r,s = 1,2$, there exists a not trivial $h \in H^1$ such that $w \geq |w-h|$ a.e. iff there exists a constant $k \neq 0$ such that $k \in L^2(\frac{1}{w}\, dt)$, that is, iff $\frac{1}{w} \in L^1$. This condition appears in a remarkable theorem of P. Koosis [12], concerning weighted quadratic means of Hilbert transforms, that can be obtained by developing the above sketched reasonement [6].

IX. SOME PROPERTIES OF CANONICAL MATRICES

Let M be a canonical matrix, so the same happens with $W = W(M)$. Then $H_W \approx L^2 \oplus L^2(p\, dt) \equiv E$, with $p = w_{11} = |w_{12}|^2/w_{22}$. So, considering H_o as a subspace of H_W and with B as in (VII.1), $B(H_o)$ has codimension one in E. Now, $(F,G) \in E \ominus B(H_o)$ iff

$$\text{(IX.1)} \qquad F\bar{w}_{12}/w_{22}^{\frac{1}{2}} + Gp = \bar{h}_1, \quad Fw_{22}^{\frac{1}{2}} = h_2, \quad \text{with } h_1, h_2 \in H^1.$$

Suppose that $p = 0$ does not hold a.e. Then there exists a set $A \subset T$, of positive measure, and positive constants a_1, a_2, such that $a_1 \geq w_{11} \geq p \geq a_2$ holds in A. Proceeding as in section 3 of [2], let a non constant $h \in H^\infty$ be such that h is real in $(T-A)$. Set $F_1 = hF$ in $(T-A)$ and $G_1 = (\bar{h}\bar{h}_1 - Fh\bar{w}_{12}/w_{22}^{\frac{1}{2}})\frac{1}{p}$ in A. Then $F_1 \in L^2$ and

$$\int_T |G_1|^2 p\, dt \leq \|h\|_\infty^2 \left[3\int_T |G|^2 p\, dt + 8\frac{a_1}{a_2}\int_T |F|^2 dt\right], \text{ so } (F_1,G_1) \in E \ominus B(H_o).$$

Moreover, (F_1,G_1) is not parallel to (F,G) if the last is not trivial, so the dimension of $E \ominus B(H_o)$ would not be one; then it must be $p = 0$ a.e., that is:

(IX.2) THEOREM. Let K be a p.d. GTK such that $\#[\mathfrak{m}(K)] > 1$, $M = (m_{rs})$ a canonical element of $\mathfrak{m}(K)$ and $dm_{rs} = w_{rs}dt + dv_{rs}$, with v_{rs} singular with respect to Lebesgue measure. Then $w_{11}w_{22} - w_{12}w_{21} = 0$ a.e.

Consequently - in the above hypothesis - $H_W \approx L^2$, and $F \in L^2 \ominus B(H_o)$ iff $F\bar{w}_{12}/w_{22}^{\frac{1}{2}} = \bar{h}_1$ and $Fw_{22}^{\frac{1}{2}} = h_2$, with h_1 and $h_2 \in H^1$. Set $h_1 = uh$, with u interior and h outer, so $Fu\bar{w}_{12}/w_{22}^{\frac{1}{2}} = \bar{h}$, $Fuw_{22}^{\frac{1}{2}} = uh_2$; then the same argument concerning the codimension employed in the last proof shows that $u \equiv 1$ and leads to the following.

(IX.3) THEOREM. In the same conditions of (IX.2) there exist two outer functions in H^1, h_1 and h_2, such that $\frac{w_{12}}{w_{22}} = \frac{h_1}{\bar{h}_2} = \frac{w_{11}}{w_{21}}$.

Note that if, moreover, $w_{11} = w_{22} = w$, then $|h_1| = |h_2|$ so there exists an outer function $h \in H^1 \cap L^2(\frac{1}{w}\,dt)$ such that $w_{12}/w = h/\bar{h}$.

The last theorems extend those of section 3 in [2].

X. APPLICATIONS RELATED TO THE HELSON-SZEGÕ THEOREM

Let $W = (w_{rs})$ be an hermitean function matrix with $w_{11} = w_{22} = w \geq 0$, the kernel $K \sim \hat{W}$ and the number s given by

$$(X.1) \qquad s = \inf\{\hat{w}(0) + \sum_{k\cdot j>0} c_j\bar{c}_k(k-j) + 2\,\mathrm{Re}\sum_{j>0} c_j\hat{w}(-j) + \sum_{j\cdot k<0} c_j\bar{c}_k\hat{w}_{12}(k-j) + 2\,\mathrm{Re}\sum_{j<0} c_j\hat{w}_{12}(-j)\},$$

where the infimum is considered on the sequences $\{c_j\}_{j\in Z}$ of finite support. It is not hard to see that K is p.d. iff $s \geq 0$; in such case $s = \mathrm{dist}(e_o, H_o)$. Consequently:

(1) The following conditions are equivalent: $W > 0$; $w \geq |w_{12}-h|$, a.e. with $h \in H^1$; K is p.d.; $s \geq 0$.

(2) There exists only one $h \in H^1$ such that $w \geq |w_{12}-h|$ a.e.

iff $s = 0$ (see (III.7)).

(3) If $s > 0$, $\exists\ h \in H^1$ such that $w = |w_{12} - h|$, a.e. (IX.2).

Let us consider an example. Let H be the Hilbert transform in L^2, given by $He_n = (-i)(\operatorname{sg} n)e_n$; let ω be a non negative and integrable function and $k > 1$ a constant. Set $s_k = s$, given by (X.1) with $w = \frac{k-1}{k+1}\omega$, $w_{12} = \omega$. Then:

(4) The following conditions are equivalent: (i) $\int_T |Hf|^2\omega\, dt \leq k \int_T |f|^2\omega\, dt$, $\forall\ f \in P$; (ii) $s_k \geq 0$; (iii) $\exists\ h \in H^1$ such that $\frac{k-1}{k+1}\omega \geq |\omega - h|$ a.e.

This result follows ([9]) from the equivalence of (i) with $W > 0$ and of (1). With the same notation, it stems from (2) and (3), respectively, that:

(5) $s_k = 0$ iff there exists only one $h \in H^1$ such that $\frac{k-1}{k+1}\omega \geq |\omega - h|$ a.e.

(6) If $s_k > 0$, $\exists\ h \in H^1$ such that $\frac{k-1}{k+1}\omega = |\omega - h|$ a.e.

When ω verifies (4i), we say that it belongs to the set $\mathcal{R}_k$; let $\mathcal{R} = \bigcup\{\mathcal{R}_k : k > 1\}$. The Helson-Szegö theorem [11] says that $\omega \in \mathcal{R}$ iff $\omega = e^{u+\tilde{v}}$, with $u,v \in L^\infty_r$, $\|v\|_\infty < \pi/2$ (and $\tilde{v}$ the harmonic conjugate of v, that is, $\tilde{v} = Hv + i\hat{v}(0)$). In [8] it was shown that (4.iii) implies the following refinement:

(7) $\omega \in \mathcal{R}_k$ iff $\omega = ce^{u+\tilde{v}}$, c a positive constant, $u,v \in L^\infty_r$, $\|v\|_\infty < \operatorname{artg}\frac{k-1}{2k^{\frac{1}{2}}}$, $|u| \leq \operatorname{arch}(\frac{k+1}{2k^{\frac{1}{2}}} v)$ a.e.

Now, if $\omega \in \mathcal{R}_k$ and $k' > k$, it follows that $s_{k'} > 0$. So the deduction that leads to (7) can start of (6) instead of (4.iii). Thus:

(X.2) THEOREM. $0 \leq \omega \leq L^1$ is such that $\int_T |Hf|^2\omega\, dt \leq k\int_T |f|^2\,\omega\, dt$ for a fixed constant $k > 1$ and every $f \in P$ iff

$\omega = ce^{u+\tilde{v}}$, where c is a positive constant, $u, v \in L^\infty_r$, $\|v\|_\infty <$ $< \operatorname{artg} \frac{k'-1}{2k'^{\frac{1}{2}}}$, $|u| = \operatorname{arch}(\frac{k'+1}{2k'^{\frac{1}{2}}} v)$ a.e., with $k' \geq k$, a constant.

Also, the equality in (6) is related to the characterization of extremal measures given in [8] which implies:

(8) ω belongs to an extremal ray of the cone $\mathcal{R}_k$ iff $s_k = 0$ and $\exists\, h \in H^1$ such that $\frac{k-1}{k+1}\omega = |\omega - h|$ a.e.

XI. EXTENSION OF ADAMJAN, AROV AND KREIN PARAMETRIZATION

In section 4 of [2] the following remarkable result is stated.

(XI.1) Let $f \in L^\infty$ be such that more than one function $u \in H^\infty$ verifies $\|f-u\|_\infty \leq 1$. Then there exists $F \in H^1$, outer and with $\|F\|_1 = 1$ such that, if $g \in H^\infty$ is given by $\frac{1+g(z)}{1-g(z)} =$ $= \frac{1}{2\pi}\int_0^{2\pi} \frac{e^{ix}+z}{e^{ix}-z}|F(e^{ix})|dx$, $|z| < 1$, the following holds:

$$\{u \in H^\infty : \|f-u\|_\infty \leq 1\} = \{F\,\frac{(1-g)(1-\varphi)}{1-g\varphi} : \varphi \in H^\infty,\ \|\varphi\|_\infty \leq 1\}.$$

Let $M > 0$ and $K \sim M^\wedge$; assume $\#[\mathfrak{m}(K)] > 1$. Set $(w_{rs}) =$ $= W(M)$ and $w = (w_{11} \cdot w_{22})^{\frac{1}{2}}$. Then $\log w \in L^1$ so there exists $\psi \in H^1$, outer, such that $w = |\psi|$ a.e. Set $H^1(M) =$ $= \{h \in H^1 : M + \begin{pmatrix} 0 & h\,dt \\ \bar{h}\,dt & 0 \end{pmatrix} \geq 0\}$ and fix $h_o \in H^1(M)$; then $w \geq |w_{12}+h_o|$, so $1 \geq |(w_{12}+h_o)/\psi|$; set $f = (w_{12}+h_o)/\psi$, so $f \in L^\infty$. Then $h \in H^1(M)$ iff $1 \geq \|(w_{12}+h)/\psi\|_\infty = \|f + \frac{h-h_o}{\psi}\|_\infty$. Let $u = -\frac{h-h_o}{\psi}$; since $f \in L^\infty$, $(h-h_o) \in H^1$ and ψ is outer, it follows that $u \in H^\infty$. Consequently $h \in H^1(M)$ iff $h = h_o - \psi u$, with $u \in H^\infty$ and $\|f-u\|_\infty \leq 1$. Then (XI.1) implies the following.

(XI.2) THEOREM. Let $M = (m_{rs})$ $r,s = 1,2$, be a weakly positive matrix, such that, if $K \sim M^\wedge$, then $\mathfrak{m}(K)$ contains more than one element. There exist $h_o, \psi, F \in H^1$, ψ and F outer functions, $\|F\|_1 = 1$, such that, if $g \in H^\infty$ is given by

$$\frac{1+g(z)}{1-g(z)} = \frac{1}{2\pi}\int_0^{2\pi} \frac{e^{ix}+z}{e^{ix}-z}\,|F(e^{ix})|\,dx, \qquad |z| < 1,$$

then M' is a positive lifting of M iff there exists $\varphi \in H^\infty$ such that $\|\varphi\|_\infty \leq 1$ and

$$M' = \begin{pmatrix} m_{11} & m_{12} + [h_o + \psi F \frac{(1-g)(1-\varphi)}{1-g\varphi}]\,dt \\ m_{21} + [\bar{h}_o + \bar{\psi}\bar{F} \frac{(1-\bar{g})(1-\bar{\varphi})}{1-\bar{g}\bar{\varphi}}]\,dt & m_{22} \end{pmatrix}$$

REFERENCES

1. V.M. ADAMJAN, D.Z. AROV and M.G. KREIN, Infinite Hankel matrices and generalized Carathéodory-Fejer and Riesz problems, Funct. Anal. Appl. 2, no. 1 (1968), 1-19.

2. V.M. ADAMJAN, D.Z. AROV and M.G. KREIN, Infinite Hankel matrices and generalized Carathéodory-Fejer and I. Schur problems, Funct. Anal. Appl. 2, no. 4 (1968), 1-17.

3. R. AROCENA, M. COTLAR and C. SADOSKY, Weighted inequalities in L^2 and lifting properties, Adv. Math., Supplementary studies, v. 74 (1981), 95-128.

4. R. AROCENA and M. COTLAR, On a lifting theorem and its relation to some approximation problems, North Holland Mathematical Studies 71 (1982), J. Barroso Ed.

5. R. AROCENA and M. COTLAR, Generalized Toeplitz kernels and Adamjan-Arov-Krein moment problems, Operator Theory: Advances and Applications, Vol. 4 (1982), 37-55.

6. R. AROCENA and M. COTLAR, Generalized Herglotz-Bochner theorem and L^2-weighted problems with finite measures, Proc. Conference in honour of Prof. A. Zygmund, Chicago, (1981), 258-269.

7. R. AROCENA and M. COTLAR, Dilation of generalized Toeplitz kernels and some vectorial moment and weighted problems, Springer L. N. in Math., 908 (1982), 169-188.

8. R. AROCENA, A refinement of the Helson-Szegö theorem and determination of the extremal measures, Studia Math. 71, 2 (1981), 203-221.

9. M. COTLAR and C. SADOSKY, On the Helson-Szegö theorem and a related class of modified Toeplitz kernels, Proc. Symp. Pure Math. AMS 35: I (1979), 383-407.

10. J. GARNETT, Bounded analytic functions, Academic Press, 1981.

11. H. HELSON and G. SZEGÖ, A problem in prediction theory, Ann. Mat. Pura Appl. 51 (1960), 107-138.

12. P. KOOSIS, Moyennes quadratiques pondérées, C.R. Acad. Sc. Paris 291 (1980), 255-256.

13. Z. NEHARI, On bounded bilinear forms, Ann. of Math. 65 (1957), 153-162.

14. V.V. PELLER and S.V. HRUSCEV, Hankel operators, Best approximations and stationary Gaussian processes, LOMI preprint E-4-81 Leningrad 1981.

ADDED IN PROOF. A self contained proof of the parametrization theorem (XI.2) can be given. (R. Arocena, "On the parametrization of Adamjan, Arov and Krein", in Publications Mathématiques d'Orsay).

Departamento de Matemáticas
Universidad Central de Venezuela

Mailing address:

Apartado Postal 47380
Caracas 1041-A
Venezuela

Functional Analysis, Holomorphy and Approximation Theory II, G.I. Zapata (ed.)

WEAKLY SEQUENTIALLY CONTINUOUS ANALYTIC FUNCTIONS ON A BANACH SPACE

Richard M. Aron and Carlos Herves

ABSTRACT

Let $H_{wsc}(E)$ be the space of complex valued analytic functions on the complex Banach space E which map weakly convergent sequences to convergent sequences. $H_{wsc}(E)$ contains, in general properly, the space $H_w(E)$ of analytic functions on E which are weakly continuous on bounded subsets of E and is, in general, properly contained in the space $H_d(E)$ of analytic functions which are bounded on weakly compact subsets of E. Here we study more closely the relation between these spaces. We show, for example, that for many Banach spaces E, a function f belongs to $H_{wsc}(E)$ if and only if its differential df belongs to the ε-product $H_d(E)\,\varepsilon\,E'$.

INTRODUCTION

For a complex Banach space E, we will be interested in the following subspaces of the space $H(E)$ of entire functions on E.

$H_d(E) = \{f \in H(E):$ f is bounded on each weakly compact subset of $E\}$

$H_{wsc}(E) = \{f \in H(E):$ f takes weakly convergent sequences in E to convergent sequences$\}$

$H_w(E) = \{f \in H(E):$ f is weakly continuous when restricted to any bounded subset of $E\}$

$H_{wu}(E) = \{f \in H(E):\ f$ is uniformly weakly continuous when restricted to any bounded subset of $E\}$

$H_b(E) = \{f \in H(E):\ f$ is bounded on bounded subsets of $E\}$.

The intersection $P(^nE) \cap H_{wsc}(E)$ is denoted $P_{wsc}(^nE)$; $P_{wu}(^nE) = P_w(^nE)$ is defined similarly (cf. Theorem 2.9 of [4]).

It is easy to see that $H_{wu}(E) \subset H_b(E)$ and that $H_{wu}(E) \subset \subset H_\omega(E) \subset H_{wsc}(E) \subset H_d(E)$. It is known (cf. Lemma 3.1 of [4]) that a function $f \in H_w(E)$ is a member of $H_{wu}(E)$ if and only if f is bounded on bounded subsets of E. Moreover, it is trivial that for reflexive Banach spaces E, $H_w(E) = H_{wu}(E)$. It is reasonable to ask whether, conversely, the equality of these spaces implies that E is reflexive. In fact, it might seem reasonable to hope that these conditions are equivalent, in light of the fact that the corresponding result holds for continuous functions by a theorem of Valdivia [23] and for differentiable functions by a result of Gil [12]. Such a hope was brutally dashed recently by Dineen [7], who showed that every function in $H_w(c_o)$ is bounded on balls of c_o and thus $H_w(c_o) = H_{wu}(c_o)$. The case $E = c_o$ is somewhat special, however, in that all of the above defined spaces coincide in this case. (It is easy to see that, for general E, $H_{wsc}(E)$, $H_d(E)$ and $H_w(E)$ are all different. On the other hand, we do not know if $H_w(E)$ is always equal to $H_{wu}(E)$).

Here we concentrate primarily on $H_{wsc}(E)$, which we endow with either the weakly-compact open topology τ_{Od} or the Nachbin weak-ported topology $\tau_{\omega d}$. Our discussion falls under three general headings. First, we examine the question of whether $H_{wsc}(E)$ is barreled. In fact, we show that $H_{wsc}(E)$ is barreled with the τ_{Od} topology if and only if E is reflexive. We also study some special situations in which $(H_{wsc}(E), \tau_{\omega d})$ is, and is not, barreled.

Next, we consider the question of completeness of $H_{wsc}(E)$, $H_d(E)$ and $H_w(E)$, and approximation by finite-type holomorphic functions. Here we make use of the recently defined concept of bounded-weak approximation property due to Gil and Llavona. Finally, it is known that a function belongs to $H_{wsc}(E)$ if and only if its restriction to each weakly compact subset of E is weakly continuous. This observation permits us to obtain a useful characterization of $H_{wsc}(E)$ (for many spaces E) in terms of $H_d(E)$, which is analogous to the one given by the first author [2] of $H_{wu}(E)$ in terms of $H_b(E)$. We conclude the paper with some specific observations and questions concerning two special Banach spaces, the James quasi-reflexive space J and the Cartesian product $c_o \times T$ where T is Tsirelson's original space.

Throughout much of our discussion, we will be assuming that our space E does not contain a copy of ℓ_1, and the special case $E = \ell_1$ will be treated separately. It would be very interesting to know the relationship of the assumption $E \not\supset \ell_1$ to the results we obtain here.

Our notation and terminology will follow the standard works on the subject by Nachbin [17] and Dineen [6].

The second author acknowledges with thanks the support he received as a postdoctoral fellow at University College Dublin, while on a grant from the Irish Department of Education.

ON THE SPACES $H_d(E)$ AND $H_{wsc}(E)$

We will endow $H_d(E)$ and its subspace $H_{wsc}(E)$ with either the topology τ_{0d} of uniform convergence on weakly compact subsets of E or the Nachbin weak-ported topology $\tau_{\omega d}$. This locally convex topology is generated by all seminorms p on $H_d(E)$ for which there is some associated weakly compact set $K \subset E$ satisfying the following condition:

For every $\varepsilon > 0$ there is a constant $c(\varepsilon) > 0$ such that for all $f \in H_d(E)$, $p(f) \leq c(\varepsilon)\cdot\sup\{\|f(x)\| : \text{dist}(x,K) < \varepsilon\}$.

The $\tau_{\omega d}$ topology has been studied by Paredes [18], who showed that it is generated by all seminorms of the form

$$p(f) = \sum_{n=0}^{\infty} \sup\{\|P_n(x)\| : \text{dist}(x,K) < \alpha_n\}$$

for all $f = \Sigma P_n \in H_d(E)$, where $K \subset E$ is weakly compact and (α_n) is a sequence of non-negative real numbers converging to 0.

One of the main reasons for our interest in $H_{wsc}(E)$ lies in the following result, which suggests that some properties of $H_{wu}(E)$ may have analogues which hold for $H_{wsc}(E)$.

PROPOSITION 1 (Proposition 3.3, [4]). $H_{wsc}(E)$ is the space of analytic functions on E which are weakly continuous on each weakly compact subset of E.

This result has been recently extended by Ferrera, Gil and Llavona [11] to the case of continuous functions.

Another reason for our interest in these spaces is that for special choices of E, $H_{wsc}(E)$ and $H_d(E)$ are of particular importance. Thus, for example, if E is reflexive then $(H_{wsc}(E),\tau) = H_{wu}(E)$ where $\tau = \tau_{0d}$ or $\tau_{\omega d}$ and $H_{wu}(E)$ has the topology of uniform convergence on bounded subsets of E. Further, for reflexive E, $(H_d(E),\tau) = H_b(E)$ where $\tau = \tau_{0d}$ or $\tau_{\omega d}$. Also, if E is a Schur space, then $(H_{wsc}(E),\tau_{0d}) = (H_d(E),\tau_{0d}) = (H(E),\tau_0)$ and $(H_{wsc}(E),\tau_{\omega d}) = (H_d(E),\tau_{\omega d}) = (H(E),\tau_\omega)$, where τ_0 and τ_ω are, respectively, the compact-open and the Nachbin ported topologies.

Note that if E is reflexive, then it is trivial that both $H_{wsc}(E)$ and $H_d(E)$ are barreled, while $(H_d(\ell_1),\tau_{\omega d})$ is also barreled [8]. On the other hand, neither $(H_d(\ell_\infty),\tau_{\omega d})$ nor

$(H_{wsc}(\ell_\infty),\tau_{\omega d})$ is barreled. To see this, recall first that the closed unit ball of c_o, $B(c_o)$, is bounding for $H(\ell_\infty)$ [16]. Thus, $\{f \in H_d(\ell_\infty) : \sup\{\|f(x)\|, x \in B(c_o)\} \leq 1\}$ is a barrel in $(H_d(\ell_\infty),\tau_{\omega d})$. If this space were barreled, then for some absolutely convex weakly compact set K in ℓ_∞, we would have that for all $\varepsilon > 0$, there is $c(\varepsilon) > 0$ such that if $f \in H_d(\ell_\infty)$
$\|f\|_{B(c_o)} = \sup\{\|f(x)\| : x\in B(c_o)\} \leq c(\varepsilon)\cdot\sup\{\|f(x)\| : \mathrm{dist}(x,K) < \varepsilon\}$.
In particular, for all $\varphi \in \ell_1 \subset \ell'_\infty$ and for all n,

$$\|\varphi^n\|_{B(c_o)} \leq c(\varepsilon)\cdot\sup\{\|\varphi^n(x)\| : \mathrm{dist}(x,K) < \varepsilon\}.$$

Taking n^{th} roots and letting $n \to \infty$, we see that $B(c_o)$ is contained in the weakly compact set K. This contradiction shows that $(H_d(\ell_\infty),\tau_{\omega d})$ isn't barreled. The proof for $H_{wsc}(\ell_\infty)$ is identical.

Thus far our discussion on barreledness of $H_d(E)$ and $H_{wsc}(E)$ has closely paralleled the analogous situation for $(H(E),\tau_\omega)$. An interesting difference follows from the following simple proposition which also holds for $H_{wsc}(E)$. (Compare this result with Ferrera [9]).

PROPOSITION 2. (1). $(H_d(E),\tau_{Od})$ is barreled if and only if E is reflexive. (2). If $(H_d(E),\tau_{\omega d})$ is barreled, then either E is reflexive or $H_b(E) \neq H_d(E)$.

PROOF. (1). Suppose that $(H_d(E),\tau_{Od})$ is barreled. Let
$A = \{f \in H_d(E) : \|df(0)\| = \sup\{|df(0)(x)|, \|x\| \leq 1\} \leq 1$. Clearly A is convex, balanced and absorbing. Also, if $f_\alpha \in A$ and $f_\alpha \to f$ for τ_{Od}, then for all unit vectors x, $f_\alpha \to f$ uniformly on $\{zx : z \in C, |z| = 1\}$. Thus, $df(0)(x) =$

$$= \frac{1}{2\pi i}\int_{|z|=1} \frac{f(zx)}{z^2}\,dz = \lim_\alpha \frac{1}{2\pi i}\int_{|z|=1} \frac{f_\alpha(zx)}{z^2}\,dz = \lim_\alpha df_\alpha(0)(x).$$

Therefore A is a barrel in $(H_d(E),\tau_{Od})$ and thus there is an absolutely convex weakly compact set $K \subset E$ and a constant $\varepsilon > 0$ such that $A \supset \{f \in H_d(E) : \|f\|_K < \varepsilon\}$. It follows that the unit ball of E, B(E), is contained in $\varepsilon^{-1}K$, and so E is reflexive. The converse is trivial.

(2). Assume that $H_b(E) = H_d(E)$. Since $(H_d(E), \tau_{\omega d})$ is barreled, an application of the open mapping theorem (see, for example, page 299 of [14]) shows that the identity mapping $H_b(E) \to (H_d(E), \tau_{\omega d})$ is a topological isomorphism. Arguing as above, we conclude that the unit ball of E must be contained in a weakly compact subset of E and therefore E must be reflexive.

Q.E.D.

Since Dineen has shown that $H_d(c_o) = H_b(c_o)$, we have the following.

COROLLARY 3. $(H_d(c_o), \tau_{\omega d})$ is not barreled.

We turn now to a review of the question of completeness of the spaces $H_w(E)$, $H_{wsc}(E)$, and $H_d(E)$. The basic known results are summarized in the following proposition. The ℓ_1 case is discussed later.

PROPOSITION 4. (1). $H_w(E)$ is complete for both τ_{Od} and $\tau_{\omega d}$, provided E is reflexive or E is separable and does not contain a copy of ℓ_1.

(2). $H_{wsc}(E)$ is complete for both τ_{Od} and $\tau_{\omega d}$.

(3). $H_d(E)$ is complete for both τ_{Od} and $\tau_{\omega d}$.

PROOF. (1). If E is reflexive, then the result is trivial since in this case $H_w(E) = H_{wu}(E)$ and the topologies τ_{Od} and $\tau_{\omega d}$ are both the topology of uniform convergence on bounded subsets of E. If E is separable and does not contain a copy of ℓ_1 then by Proposition 3.3 of [4], $H_{wsc}(E) = H_w(E)$. Thus to complete the proof of (1), it suffices to prove (2).

(2). Let (f_α) be a $\tau_{\omega d}$-Cauchy net in $H_{wsc}(E)$. Then there is a function $f \in H(E)$ such that $f_\alpha \to f$ for $\tau_{\omega d}$, by a routine argument using the completeness of $(H(E),\tau_\omega)$. We claim that $f \in H_{wsc}(E)$. To see this, let (x_n) converge weakly to x in E. Since the set $K = \{x\} \cup \bigcup_n \{x_n\}$ is weakly compact and since $f_\alpha \to f$ for $\tau_{\omega d}$ and hence for τ_{Od}, it is easy to see that $f(x_n) \to f(x)$. The proof for the τ_{Od} case is omitted, as is the proof of (3) which is due to Paredes [18]. Q.E.D.

The assumption in Proposition 4.1 seems to be necessary, as the following example shows.

EXAMPLE 5. Let $f(x) = \Sigma_n\ x_n^n \in H(\ell_1)$ where $x = (x_n) \in \ell_1$. It is well-known that the finite sums of a Taylor series expansion of an entire function converge to the function for τ_ω. Thus $f \in \widehat{H_{wu}(\ell_1)}^{\tau_\omega}$, since the finite Taylor series terms of f are in $H_{wu}(\ell_1)$. However, $f \notin H_w(\ell_1)$. Indeed, let $B_4 = \{x \in \ell_1 : \|x\|_1 \leq 4\}$, and suppose that there is $\varepsilon > 0$ and a finite set $\{\varphi^1,\ldots,\varphi^k\}$ in ℓ_∞ such that if $x \in B_4$ satisfies $|\varphi^i(x)| < \varepsilon$ $(1 \leq i \leq k)$ then $|f(x)| \leq 1$. It is easy to find an infinite subset N' of the natural numbers such that $|\varphi^i_j - \varphi^i_{j'}| \to 0$ as $j,j' \to \infty$ $(j,j' \in N')$, for each $i = 1,\ldots,k$. Taking $x = 2(e_j - e_{j'})$ we get that $x \in B_4$ and for sufficiently well-chosen j and j', $|\varphi^i(x)| < \varepsilon$ $(1 \leq i \leq k)$ and yet $|f(x)|$ can be arbitrarily large.

There are two interesting observations concerning the relationship of the completeness of $H_w(E)$ to whether E contains ℓ_1. First, we do not know whether the above argument can be extended to the case of arbitrary Banach spaces E containing ℓ_1, to show that in this situation $H_w(E)$ is not complete with respect to either topology τ_{Od} or $\tau_{\omega d}$. Second, in proving that $H_w(E)$ is complete when E is separable and does not contain ℓ_1, we use the fact that in this case $H_{wsc}(E) = H_w(E)$. This makes use of a result of Ferrera [10], which we now describe. Let $X_n = B_n(E)$ with the relative $\sigma(E,E')$ topology, and let $X = \lim_n X_n$. Then $C(X)$ consists of those functions f on E whose restrictions are in $C(X_n)$ for each n; thus $H_w(E) = H(E) \cap C(X)$. Ferrera's result is that if E is separable and does not contain ℓ_1, then X is a k-space. The proof of this result makes use, in turn, of a result of Rosenthal [20], that for a separable Banach space E, E does not contain a copy of ℓ_1 if and only if every bounded subset of E is weakly sequentially dense in its weak closure.

We turn now to the question of describing some dense subspaces of the above spaces. Recall that $\mathcal{P}_f(E) = \sum_n \mathcal{P}_f({}^nE)$, where $\mathcal{P}_f({}^nE)$ is the subspace of $\mathcal{P}({}^nE)$ generated by $\{\varphi^n : \varphi \in E'\}$. The following concept, due to Gil and Llavona [13], will be useful in our discussion. We say that a Banach space E has the bounded weak approximation property if for any weakly compact set $K \subset E$, there is a net $(u_\alpha): E \to E$ of finite rank operators with the properties:

(a) (u_α) converges to the identity uniformly on K with respect to the weak topology, and

(b) $\bigcup_\alpha u_\alpha(K)$ is a bounded subset of E.

Gil and Llavona have shown that if E' has the bounded approximation property then E has the bounded weak approximation property, and that a reflexive Banach space has the approximation property if and only if it has the bounded weak approximation property.

PROPOSITION 6. (1) Assume that E' has the approximation property and that E does not contain ℓ_1. Then the completion of $\mathcal{P}_f(E)$ is $H_{wsc}(E)$ with respect to either τ_{0d} or $\tau_{\omega d}$.

(2) If E has the bounded weak approximation property, then $\mathcal{P}_f(E)$ is τ_{0d}-dense in $H_w(E)$. If E' has the approximation pro-

perty, then $\mathcal{P}_f(E)$ is $\tau_{\omega d}$-dense in $H_w(E)$.

PROOF. (1) Let $f \in H_{wsc}(E)$, with Taylor series expansion $f = \Sigma P_n$. By [18, p.64] this series converges to f for the topology $\tau_{\omega d}$. Since E does not contain ℓ_1 and since each $P_n \in \mathcal{P}_{wsc}({}^nE)$ [2], it follows by Proposition 2.12 of [4] that each $P_n \in \overline{\mathcal{P}_f({}^nE)}$, closure with respect to the norm topology of $\mathcal{P}({}^nE)$. The proof in this case is completed by noting that on $\mathcal{P}({}^nE)$ the norm and the relative $\tau_{\omega d}$ topologies coincide. The proof for the τ_{0d} case follows from the above argument and from Proposition 4 above.

(2) (cf [13]). Assume that E has the bounded weak approximation property and that $f \in H_w(E)$. Let K be an arbitrary weakly compact subset of E, $\varepsilon > 0$, and let (u_α) be the associated net of finite rank operators. Let $B_r(E)$ be a ball containing $K \cup \bigcup_\alpha u_\alpha(K)$. Since $f \in H_w(E)$, a simple compactness argument shows that there is a weak neighborhood V of 0 in E such that if $x \in K$, $y \in B_r(E)$ and $x-y \in V$, then $|f(x)-f(y)| < \varepsilon$. By our assumption, there is α such that $x - u_\alpha(x) \in V$ for all $x \in K$. Hence $|f(x) - f\circ u_\alpha(x)| < \varepsilon$ for all $x \in K$. Since it is easy to see that $f\circ u \in \overline{\mathcal{P}_f(E)}^{\tau_{0d}}$, the proof of the first part is complete. The second part is proved as in (1) above. Q.E.D.

As before, the ℓ_1 case is somewhat different.

EXAMPLE 7. (1) $\overline{\mathcal{P}_f(\ell_1)}^{\tau_{0d}} = (H_{wsc}(\ell_1), \tau_{0d}) = (H(\ell_1), \tau_0)$.

(2) $\overline{\mathcal{P}_f(\ell_1)}^{\tau_{\omega d}} = (H_c(\ell_1), \tau_\omega)$, where $H_c(\ell_1)$ is the space of entire functions $f = \Sigma P_n$ such that for each n, $P_n \in \overline{\mathcal{P}_f({}^nE)}$, closure in the norm topology.

PROOF. The basic fact we make use of is that ℓ_1 is a Schur space, so that weakly compact sets are compact. From this it follows that $H_{wsc}(\ell_1) = H(\ell_1)$ and that both $\tau_0 = \tau_{0d}$ and $\tau_\omega = \tau_{\omega d}$. Thus (1) follows from the well-known fact that if a Banach space E has the approximation property, then $\mathcal{P}_f(E)$ is τ_0-dense in $H(E)$. The proof of (2) is easy. Q.E.D.

From the above proposition and example, we see that for many spaces E, $\overline{\mathcal{P}_f(E)}^{\tau_{0d}} = (H_{wsc}(E), \tau_{0d})$. We do not know whether this holds in general (i.e. for arbitrary Banach spaces E containing ℓ_1), even if one assumes that E and E′ have some type of approximation property.

$H_{wsc}(E)$ has played an important role in the preceeding discussion. As we now show, this space has several additional interesting properties.

PROPOSITION 8. Assume that $H_{wsc}(E) = H_{wu}(E)$ for a Banach space E and that $\ell_1 \not\subset E$. Then for all Banach spaces F, $H_{wsc}(E;F) = H_{wu}(E;F)$.

Note that the assumptions are satisfied if E is reflexive or if $E = c_0$.

PROOF. Let $f = \Sigma P_n \in H_{wsc}(E,F)$. By [2], it follows that $P_n \in \mathcal{P}_{wsc}(^nE;F)$ for each n, and so by Proposition 2.12 of [4], each $P_n \in \mathcal{P}_{wu}(^nE;F)$. Also, f is bounded on bounded subsets of E. Indeed if this were not the case, then there would be an element $\varphi \in F'$ such that $\varphi \circ f$ is unbounded on some ball of E. Since the function $\varphi \circ f$ is then in $H_{wsc}(E) \setminus H_{wu}(E)$, we have a contradiction. The result follows by Lemma 3.1 of [4]. Q.E.D.

The following result gives an interesting characterization of $H_{wsc}(E)$ for many spaces E, which is analogous to the characterization of $H_{wu}(E)$ given in [2].

THEOREM 9. Assume that E does not contain a copy of ℓ_1. Then $H_{wsc}(E) = \{f \in H(E) :$ for each weakly compact set K in E, $df(K) \subset E'$ is relatively compact$\}$.

The proof makes use of the following lemma, which is similar to one given in [3].

LEMMA 10. Let $P \in \mathcal{P}(^nE)$. If $dP(K)$ is relatively compact in E' for every weakly compact set K in E, then $P \in \mathcal{P}_{wsc}(^nE)$. Conversely, if $E \not\supset \ell_1$ and $P \in \mathcal{P}_{wsc}(^nE)$, then for every weakly compact subset K in E, $dP(K) \subset E'$ is relatively compact.

PROOF OF THE LEMMA. Assume that dP takes weakly compact subsets K of E to relatively compact subsets in E'. Using Proposition 2.13 of [4], to show $P \in \mathcal{P}_{wsc}(^nE)$ it is equivalent to show that P is weakly continuous on each such K. Arguing as in [3,5], we first observe that the set $\{A(x_1,\ldots,x_{n-1},\cdot) : x_i \in K,\ 1 \leq i \leq n-1\}$ is a relatively compact subset of E', where A is the symmetric n-linear mapping associated to P. Thus given $\varepsilon > 0$, there is a finite set $\{\varphi_1,\ldots,\varphi_k\} \subset E'$ such that for all $x_1,\ldots,x_{n-1} \in K$, there is φ_j satisfying $\|A(x_1,\ldots,x_{n-1},\cdot) - \varphi_j\| < \varepsilon/4n\,\|K\|$ (where $\|K\| = \sup\{\|x\| : x \in K\}$). Now, let $x,y \in K$ be such that $|\varphi_j(x-y)| < \varepsilon/2n$ for all $j = 1,\ldots,k$. Then $|P(x)-P(y)| =$ $= |A(x,\ldots,x) - A(y,\ldots,y)| \leq |A(x-y,x,\ldots,x)| + |A(y,x-y,x,\ldots,x)| +$ $+\ldots+ |A(y,\ldots,y,x-y)|$. Using the symmetry of A, we see that each of the above terms is of the form $|A(x,\ldots,x,y,\ldots,y,x-y)| \leq$ $\leq |A(x,\ldots,x,y,\ldots,y,x-y) - \varphi_j(x-y)| + |\varphi_j(x-y)| < \varepsilon/n$. Hence $|P(x)-P(y)| < \varepsilon$ and so $P \in \mathcal{P}_{wsc}(^nE)$.

Conversely, let $P \in \mathcal{P}_{wsc}(^nE)$ where $E \not\supset \ell_1$ and let K be weakly compact in E. For simplicity of notation, we will not use additional subscripts to denote subsequences in the following argument. To show that $dP(K)$ is relatively compact, it suffices to show that each infinite sequence $\{dP(x_j) : x_j \in K\}$ has a convergent subsequence. Using the weak compactness of K, there is a point $x \in K$ to which a subsequence of (x_j) converges weakly. Thus it suffices to show that $\|dP(x_j)-dP(x)\| \to 0$ as $j \to \infty$. That is, we must show that $\sup\{|A(x_j,\ldots,x_j,y) - A(x,\ldots,x,y)| : \|y\|=1\} \to 0$ as $j \to \infty$. If this fails, then there is $\varepsilon > 0$ such that for each point of a subsequence of (x_j) there corresponds a unit vector y_j for which $|A(x_j,\ldots,x_j,y_j) - A(x,\ldots,x,y_j)| > \varepsilon$. We now use the fact that $\ell_1 \not\subset E$ to conclude that there is a weakly Cauchy subsequence of the sequence (y_j). Finally, we may write $A(x_j,\ldots,x_j,y_j) - A(x,\ldots,x,y_j) = A(x_j-x,x_j,\ldots,x_j,y_j) + A(x,x_j-x,x_j,\ldots,x_j,y_j) +\ldots+ A(x,\ldots,x_j-x,y_j)$. In each of these terms, one coordinate tends weakly to 0 while the ramaining coordinates are weakly Cauchy. Thus by Lemma 2.4 of [4], each of the above terms tends to 0 as $j \to \infty$, which contradicts our assumption. Hence $dP(K)$ is relatively compact in E', and the proof is complete. Q.E.D.

PROOF OF THEOREM. Let $f = \Sigma P_n \in H_{wsc}(E)$ and let $K \subset E$ be weakly compact. Since f is bounded on each weakly compact set, $f = \Sigma P_n$ uniformly on K, and thus $df = \Sigma dP_n$ uniformly on K. By [2], each $P_n \in \mathcal{P}_{wsc}(^nE)$, and the preceeding lemma tells us that $dP_n(K)$ is relatively compact in E' for all n. Therefore $df(K)$ is relatively compact in E'.

Conversely, suppose that $df(K)$ is compact in E' for each weakly compact set $K \subset E$. Thus $df = \Sigma dP_n$ uniformly on each such K and so $f = \Sigma P_n$ uniformly on K. Since $dP_n(K)$ is contained in the absolutely convex hull of $f(K)$ and is therefore relatively compact, another application of the lemma shows that each P_n is weakly continuous on K. Finally the uniform convergence of the series on K gives the result. Q.E.D.

We remark that we have no example of a Banach space E for which the lemma fails. Thus, we do not know of any situation in which the conclusion of the theorem fails. Note that if $E = \ell_1$ then the lemma and hence the theorem are trivially true. In any case, the above proof shows the following.

COROLLARY 11. Let $f \in H(E)$ where E is an arbitrary Banach space. If $df(K)$ is relatively compact in E' for each weakly compact set K in E, then $f \in H_{wsc}(E)$.

Note that if E is separable and does not contain ℓ_1, then $H_{wsc}(E) = H_w(E)$ (Proposition 3.3, [4]), and it is unknown if $H_w(E) = H_{wu}(E)$ for all E. In other words, using [2], it is unknown if whenever $f \in H(E)$ is such that $\overline{df(K)}$ is compact in E' for every weakly compact subset $K \subset E$ it necessarily follows that $\overline{df(B)}$ is compact in E' for every bounded subset $B \subset E$.

Theorem 10 and its corollary are interesting, in light of the following ε-product representation theorem (cf [21]), whose proof is omitted.

THEOREM 12. For arbitrary Banach spaces E and F, $(H_d(E),\tau_{0d})\ \varepsilon\ F = \{f \in H(E;F) : f(K)$ is relatively compact in F for all weakly compact subsets K in $E\}$, with the topology of uniform convergence on weakly compact subsets of E. Therefore, if $E \not\supset \ell_1$ (or if $E = \ell_1$), $f \in H_{wsc}(E)$ if and only if $df \in H_d(E)\ \varepsilon\ E'$.

Finally, we mention some related open problems in connection with two special Banach spaces. First, consider the quasi-reflexive James space J [15], which is a non-reflexive Banach space with basis which is isomorphic to its second dual. Thus, since J is separable and does not contain ℓ_1, $P_{wsc}({}^nJ) = \overline{P_f({}^nJ)}$ for all n, $H_{wsc}(J) = H_w(J)$ and every entire function which takes weak Cauchy sequences in J to convergent sequences of scalars is an element of $H_{wu}(J)$ (cf. Proposition 3.3 of [4]). However, we do not know whether $H_w(J) = H_{wu}(J)$, although $H_w(J) \neq H_d(J)$. To see this, take the polynomial $P(x) = \sum_{i=1}^{\infty} (x_{2i-1} - x_{2i})^2$ in $P({}^2J)$.

Then $P \notin \mathcal{P}_{wsc}({}^2J)$, since $P(e_n) = 1$ for all n.

Second, let $E = c_o \times T$ where T is Tsirelson's (original) space [22], a reflexive Banach space with unconditional basis which contains no ℓ_p, $1 \leq p \leq \infty$. In [1], it is shown that both the space of continuous n-linear forms on T, $\mathcal{L}({}^nT)$, and $\mathcal{P}({}^nT)$ are reflexive and, in fact, that $\mathcal{P}({}^nT) = \overline{\mathcal{P}_f({}^nT)}$. We do not know if $H_b(c_o \times T) = H_{wu}(c_o \times T) = H_d(c_o \times T)$, although we do know that $\mathcal{P}({}^nc_o \times T) = \overline{\mathcal{P}_f({}^nc_o \times T)}$ for all n. Indeed, let $P \in \mathcal{P}({}^nc_o \times T)$ and let A be the symmetric n-linear mapping corresponding to P. Then the action of P can be expressed as a finite sum of the form $x \in c_o \mapsto A(x^k) \in \mathcal{P}({}^{n-k}T)$, where $A(x^k)(y) = A(\underbrace{x,\ldots,x}_{k},\underbrace{y,\ldots,y}_{n-k})$ for $y \in T$. That is, the action of P can be expressed as a finite sum of elements each of which is in some $\mathcal{P}({}^kc_o, \mathcal{P}({}^{n-k}T))$. Since $c_o \not\subset \mathcal{P}({}^{n-k}T)$, it follows by [19] that $\mathcal{P}({}^kc_o, \mathcal{P}({}^{n-k}T)) = \overline{\mathcal{P}_f({}^kc_o, \mathcal{P}({}^{n-k}T))}$. Moreover, since $\mathcal{P}({}^{n-k}T) = \overline{\mathcal{P}_f({}^{n-k}T)}$, we see that each element of $\mathcal{P}({}^kc_o, \mathcal{P}({}^{n-k}T))$ is a limit of elements of $\mathcal{P}_f({}^kc_o, \mathcal{P}_f({}^{n-k}T))$, which completes the proof.

REFERENCES

1. R. ALENCAR, R. ARON, and S. DINEEN, A reflexive space of holomorphic functions in infinitely many variables, to appear in Proc. A.M.S.

2. R. ARON, Weakly uniformly continuous and weakly sequentially continuous entire functions, in Adv. in Hol. (J.A. Barroso, ed.), North-Holland Math. Studies 34 (1979), 47-66.

3. R. ARON, Polynomial approximation and a question of G.E. Shilov, in Approx. Theory and Funct. Anal. (J.B. Prolla, ed.), North-Holland Math. Studies (1979), 1-12.

4. R. ARON, C. HERVES, and M. VALDIVIA, Weakly continuous mappings on a Banach space, to appear in J. Funct. Anal.

5. R. ARON and J.B. PROLLA, Polynomial approximation of differentiable functions on Banach spaces, J. reine angew. Math. (Crelle), 313, (1980), 195-216.

6. S. DINEEN, Complex analysis in locally convex spaces, North-Holland Math. Studies 57 (1981).

7. S. DINEEN, Entire functions on c_o, to appear in J. Funct. Anal.

8. S. DINEEN, Holomorphic functions on (c_o, X_b) modules, Math. Ann., 196 (1972), 106-116.

9. J. FERRERA, Spaces of weakly continuous functions, Pac. J. Math., 102 (2), (1982), 285-291.

10. J. FERRERA, Espacios de funciones debilmente continuas sobre espacios de Banach, thesis, Univ. Complutense de Madrid (1980).

11. J. FERRERA, J. GOMEZ GIL and J.G. LLAVONA, On completion of spaces of weakly continuous functions, Bull. L.M.S. 15, part 3, no. 54 (1983), 260-264.

12. J. GOMES GIL, Espacios de funciones debilmente diferenciables, thesis, Univ. Complutense de Madrid (1981).

13. J. GOMEZ GIL and J.L.G. LLAVONA, Polynomial approximation of weakly differentiable functions on Banach spaces, Proc. Royal Irish Acad., 82 A, 2 (1982), 141-150.

14. J. HORVATH, Topological vector spaces and distributions, vol. 1, Addison-Wesley, Mass. (1966).

15. R.C. JAMES, Bases and reflexivity of Banach spaces, Ann. Math. 52, no. 3 (1950), 518-527.

16. B. JOSEFSON, Bounding subsets of $\ell^\infty(A)$, thesis, Univ. of Uppsala (1975), and J. Math. Pures et Appl. 57 (1978), 397-421.

17. L. NACHBIN, Topology on spaces of holomorphic mappings, Erg. der Math. 47, Springer-Verlag (1969).

18. J. PAREDES ALVAREZ, Estudio de algunos espacios de funciones holomorfas, thesis, Univ. de Santiago, Spain (1981).

19. A. PELCZYNSKI, A theorem of Dunford-Pettis type for polynomial operators, Bull. Acad. Pol. Sc. XI, 6 (1963), 379-386.

20. H. ROSENTHAL, Some recent discoveries on the isomorphic theory of Banach spaces, Bull. A.M.S. 84, no. 5 (1978), 803-831.

21. L. SCHWARTZ, Theorie des distributions a valeurs vectorielles I, Ann. Inst. Fourier 7 (1957), 1-142.

22. B.S. TSIREL'SON, Not every Banach space contains an imbedding of ℓ_p or c_o, Funct. Anal. and Appl. 8 (1974), 138-141.

23. M. VALDIVIA, Some new results on weak compactness, J. Funct. Anal. 24 (1977), 1-10.

R.M. ARON

Trinity College, Dublin, Ireland

Current address: Dept. of Mathematics
Kent State University
Kent, Ohio 44242
U.S.A.

C. HERVES

Colegio Universitario de Vigo
Apartado 14
Vigo, Spain

Functional Analysis, Holomorphy and
Approximation Theory II, G.I. Zapata (ed.)

THE PRECOMPACTNESS-LEMMA FOR SETS OF OPERATORS

Andreas Defant and Klaus Floret

SUMMARY

Grothendieck's lemma on precompactness in dual systems of vector-spaces is generalized to sets of operators. We want to focus the reader's attention to the method of proving compactness results by using "individual" duality statements on precompactness. In particular, certain localization results in spaces of semi-precompact operators are deduced in this manner; for the special case of Schwartz' ε-product $G \varepsilon F$ these localization properties show when a kind of lifting holds true, namely: For every compact $H \subset G \varepsilon F$ there are compact sets $K \subset G$ and $L \subset F$ such that $H \subset (K^o \otimes L^o)^o$.

1. INTRODUCTION

For an absolutely convex subset C of a (real or complex) vector space E the Minkowski-functional m_C

$$m_C(x) := \inf\{\lambda > 0 \mid x \in \lambda C\} \in [0,\infty]$$

is a semi-norm on the linear hull $\operatorname{span} C = \{x \in E \mid m_C(x) < \infty\}$. A subset $A \subset E$ is called m_C-precompact or C-precompact, if it is precompact in the semi-normed space $[\![C]\!] := (\operatorname{span} C, m_C)$, i.e. $A \subset \operatorname{span} C$ and for every $\varepsilon > 0$ there is a finite set $A_\varepsilon \subset \operatorname{span} C$ such that $A \subset A_\varepsilon + \varepsilon C$; the set A_ε can always be chosen as a subset of A. If $\langle E_1, E_2\rangle$ is a (not necessarily separating) dual system of vector spaces, $A \subset E_1$ any subset and

$$A^{o} := \{y \in E_2 \mid |\langle a,y\rangle| \leq 1 \text{ for all } a \in A\}$$

the absolute polar of A, then the relation

$$m_{A^o}(y) = \sup_{a \in A} |\langle a,y\rangle|$$

holds for all $y \in E_2$. The following lemma is due to A. Grothendieck ([6], p.133).

PRECOMPACTNESS-LEMMA. Let $\langle E_1,E_2\rangle$ be a dual system, $A \subset E_1$ and $B \subset E_2$. Then A is B^o-precompact if and only if B is A^o-precompact.

The core of Grothendieck's original proof is the immediate fact that

$$B \subset \mathcal{C}((A,m_{B^o}))$$

(continuous scalar-valued functions, every $b \in B$ is operating on A by the duality bracket) is always uniformly equicontinuous; if (A,m_{B^o}) is a precompact space the Arzelà-Ascoli theorem applies. This idea will be used later on. There is also an easy direct proof by polarity calculations (see e.g. [4], where the result is called "théorème de précompacité réciproque").

Unfortunately, the precompactness-lemma has not yet found the attention it deserves by its structural elegance and its usefulness. It is basic for the duality theory of locally convex spaces; it can (and should) be an early part of lectures on this topic. We observed that having this lemma in mind many questions are easier to investigate. The purpose of this note is to attack the problem of compactness of sets of operators within the spirit of the precompactness-lemma.

2. SETS OF OPERATORS

Let $\langle E_1,E_2\rangle$ and $\langle F_1,F_2\rangle$ be two dual systems. $L(\langle E_1,E_2\rangle, \langle F_1,F_2\rangle)$ is the set of linear operators $T: E_1 \to F_1$ having a dual operator $T': F_2 \to E_2$, i.e. the set of $\sigma(E_1,E_2)$-$\sigma(F_1,F_2)$-continuous operators. An immediate well-known consequence of the precompactness-lemma is the following

COROLLARY. For $T \in L(\langle E_1,E_2\rangle,\langle F_1,F_2\rangle)$, $A \subset E_1$ and $B \subset F_2$ the set $T(A)$ is B^o-precompact if and only if $T'(B)$ is A^o-precompact.

If E_1 and F_1 are normed spaces, $A \subset E_1$ and $B \subset F_2 := F_1'$ the unit balls, $E_2 := E_1'$, this is just Schauder's theorem.

For $A \subset E_1$ and $V \subset F_1$ denote

$$N(A,V) := \{T \in L(\langle E_1,E_2\rangle,\langle F_1,F_2\rangle) \mid T(A) \subset V\}$$

and for $H \subset L(\langle E_1,E_2\rangle,\langle F_1,F_2\rangle)$

$$H(A) := \{T(a) \mid T \in H,\ a \in A\}$$

$$H' := \{T' \in L(\langle F_2,F_1\rangle,\langle E_2,E_1\rangle) \mid T \in H\}\ .$$

The basic result reads now as follows:

THEOREM (Individual precompactness-theorem for sets of operators). Let $\langle E_1,E_2\rangle$ and $\langle F_1,F_2\rangle$ be two dual systems, $A \subset E_1$ and $B \subset F_2$ subsets. For every $H \subset L(\langle E_1,E_2\rangle,\langle F_1,F_2\rangle)$ the following statements (1)-(4) are equivalent:

(1) (a) $H(A)$ is B^o-precompact

(b) $H'(b)$ is A^o-precompact for all $b \in B$

(2) (a) $H'(B)$ is A^o-precompact

(b) $H(a)$ is B^o-precompact for all $a \in A$

(3) (a) H is $N(A,B^o)$-precompact

(b) $T(A)$ is B^o-precompact for all $T \in H$

(4) (a) H' is $N(B,A^{o})$-precompact

(b) $T'(B)$ is A^{o}-precompact for all $T \in H$.

PROOF. The implications (1a) → (2b), (2a) → (1b), (1a) → (3b) and (2a) → (4b) are obvious; (3b) ↔ (4b) holds by the corollary. Since

$$m_{N(A,B^{o})}(T) = \inf\{\lambda > 0 \mid T(A) \subset \lambda B^{o}\} = \sup_{a\in A} \sup_{b\in B} |\langle T(a),b\rangle| =$$

$$= \sup_{b\in B} \sup_{a\in A} |\langle a,T'(b)\rangle| = \ldots = m_{N(B,A^{o})}(T')$$

the statements (3a) and (4a) are equivalent, whence (3) ↔ (4).

(1) → (2a): By the precompactness-lemma B is $H(A)^{o}$-precompact; therefore for every $\varepsilon > 0$ there are $b_1,\ldots,b_n \in B$ with

$$B \subset \{b_1,\ldots,b_n\} + \varepsilon H(A)^{o}.$$

Consequently

$$H'(B) \subset \bigcup_{i=1}^{n} H'(b_i) + \varepsilon H'(H(A)^{o});$$

since all $H'(b_i)$ are A^{o}-precompact by (1b) and $H'(H(A)^{o}) \subset A^{o}$ it is immediate that $H'(B)$ is A^{o}-precompact.

Therefore (1) → (2) and, by symmetry, (2) → (1).

(3) → (1a): For every $\varepsilon > 0$ there are $T_1,\ldots,T_n \in H$ with

$$H \subset \{T_1,\ldots,T_n\} + \varepsilon N(A,B^{o})$$

whence

$$H(A) \subset \bigcup_{i=1}^{n} T_i(A) + \varepsilon B^{o}.$$

Since all $T_i(A)$ are B^{o}-precompact (1a) follows.

By symmetry the implication (4) → (2a) is true and therefore (3) ↔ (4) → (2a) → (1b). To finish the proof it is enough to show that

(2) → (3a): By (2b) every $T \in H$ defines a function

$$f_T\colon A \to [\![B^{o}]\!]$$
$$a \rightsquigarrow T(a)$$

and it is easy to see that $\{f_T \mid T \in H\}$ is a uniformly equicontinuous set of functions from $(A, m_{H'(B)^o})$ into $[\![B^o]\!]$. The precompactness-lemma applied to (2a) implies, that $(A, m_{H'(B)^o})$ is a precompact semi-metric space; since by (2b) for each $a \in A$ the set $\{f_T(a) \mid T \in H\} = H(a)$ is precompact in the semi-normed space $[\![B^o]\!]$, the Arzelà-Ascoli-theorem for the vector-valued setting ascertains that $\{f_T \mid T \in H\}$ is precompact in the space $\mathcal{C}((A, m_{H'(B)^o}), [\![B^o]\!])$ with respect to the topology of uniform convergence. It was already mentioned that the equation

$$\sup_{a \in A} m_{B^o}(f_T(a)) = \sup_{a \in A} \sup_{b \in B} |\langle Ta, b\rangle| = m_{N(A,B^o)}(T)$$

holds which implies that precompact with respect to the uniform convergence just means $N(A,B^o)$-precompact. ■

For Banach spaces E and F, the sets A and B being the unit balls of E and F' respectively, this is a well-known result on sets of collectively compact operators, [9]; from this special case it is also clear that the extra-conditions (1b) and (2b) are not superfluous.

3. SETS OF Σ-PRECOMPACT OPERATORS

Let E and F be not necessarily separated locally convex spaces and Σ a cover of E by bounded sets. The space

$$P_\Sigma(E,F) := \{T \in L(\langle E,E'\rangle, \langle F,F'\rangle) \mid T(A) \text{ is precompact in } F \text{ for all } A \in \Sigma\}$$

of all Σ-<u>precompact</u> operators is equipped with the topology of uniform convergence on all $A \in \Sigma$; a basis of continuous semi-norms on $P_\Sigma(E,F)$ is given by all $m_{N(A,W)}$, where A is running through all of Σ and W through a basis $\mathfrak{U}_F(0)$ of absolutely convex closed zero-neighbourhoods of F. The symbol $L_\Sigma(E,F)$ stands for the space $L(E,F)$ of all continuous linear operators $E \to F$ endowed as well with the topology of uniform convergence on all $A \in \Sigma$. If F is the scalar field $\mathbb{K}$ the notation $E'_\Sigma := P_\Sigma(E,\mathbb{K}) = L_\Sigma(E,\mathbb{K})$ will be used. The following particular cases for Σ are important:

b := set of all bounded sets

pc := set of all precompact sets

co := set of all absolutely convex, relatively compact sets

e := set of all equicontinuous sets (if E is a dual space).

Just adding quantifiers over all $A \in \Sigma$ and all $B = W^o$, $W \in \mathfrak{U}_F(0)$, the "individual" theorem has the following "global precompactness theorem" as a

COROLLARY. For $H \subset L(\langle E,E'\rangle,\langle F,F'\rangle)$ the following statements (1)-(4) are equivalent:

(1) (a) $H(A)$ is precompact in F for all $A \in \Sigma$

(b) $H'(\varphi)$ is precompact in E'_Σ for all $\varphi \in F'$

(2) (a) $H'(W^o)$ is precompact in E'_Σ for all $W \in \mathfrak{U}_F(0)$

(b) $H(x)$ is precompact in F for all $x \in E$

(3) H is precompact in $P_\Sigma(E,F)$

(4) H' is precompact in $P_e(F'_{pc},E'_\Sigma)$.

(In (4) the topology of F' can be replaced by any locally convex topology α such that all equicontinuous sets are α-bounded and all $T' \in H'$ are weakly continuous $(F',\alpha) \to E'_\Sigma$.) For special cases this corollary is related to results of A. Geue [5] and W. Ruess [10]; see also [1]. It is worthwhile to note that by a simple argument (1a) is equivalent to

(1a') H' is equicontinuous in $L(F'_{pc},E'_\Sigma)$;

denoting by δ the topology on E which is induced by $(E'_\Sigma)'_{pc}$, the statement (2a) is equivalent to

(2a') H is equicontinuous in $L((E,\delta),F)$.

If every precompact set in E'_Σ is equicontinuous then $E \hookrightarrow (E,\delta)$ is continuous and $H \subset L(E,F)$ is equicontinuous. In particular, this holds true in the following cases (F arbitrary):

(i) E is barrelled, Σ arbitrary

(ii) E is quasibarrelled, $\Sigma \supset pc$

(iii) E has a countable basis of bounded sets and every zero-sequence in E'_b is equicontinuous (e.g. E σ-locally topological = gDF), $\Sigma = b$.

For Schwartz' ε-product of two quasicomplete locally convex separated spaces G and F

$$G \,\varepsilon\, F := L_e(G'_{co},F) = P_e(G'_{co},F)$$

the corollary (with $E := G'_{co}$, whence $E'_e = G$ and $(E,\delta) = G'_{co}$) is a refinement of L. Schwartz' characterization of relatively compact sets in $G \,\varepsilon\, F$ (see [8], §44) which was also obtained by W. Ruess [10].

4. LOCALIZATION

By definition, a subset $H \subset L(\langle E,E'\rangle,\langle F,F'\rangle)$ Σ-<u>localizes</u> if there is a precompact set $K \subset E'_\Sigma$ such that $H(K^o)$ is precompact in F; if the precompact set $K \subset E'_\Sigma$ can be chosen to be equicontinuous H Σ-<u>localizes fully</u>. The study of this property was initiated by some results of W. Ruess, e.g. in [10].

The difference between "Σ-localization" and "full Σ-localization" is of a more technical nature; in most of the applications every precompact set in E'_Σ is already equicontinuous (compare the remarks after property (2a') in Section 3.).

PROPOSITION 1.

(1) If $H \subset L(\langle E,E'\rangle,\langle F,F'\rangle)$ Σ-localizes, then H is precompact in $P_\Sigma(E,F)$.

(2) If G and F are quasicomplete, then $H \subset L_e(G'_{co},F) = G\varepsilon F$ e-localizes (= e-localizes fully) if and only if there are compact sets $K \subset G$ and $L \subset F$ such that

$$H \subset (K^o \otimes L^o)^o.$$

(Here $G' \otimes F' \subset (G \varepsilon F)'$ is the natural embedding).

PROOF. For (1) check conditions (1) (a) and (b) of the global precompactness result, (2) is as easy by taking $L := \overline{H(K^o)}$. ∎

Our concern is now to find conditions under which all precompact sets in $P_\Sigma(E,F)$ Σ-localize, i.e. when (1)-(4) of the global precompactness-theorem is equivalent to

(5) H Σ-localizes.

Certainly this includes the problem under which circumstances each b-precompact operator is precompact (:= there is a zero-neighbourhood whose image is precompact) which in general is not true. In the setting of ε-products the question involves finding out when all compact sets $H \subset G \varepsilon F$ can be "lifted" in the sense expressed in Proposition 1 (2). Note, that this is equivalent to the continuity of the natural embedding*)

$$G'_{co} \otimes_\pi F'_{co} \hookrightarrow (G \varepsilon F)'_{co} .$$

The analysis will be split up in essentially two parts: When Σ-localizes every precompact set $H \subset P_\Sigma(E,F)$ which is equibounded (i.e. there is a $U \in \mathfrak{U}_E(0)$ such that $H(U)$ is bounded)? When are

*) There is a close relationship between the notion of e-localization and the duality of ε- and π-topologies on tensorproducts. We shall deal with this question in another paper.

precompact sets in $P_\Sigma(E,F)$ equibounded?

For the first question some further notation is helpful: A family Ω of bounded sets of a locally convex space E satisfies the Mackey-condition (resp. the strict Mackey-condition) if for every $A \in \Omega$ there is an absolutely convex $B \in \Omega$ such that $A \subset B$ and every (in E) precompact subset of A is B-precompact (resp. A is B-precompact). By the precompactness-lemma pc satisfies the strict Mackey-condition if and only if F'_{pc} is a Schwartz-space.

One answer to the first question is given by the

LEMMA 1

(1) If the family e in E'_Σ satisfies the Mackey-condition or if $b = pc$ in F, then every equibounded precompact subset $H \subset P_\Sigma(E,F)$ is equiprecompact, i.e. there is a $U \in \mathfrak{U}_E(0)$ such that $H(U)$ is precompact.

(2) If F'_{pc} is Schwartz or $\Sigma \subset pc$ in E then every precompact subset $H \subset P_\Sigma(E,F)$ which is equiprecompact Σ-localizes fully.

PROOF. (1) If $b = pc$ in F nothing has to be shown. In the other case take $U \in \mathfrak{U}_E(0)$ such that $H(U)$ is bounded. If B is an equicontinuous set in E' which is chosen to U^o according to the Mackey-condition it is enough to show that $H(B^o)$ is W-precompact for every $W \in \mathfrak{U}_F(0)$. Since (by the precompactness of $H \subset P_\Sigma(E,F)$) all $H(x)$, $x \in E$, are W-precompact it suffices by the individual theorem to check that $H'(W^o)$ is B-precompact. But (by the global result) $H'(W^o)$ is precompact in E'_Σ and $H'(W^o)$ is absorbed by U^o, the Mackey-condition gives that $H'(W^o)$ is B-precompact.

(2) Take $U \in \mathfrak{U}_E(0)$ such that $H(U)$ is precompact. If $\Sigma \subset pc$ every equicontinuous set in E'_Σ is precompact by the precompactness-lemma, whence H Σ-localizes fully.

The other case runs as follows: According to the strict Mackey-condition of pc in F there is a precompact set D such that $H(U) \subset D$ and $H(U)$ is D-precompact. Take $K := H'(D^o) \subset U^o$. Obviously $H(K^o)$ is contained in D^{oo} whence precompact. So it remains to show that $K = H'(D^o)$ is A^o-precompact for every $A \in \Sigma$: to apply the individual theorem note that U absorbs A whence $H(A)$ is D-precompact. Again the individual theorem (and the precompactness of H in $P_\Sigma(E,F)$) shows that $H'(D^o)$ is A^o-precompact. ∎

With rather the same arguments the following lemma can be shown:

LEMMA 2

(1) If b in F satisifes the Mackey-condition or if $\Sigma \subset$ pc in E then every equibounded precompact set $H \subset P_\Sigma(E,F)$ has the following property (*): There is an equicontinuous precompact set K in E'_Σ such that $H(K^o)$ is bounded.

(2) If b = pc in F or if the family $e \cap pc$ of all equicontinuous precompact sets in E'_Σ satisfies the strict Mackey-condition then every precompact subset $H \subset P_\Sigma(E,F)$ with (*) Σ-localizes fully.

Collecting the results of the two lemmata gives the

PROPOSITION 2. In each of the following cases (a)-(e) every equibounded precompact subset of $P_\Sigma(E,F)$ Σ-localizes fully:

(a) $\Sigma \subset$ pc in E and b = pc in F.

(b) F'_b is a Schwartz-space.

(c) e in E'_Σ satisfies the Mackey-condition and F'_{pc} is Schwartz.

(d) E is a Schwartz-space.

(e) $e \cap pc$ in E'_Σ satisfies the strict Mackey-condition and b in F satisfies the Mackey-condition.

PROOF. (a) and (c) follows from Lemma 1 as well as (b) by observing the following fact: F'_b is Schwartz if and only if F'_{pc} is Schwartz and $b = pc$ in F. The statements (e) and (d) come from Lemma 2 noticing for the latter that the following holds by the precompactness-lemma: E is Schwartz if and only if $e \cap pc$ in E'_Σ satisfies the strict Mackey-condition and $\Sigma \subset pc$ in E. ■

Later on it turns out that the assumption of one of the spaces being Schwartz is not at all artificial.

Since $H \subset P_\Sigma(E,F)$ Σ-localizes if and only if $H' \subset P_e(F'_{pc}, E'_\Sigma)$ e-localizes the following result is a corollary of Proposition 2 (d) and (a) (and, of course, the global theorem):

PROPOSITION 3. Let F'_{pc} be Schwartz or $b = pc$ in E'_Σ. Then every precompact subset $H \subset P_\Sigma(E,F)$ such that $H' \subset L(F'_{pc}, E'_\Sigma)$ is equibounded, Σ-localizes.

Coming back to the original question "When Σ-localizes every precompact set $H \subset P_\Sigma(E,F)$?" note first that according to condition (1a') of the global theorem $H' \subset L(F'_{pc}, E'_\Sigma)$ is equicontinuous and by (2a') the set H itself is equicontinuous in $L((E,\delta),F)$ which in most cases implies that it is equicontinuous in $L(E,F)$. In view of Propositions 2 and 3 it is therefore reasonable to investigate under which circumstances a given pair (M,N) of locally convex spaces satisfies the following <u>localization principle</u>: Every equicontinuous subset of $L(M,N)$ is equibounded.

PROPOSITION 4. (M,N) satisfies the localization principle in each of the following five cases:

(a) M or N is normed.

(b) M has the countable-neighbourhoods-property (i.e. for every sequence (U_n) in $\mathfrak{U}_M(0)$ there is a $U \in \mathfrak{U}_M(0)$ which is absorbed by each U_n) and N is metrizable.

(c) M is Baire-like [11] and N has a countable basis of bounded sets.

(d) M is metrizable and N'_b (or even the completion $\widetilde{N'_b}$) is Baire-like.

(e) M is metrizable and N has a countable basis of bounded sets.

PROOF. (a) is obvious.

(b) If (W_n) is a neighbourhood basis of N and $H \subset L(M,N)$ equicontinuous take $U_n := \bigcap_{T \in H} T^{-1}(W_n)$ and a neighbourhood U according to the definition; then $H(U)$ is bounded.

(c) If (B_n) is a basis of closed, absolutely convex, bounded sets, consider

$$D_n := \bigcap_{T \in H} T^{-1}(B_n).$$

(d) If $H \subset L(M,N)$ is equicontinuous then $H' \subset L(N'_b, M'_b)$ is equicontinuous; since M'_b is a complete (DF)-space the set $\widetilde{H}'$ of all extensions $\widetilde{T}'$ is equicontinuous in $L(\widetilde{N'_b}, M'_b)$.

By (c) there is in both cases a bounded set $A \subset N$ such that $H'(A^o)$ is bounded = equicontinuous in M'_b. By dualizing follows that H is equibounded.

(e) is a special case of (d) (an alternative proof can be found in [3], 4.2.). ■

There are even pairs (M,N) of <u>Fréchet-spaces</u> which satisfy the localization principle: e.g. M a power sequence space of finite type and N one of infinite type (see V.P. Zahariuta [13], p 208, 209). The localization principle for pairs of Fréchet-spaces was recently charactericed by D. Vogt [12].

It is not too difficult to see, that for an arbitrary locally convex space E and a quasibarrelled space F the pair (E,F'_b) satisfies the localization principle if and only if on $E \otimes F$ the projective and the (b,b)-hypocontinuous topologies coincide.

5. APPLICATIONS

It was shown

(A) If (E,F) satisfies the localization principle then in any of the cases of Proposition 2 every equicontinuous precompact subset of $P_\Sigma(E,F)$ Σ-localizes fully.

(For the equicontinuity recall the remarks at the end of 3.)

(B) If (F'_{pc},E'_Σ) satisfies the localization principle then in both cases of Proposition 3 every precompact subset of $P_\Sigma(E,F)$ Σ-localizes.

For the following examples note first, that for metrizable F standard manipulations with precompact sets show that F'_{pc} is Schwartz and has the countable-neighbourhoods-property. Moreover, for every quasinormable space E the family $e \cap pc$ in E'_b satisfies both the Mackey-condition and the strict one.

The assumptions of (A) hold true in the following cases:

(a) E is quasinormable, F normed, $\Sigma = b$ (Prop. 4(a), Prop. 2(e)).

(b) E is Schwartz, F normed, Σ arbitrary (Prop. 4(a), Prop. 2(d)).

(c) E has the countable-neighbourhoods-property (e.g. E is σ-locally topological), F is metrizable, $b = pc$ in F, Σ arbitrary (Prop. 4(b), Prop. 2(b)).

(d) E has the countable-neighbourhoods-property and is Schwartz, F metrizable, Σ arbitrary (Prop. 4(b), Prop. 2(d)).

Since σ-locally topological spaces are quasinormable (see [7], p. 260)

(d) includes the case that E is σ-locally topological, $b = pc$ in E, F metrizable and Σ arbitrary.

(e) E is metrizable and Schwartz, F has the countable basis of bounded sets, Σ is arbitrary (Prop. 4 (b), Prop. 2(d)).

The assumptions of (B) hold true for

(f) E has a countable basis of bounded sets, F is metrizable, $\Sigma = b$ (Prop. 4(b), Prop. 3).

(g) E is metrizable, F'_b Schwartz, F has a countable basis of bounded sets (e.g. F an (LS)-space), Σ is arbitrary. (Prop. 4(e), Prop. 3 since F'_b Schwartz implies F'_{pc} Schwartz.)

For ε-products $G \varepsilon F = P_e(G'_{co}, F)$ the principle (B) gives

PROPOSITION 5. Let G and F be quasicomplete separated locally convex spaces such that (F'_{co}, G) satisfies the localization principle and: G is semi-Montel or F'_{co} is Schwartz then for every subset $H \subset G \varepsilon F$ the following two statements are equivalent:

(1) H is relatively compact.

(2) There are compact sets $K \subset G$ and $L \subset F$ such that $H \subset (K^o \otimes L^o)^o$.

The assumptions of this result are satisfied e.g. in the following cases:

(a) G is Banach and F'_{co} Schwartz.

(b) G and F are Fréchet-spaces.

(c) G has a countable basis of bounded sets and F is an (LS)-space.

(d) G and F both have a countable basis of compact sets (this implies that G and F are semi-Montel, see [7], p. 266).

((b) was already treated in [8], §44). Note that $G \varepsilon F = F \varepsilon G$, i.e. the assumptions on G and F can be interchanged. Note that (d) comprises the case that both spaces G and F are semi-Montel σ-locally topological spaces. For an illustration of (a) take a compact set X and an (LS)-space F, then for every compact set $H \subset \mathcal{C}(X) \varepsilon F = \mathcal{C}(X,F)$ there is a compact set $K \subset \mathcal{C}(X)$ such that

$$\{ \int f d\mu \mid f \in H, \mu \in K^{o} \}$$

is relatively compact in F.

6. NECESSITY RESULTS

By the very nature of Σ-localization it is clear that once it holds in $P_{\Sigma}(E,F)$ certain sets have to be equibounded in $L((E,\delta),F)$ and $L(F'_{pc},E'_{\Sigma})$. However it is surprising that the assumption of E or F'_{pc} being Schwartz which appears frequently is sometimes even necessary.

The key for the following results is the external characterization of Schwartz-spaces ([2], 12.4.): A locally convex space E is Schwartz if and only if for every Banach-space G all continuous linear mappings $E \to G$ are compact.

PROPOSITION 6. (1) If $\Sigma \subset pc$ in E and for every Banach-space G every one-point set in $P_{\Sigma}(E,G)$ Σ-localizes fully, then E is a Schwartz-space.
(2) Let F be a semi-Montel-space such that for all Banach-spaces G every one-point set in $P_{b}(G,F)$ b-localizes, then F'_{b} is Schwartz.

PROOF. (1) directly by the external characterization, (2) with a simple additional duality argument. ■

Since there is a widely known Fréchet-Montel-space which is not

Schwartz this result shows that there is a Banach-space G and a Montel-(LB)-space F such that not every precompact set in $P_b(G,F)$ b-localizes.

In the setting of the ε-product (1) implies the following

COROLLARY. If a quasicomplete separated locally convex space F has the property that for every Banach-space G and for every $T \in F \varepsilon G$ there are compact sets $K \subset F$ and $L \subset G$ with $T \in (K^o \otimes L^o)^o$ then F'_{co} is a Schwartz-space.

Together with the above-mentioned example the last result shows that the lifting-property of $G \varepsilon F$ may be false in the case "F a Montel (LB)-space and G a Banach-space" (compare Proposition 5(a) and (c)). Since it is immediately clear that in the external characterization of Schwartz-spaces only spaces of the form $G = \mathcal{C}(X)$ (with an arbitrary compact set X) have to be checked this also means, that the property for $\mathcal{C}(X,F)$ stated at the end of 5. does in general not hold if X is compact and F is only a Montel (LB)-space.

However: if $H \subset F \varepsilon G = P_e(F'_{co},G)$ is precompact it follows (e.g. by (2a') of the global precompactness theorem) that H is equicontinuous in $L(F'_{co},G)$. If (F'_{co},G) satisfies the localization principle (which is obviously true in the case just mentioned) H is equibounded which readily means there is a compact $K \subset F$ and a bounded $B \subset G$ such that

$$H \subset (K^o \otimes B^o)^o.$$

BIBLIOGRAPHY

1. A. DEFANT, Zur Analysis des Raumes der stetigen linearen Abbildungen zwischen zwei lokalkonvexen Räumen; Dissertation Kiel 1980

2. K. FLORET, Lokalkonvexe Sequenzen mit kompakten Abbildungen; J. reine angew. Math. 247 (1971) 155-195

3. K. FLORET, Folgenretraktive Sequenzen lokalkonvexer Räume; J. reine angew. Math. 259 (1973) 65-85

4. H.G. GARNIR, M. de WILDE, J. SCHMETS, Analyse fonctionnelle, Tome I; Birkhäuser 1968

5. A.S. GEUE, Precompact and Collectively Semi-Precompact Sets of Semi-Precompact Continuous Linear Operators; Pacific J. Math. 52 (1974), 377-401

6. A. GROTHENDIECK, Sur les applications linéaires faiblement compactes d'espaces du type C(K); Canadian J. Math. 5 (1953) 129-173

7. H. JARCHOW, Locally Convex Spaces; Teubner 1981

8. G. KÖTHE, Topological Vector Spaces I and II; Springer 1969 and 1979

9. T.W. PALMER, Totally Bounded Sets of Precompact Linear Operators; Proc. Am. Math. Soc. 20 (1969) 101-106

10. W. RUESS, Compactness and Collective Compactness in Spaces of Compact Operators; J. Math. Anal. Appl. 84 (1981) 400-417

11. S.A. SAXON, Nuclear and Product Spaces, Baire-like Spaces and the Strongest Locally Convex Topology; Math. Ann. 197 (1972) 87-106

12. D. VOGT, Fréchetäume, zwischen denen jede stetige lineare Abbilding beschränkt ist; preprint 1981

13. V.P. ZAHARIUTA, On the Isomorphism of Cartesian Products of Locally Convex Spaces; Studia Math. 46 (1975), 201-221

Universität Oldenburg
Fachbereich Mathematik
2900 Oldenburg
Fed. Rep. Germany

Functional Analysis, Holomorphy and
Approximation Theory II, G.I. Zapata (ed.)

ON LIPSCHITZ CLASSES AND DERIVATIVE INEQUALITIES IN VARIOUS BANACH SPACES

Z. Ditzian

1. INTRODUCTION

In a series of papers see [1], [2] and [3] it was shown that certain results on derivative inequalities, best approximation and convolution approximation can be extended from C (the space of continuous functions) to other Banach spaces for which translation is an isometry or contraction and for which translation is a continuous operator. In this paper we shall survey the results of those papers and extend them to some Banach spaces for which those theorems were not applicable. This group of Banach spaces will include L_∞, B.V. (functions of bounded variation) and duals to Sobolev or Besov spaces.

2. THE LANDAU-KOLMOGOROV AND SCHOENBERG-CAVARETTA INEQUALITIES

In [4] Kolmogorov has shown that for $f \in C^n(-\infty,\infty)$

$$\|f^{(k)}\| \le K(n,k)\|f^{(n)}\|^{\frac{k}{n}}\,\|f\|^{1-\frac{k}{n}} \tag{2.1}$$

where $\|g\| \equiv \sup_x |g(x)|$ and $1 \le k \le n-1$,

and calculated the best constants $K(n,k)$. For $n = 2$ and $k = 1$ the result was proved earlier by Landau. In [5] Schoenberg and Cavaretta developed a method to calculate the best constants of (2.1) for $f \in C^n(0,\infty)$. It was shown in [1, p.150] that if $T(t)$

Supported by NSERC grant A-4816 of Canada.

is a C_o contraction semigroup on a Banach space B and $f \in D(A^n)$ where $Af = \lim_{t \to 0+} \frac{T(t)f-f}{t}$ in B,

$$(2.2) \qquad \|A^k f\| \le K(n,k)\|A^n f\|^{\frac{k}{n}} \|f\|^{1-\frac{k}{n}}$$

where $K(n,k)$ are those calculated by Schoenberg and Cavaretta. (The $K(n,k)$ mentioned above are best possible in general, i.e. for all spaces, but for a particular space B and semigroup $T(t)$ it is possible that smaller constants are valid.) Moreover, if $T(t)$ $-\infty < t < \infty$ is a C_o group of isometries on B, then (2.2) is valied with $K(n,k)$ being Kolmogorov's constants. As an example of the use of the generalization one has for B a Banach space of functions on $(-\infty,\infty)$ or on $(0,\infty)$ for which $\|f(\cdot+t)\| = \|f(\cdot)\|$ or $\|f(\cdot+t)\| \le \|f(\cdot)\|$ and $\|f(\cdot+t)-f(\cdot)\| = o(1)$ $t \to 0$, then for $f^{(\ell)}(\cdot)$, the strong derivative of $f(\cdot)$, (2.1) is valid with the Kolmogorov or the Schoenberg-Cavaretta constants respectively.

One can also prove the following somewhat more general result.

For a Banach space B and a semigroup $T(t)$ we can define A_ω by $\langle A_\omega f,g\rangle = \langle h,g\rangle = \lim_{t\to 0} \langle \frac{T(t)-I}{t} f,g\rangle$ for all $g \in B^*$, and for those f for which that limit exists we say $f \in D(A_\omega)$.

If a Banach space is the dual of another Banach space X ($X^*=B$), then we can define $A_{\omega*}$ by $\langle A_{\omega*} f,g\rangle = \langle h,g\rangle = \lim_{t\to 0} \langle \frac{T(t)-I}{t} f,g\rangle$ for all $g \in X$, and if for f the limit exists, we say $f \in D(A_{\omega*})$. It can be noted that $D(A) \subset D(A_\omega)$ and $D(A) \subset D(A_{\omega*})$ if $X^* = B$.

THEOREM 2.1. Suppose $T(t)$, $0 \le t < \infty$ is a contraction semigroup on B; and f satisfying $f \in D(A_\omega^n)$ or $f \in D(A_{\omega*}^n)$ is such that $T(t)(A_\omega)^n f$ tends to $(A_\omega)^n f$ weakly or $T(t)(A_{\omega*})^n f$ tends to $(A_{\omega*})^n f$ in a weak* way, we have

$$(2.1) \qquad \|A_\omega^k f\| \le K(n,k)\|A_\omega^n f\|^{\frac{k}{n}} \|f\|^{1-\frac{k}{n}}$$

and

$$(2.2) \qquad \|A^k_{\omega_*} f\| \le K(n,k)\|A^n_{\omega_*} f\|^{\frac{k}{n}} \|f\|^{1-\frac{k}{n}}$$

with $K(n,k)$ the Schoenberg-Cavaretta constants.

If $T(t)$ $-\infty < t < \infty$ is in addition to the above a group of isometries, then the inequalities (2.1) and (2.2) are valid with $K(n,k)$ the Kolmogorov constants.

COROLLARY 2.2. If $T(t)$ is that of Theorem 2.1, $f \in D(A^n)$ and $T(t)A^n f \to A^n f$ either weakly or weakly*, then

$$(2.3) \qquad \|A^k f\| \le K(n,k)\|A^n f\|^{\frac{k}{n}} \|f\|^{1-\frac{k}{n}}$$

with $K(n,k)$ as in Theorem 2.1.

Note that this is slightly more general than the result in [1].

REMARK 2.3. (a) Let $T^*(t)$ be defined by $\langle T(t)f,g\rangle = \langle f,T^*(t)g\rangle$ for all $f \in B$ and $g \in B^*$, and if $\|T^*(t)g-g\|_{B^*} \to 0$ for all $g \in B^*$, the assumption $T(t)(A_\omega)^n f \to (A_\omega)^n f$ weakly can be dropped.

(b) Let $T'(t)$ be defined by $\langle T(t)f,g\rangle = \langle f,T'(t)g\rangle$ for all $f \in B$ and $g \in X$ $(X^*=B)$, and if $\|T'(t)g-g\|_X \to 0$ for all $g \in X$, then the assumption $T(t)(A_{\omega_*})^n f \to (A_{\omega_*})^n f$ in weak* fashion can be dropped.

Many examples of applications of this remark will be shown here.

PROOF OF THEOREM 2.1. To prove (2.1) and (2.2) we observe first that $T(t)A^\ell_\omega f = A^\ell_\omega T(t)f$ for $f \in D(A^\ell_\omega)$ and $T(t)A^\ell_{\omega_*} f = A^\ell_{\omega_*} T(t)f$ for $f \in D(A^\ell_{\omega_*})$ respectively. (The right hand side can be defined when the left hand side is not if we do not assume $f \in D(A^\ell_\omega)$ and $f \in D(A^\ell_{\omega_*})$ respectively.) Obviously $f \in D(A^n_\omega)$ implies $f \in D(A^\ell_\omega)$ $\ell < n$, and $f \in D(A^n_{\omega_*})$ implies $f \in D(A^\ell_{\omega_*})$ $\ell < n$. We choose g

such that $g \in B^*$ or $g \in X$ with $X^* = B$ with norm 1, and $F(t) = \langle T(t)f, g\rangle$ is in $C^n[0,\infty)$ when $T(t)$ is a semigroup of contraction and in $C^n(-\infty,\infty)$ when $T(t)$ is a group of isometries. Moreover, $F^{(\ell)}(t) = \langle T(t)A_\omega^\ell f, g\rangle$ or $F^{(\ell)}(t) = \langle T(t)A_{\omega*}^\ell f, g\rangle$ for $\ell \le n$. Using [4] and [5], we obtain

$$\|F^{(k)}(t)\|_C \le K(n,k)\|F^{(n)}(t)\|^{\frac{k}{n}}\|F(t)\|^{1-\frac{k}{n}}$$

where $K(n,k)$ are the Kolmogorov constants in case $T(t)$, $-\infty < t < \infty$, is a group of isometries and the Schoenberg-Cavaretta constants for $T(t)$, $0 \le t < \infty$, a semigroup of contractions. The proof can be completed now by choosing g_ε in B^* or X such that

$$\langle A_\omega^k f, g_\varepsilon\rangle \ge \|A_\omega^k f\| - \varepsilon \quad \text{or} \quad \langle A_{\omega*}^k f, g_\varepsilon\rangle \ge \|A_{\omega*}^k f\| - \varepsilon$$

respectively, and observing for $g_\varepsilon \in B^*$, $\|A_\omega^k f\| - \varepsilon \le \langle A_\omega^k f, g_\varepsilon\rangle =$
$= |F_\varepsilon^{(k)}(0)| \le |F_\varepsilon^{(k)}(t)| \le K(n,k)\|F_\varepsilon^{(n)}(t)\|^{k/n} \le K(n,k)\|A_\omega^n f\|^{k/n}\|f\|^{1-k/n}$
and the similar relation for $g_\varepsilon \in X$ and $A_{\omega*}$.

In most applications $T(t)f(x) = f(x+t)$ in some sense and we can summarize our result in the following theorem.

THEOREM 2.4. Let B be a Banach space and $B \subset S'(A)$ (B being continuously imbedded in the Schwartz distribution over A) and let A be $[0,\infty)$, $(-\infty,\infty)$ or $T = [-\pi,\pi]$ and periodic. Let $T(t)f(\cdot) = f(t+\cdot)$ be a contraction for $A = [0,\infty)$ and an isometry for $A = (-\infty,\infty)$ or $A = T$. Then for $f^{(n)}(\cdot)$, a strong derivative, a weak derivative or a weak* derivative of $f(\cdot)$, and $f^{(n)}(t+\cdot) \to f^{(n)}(\cdot)$ in weak or weak* mode, we have

$$\|f^{(k)}(\cdot)\| \le K(n,k)\|f^{(n)}(\cdot)\|^{\frac{k}{n}}\|f(\cdot)\|^{1-\frac{k}{n}} \tag{2.3}$$

where $K(n,k)$ are the Kolmogorov constants for $A = T$ or $A = (-\infty,\infty)$ and $\|f(t+\cdot)\| = \|f(\cdot)\|$, and are the Schoenberg-Cavaretta constants for $A = [0,\infty)$ and $\|f(t+\cdot)\| \le \|f(\cdot)\|$.

PROOF. If $\|f(t+\cdot)-f(\cdot)\|_B = o(1)$ $t \to 0$ for all $f \in B$, we have $f^{(n)}(t+\cdot) \to f^{(n)}(\cdot)$, and if $\|g(t+\cdot)-g(\cdot)\|_{B^*} = o(1)$, $t \to 0$, for all $g \in B^*$, we have $f^{(n)}(t+\cdot) \to f^{(n)}(\cdot)$ weakly whenever $f^{(n)}(\cdot)$ exists in a weak or a strong sense, and if $\|g(t+\cdot)-g(\cdot)\|_X = o(1)$, $t \to 0$, for all $g \in X$, $X^* = B$, we have $f^{(n)}(t+\cdot) \to f^{(n)}(\cdot)$ in weak* mode, whenever $f^{(n)}$ exists in a weak* sense and therefore this is a corollary of Theorem 2.1 where $T(t)f(\cdot) = f(t+\cdot)$, and where we use Remark 2.3.

It is well known that inequality (2.3) is valid for $L_p(R)$ [6] (where the idea seems to originate), $L_p(R^+)$ [1], Orlicz space on R or R^+ [1], and $S_p(\|f\| \equiv \sup_x (\int_x^{x+1} |f(x)|^p dx)^{1/p})$ [1]. Inequality (2.3) is valid also for, function in the Sobolev space on R or R^+ with the norm $\|f\| = \max_{0 \le x \le m} \|f^{(i)}\|_p$ or $\|f\| = \sum_{i=1}^{m} \|f^{(i)}\|_p$ and for functions in Besov or Lipschitz space on R or R^+ with certain norms (the norms for which translation is isometry or contraction).

Using strong derivatives and an earlier result [1, p.150], we can see for $T = [-\pi,\pi]$ and $-\pi$ identified with π that the Kolmogorov inequality is valid for $C(T)$, $L_p(T)$ and Orlicz space on T as well as for the Sobolev, Besov and Lipschitz space on T with appropriate norm, since $T(t)f(\cdot) = f(\cdot+t)$ satisfies the conditions imposed in [1] already.

The result here implies Kolmogorov's inequality (with $K(n,k)$ being either Kolmogorov's or those of Schoenberg and Cavaretta) for many spaces for which earlier results are not applicable. The following spaces with the appropriate derivatives will satisfy (2.3)

A. $L_\infty(R)$, $L_\infty(T)$ or $L_\infty(R^+)$ with weak* derivatives (and actually derivatives in S' which are in L_∞).

B. B.V.(R), B.V.(T) or B.V.(R^+), the functions of bounded variation, being dual to the space of continuous functions. The weak* derivatives (which are also derivatives in S' that belong to B.V.) satisfy (2.3).

C. The dual of the Sobolev space over R, T or R^+ with the regular norm and weak* derivatives.

D. The dual of the Lipschitz or Besov spaces over R, T, or R^+ with norm induced by a norm of the Lipschitz or Besov space under which translation is either isometry or contraction and with weak* derivatives. (Most of the norms given in the literature for these spaces satisfy this requirement.)

One should note that in order for (2.3) to be valid in the spaces mentioned in A, B, C and D the nth weak* derivative must exist in the given space.

We remark that the constants of Kolmogorov and Schoenberg-Cavaretta always apply but are not necessarily the best. The constants in (2.3) dependon the norm and whatever constant is valid for one norm may not be valid even for an equivalent norm. The above discussion provides an upper bound for the constants in case the norm used satisfies $\|f(\cdot+a)\|_B \le \|f(\cdot)\|_B$.

3. BEST TRIGONOMETRIC APPROXIMATION

For a Banach space B inside $S'(T)$, the Schartz space of distribution on the circle $T([-\pi,\pi]$ where $-\pi$ is identified with π) we can define the best trigonometric approximation $E_n(f,B) = \inf_{T_n \in B} \|f-T_n\|_B$ where T_n is a trigonometric polynomial. We define also $\Delta_h f(\cdot) = f(\cdot+h) - f(\cdot)$ and $\Delta_h^r f = \Delta_h(\Delta_h^{r-1} f)$. Following earlier results, we have:

THEOREM 3.1. For a Banach space $B \subset S'(T)$ for which translation is an isometry, that is $\|f(\cdot+a)\| = \|f(\cdot)\|$ for all $a \in T$, the following are equivalent:

(a) $\|\Delta_h^r f\|_B = O(h^\alpha)$ $h \to 0+$

(b) $E_n(f,B) = O(n^{-\alpha})$ $n \to \infty$.

Moreover, $\sup_{\eta \le h} \|\Delta_h^r f\|_B \le M(r) h^r \sum_{n \le h^{-1}} (n+1)^{r-1} E_n(f,B)$, and if $\sum n^{r-1} E_n(f,B)$ converges, $f^{(r)}$ exists as a strong derivative in B and $E_n(f^{(r)},B) \le M(r) \Sigma_{[n/2]}^{\infty} k^{r-1} E_k(f,B)$.

If $\|\Delta_h f\|_B = o(1)$ for all $f \in B$, then

(c) $\|T_n^{(r)}(f)\|_B = O(n^{r-\alpha})$ (where T_n is the best trigonometric approximation to f in B) is also equivalent to (a) and (b).

PROOF. This theorem was essentially proved in [2]. Using $g \in B^*$ and defining $F(x) = f * g = \langle f(x+\cdot), g(\cdot) \rangle$ which always yields a bounded function, we see that if either $\|\Delta_h f\|_B = o(1)$ or $E_n(f,B) = o(1)$, which are equivalent as we shall show, $F(x)$ is continuous and the theorem is derived from the analogue for continuous functions as in [2]. To prove that (c) also implies (a) we follow directly Sunouchi's result [7] but there $E_n(f,L_p) = o(1)$ is used and therefore $E_n(f,B) = o(1)$ is needed which follows from $\|\Delta_h f\|_B = o(1)$, $h \to 0$, that is imposed here. (In [2] it is used but omitted erroneously.) If $\|\Delta_h f\|_B = o(1)$, $h \to 0$, then $F_n(f) = \frac{1}{2\pi n} \int f(\cdot+t) (\frac{\sin nt/2}{\sin t/2})^2 dt$ is defined (since $f(\cdot+t)$ is continuous in the norm topology) and is a trigonometric polynomial, and therefore

$$E_n(f,B) \le \|F_n(f)-f\|_B \le \|\Delta_h f\| + 2\|f\| \frac{1}{2\pi n} \int_{|t|>h} \left(\frac{\sin \frac{nt}{2}}{\sin \frac{t}{2}} \right)^2 dt$$

which tends to zero. If $E_n(f,B) = o(1)$, $n \to \infty$, we have

$$\begin{aligned} \|\Delta_h f\| &= \|\Delta_h (f-T_n) + \Delta_h T_n\| \le \|\Delta_h (f-T_n)\| + \|\Delta_h T_n\| \\ &\le 2E_n(f,B) + h\|T_n'\| \le 2E_n(f,B) + hn\|T_n\| \\ &\le 2E_n(f,B) + hn(\|f\|+1) \end{aligned}$$

which tends to zero if $h = 1/n^2$ and $n \to \infty$ for example.

REMARK. We may observe that in proving the equivalence of (c) $\|\Delta_h f\| = o(1)$ for all $f \in B$ cannot be replaced by conditions like $\|\Delta_h g\| = o(1)$ for all $g \in X$ $X^* = B$, as for $B = B.V.(T)$ the function $f(x) = 0$ for $(0,\pi)$ and 1 for $(-\pi,0)$ has $T_n = 0$ or $T_n = 1$ whose derivatives are equal to zero and therefore satisfy $\|T_n^{(r)}\| = O(n^{r-\alpha})$ but not $\|\Delta_h f\| = o(1)$. All other parts of the theorem do not require $\|\Delta_h f\|_B = o(1)$ for all $f \in B$ and for instance are applicable to $B = B.V.(T)$.

4. INVERSE RESULTS IN APPROXIMATION THEORY

In [2] it was shown that direct and inverse theorems for a convolution approximation process that are valid for continuous functions on A $(R, R^+, T$ or a cartesian product of them) implies that those theorems are also valid for Banach space B of generalized functions, for which translation is a contraction, that is $\|f(\cdot+a)\| \le \|f(\cdot)\|$, and for which translation is also continuous, that is, $\|\Delta_h f(\cdot)\|_B = o(1)$, $h \to 0+$, for all f in B. We will relax here the condition on continuity of all elements of the space.

Let B be a Banach space such that $\|f(\cdot+a)\|_B \le \|f(\cdot)\|_B$ for all $a \in A$ and

(I) $\|\Delta_h f\|_B = o(1)$, $h \to 0$, for all $f \in B$ or

(II) $\langle \Delta_h f, g\rangle = o(1)$, $h \to 0+$, for all $f \in B$ and $g \in B^*$ or

(III) $\langle \Delta_h f, g\rangle = O(1)$, $h \to 0+$, for all $f \in B$ and $g \in X$ such that $X^* = B$.

Then we can define for any finite measure on A, $\mu_n(\cdot)$, $f*\mu_n = \int f(\cdot+x)d\mu_n(\cdot)$, define for (I) since f is continuous in norm topology, and for (II) and (III) by $\langle f*\mu_n, g\rangle \equiv \int \langle f(t+\cdot), g(\cdot)\rangle d\mu_n(t)$

for the appropriate g. (g in B^* or in X, $X^* = B$.)

THEOREM 4.1. For B and $f*\mu_n$ described above and $L_n(f) \equiv f*\mu_n$ the inverse theorem for continuous functions

(a) $\|L_n(\phi,\cdot)-\phi(\cdot)\|_{C(A)} \le M\sigma_n^\alpha$ implies $\|\Delta_h^r f\|_{C(A)} \le M_1 h^\alpha$ for $r > \alpha$ implies the inverse theorem on B, that is

(b) $\|L_n(f,\cdot)-f(\cdot)\|_B \le M\sigma_n^\alpha$ implies $\|\Delta_h^r f\|_B \le M_1 h^\alpha$ for $r > \alpha$.

PROOF. We observe that $L_n(f, \ )*g = L_n(f*g, \ )$ where $g \in B^*$ in case conditions (I) or (II) are satisfied and where $g \in X$ ($X^*=B$) in case of condition (III). Since $f*g = F \in C(A)$, we choose g of norm 1 and applying the theorem for continuous functions, we have $\|\Delta_h^r F\|_{C(A)} = M_1 h^\alpha$. Choosing g_ε of norm 1 in B^* or X such that $|\langle \Delta_h^r f(\cdot), g_\varepsilon(\cdot)\rangle| \ge \|\Delta_h^r f\| - \varepsilon$, and recalling that $\|\Delta_h^r f\| - \varepsilon \le |\langle \Delta_h^r f(\cdot), g_\varepsilon(\cdot)\rangle| = |\Delta_h^r F(0)| \le \|\Delta_h^r F\|_{C(A)} \le M_1 h^\alpha$ for all ε, we complete the proof.

REMARK. Our result now extends inverse results to spaces like B.V.(A), dual of Sobolev and Besov spaces on A and $L_\infty(A)$. Of course even the result in [2] is applicable to L_p spaces, Sobolev and Besov spaces, Orlicz and others.

5. AN EQUIVALENT CONDITION ON DERIVATIVES

In [3] we proved the equivalence of some asymptotic relations and we required there that $\|\Delta_h f\|_B = o(1)$ for all $f \in B$. This last condition can be relaxed in a way similar to that used in earlier sections of this paper.

Let B be a Banach space of distribution over R or T such that $\|f(\cdot+a)\| = \|f(\cdot)\|$ and

(I) $\|\Delta_h f\| = o(1)$ for all $f \in B$, or

(II) $\langle \Delta_h f, g\rangle = o(1)$ $h \to 0$ for all $f \in B$ and $g \in B^*$, or

(III) $\langle \Delta_h f, g\rangle = o(1)$ $h \to 0$ for all $f \in B$ and $g \in X$ $(X^* = B)$.

Define $A_n f$ for $f \in B$ and $G_n \in L_1$ by $A_n f = \int f(t+\cdot)G_n(t)dt$ as in section 4, while $A_n^k f = A_n(A_n^{k-1} f)$.

We have the following theorem:

THEOREM 5.1. For $f \in B$, $A_n f$ and G_n as above and $\int G_n(y)dy = 1$, $\int |G_n(y)|dy \le M$, $\|G_n^{(r-1)}\|_{B.V.} \le M_r \sigma_n^{-r}$ for some r (derivatives of L_1 function could be understood in the S' sense) and for some $\beta > 0$ $\int |y|^\beta |G_n(y)|dy \le M\sigma_n^\beta$ for σ_n satisfying $\sigma_n = o(1)$ $n \to \infty$ and $1 \le \sigma_n/\sigma_{n+1} \le M$, then for $\alpha > 0$ the following are equivalent:

(A) $\|\Delta_h^r f\|_B \le Mh^\alpha$ for any integer r, $r > \alpha$.

(B) $\|(\frac{d}{dx})^{rk} A_n^k(f,X)\|_B = O(\sigma_n^{-rk+\alpha})$ for any integers r, k such that $rk > \alpha$.

(C) $\|(A_n - I)^\ell f\|_B = O(\sigma_n^\alpha)$ for any integer ℓ such that $\ell > \alpha/\min(\beta,1)$.

(If G_n is even, $\ell > \alpha/\min(\beta,2)$ is sufficient).

REMARK 5.2. It is clear that for the f in question $\|\Delta_h f\|_B = o(1)$ $h \to 0$ since (A) is much stronger. The advantage in this theorem is that we do not have to assume $\|\Delta_h f\|_B = o(1)$ on the whole space B and therefore our theorem applies now to many spaces for which it was not valid before. For example B.V., L_∞ and duals of Sobolev and Besov spaces.

PROOF. A combination of [3] with the simple idea used in section 4 will constitute a proof of our theorem.

REFERENCES

1. Z. DITZIAN, Some remarks on inequalities of Landau and Kolmogorov, Aequationes Math., 12, 1975, 145-151.

2. Z. DITZIAN, Some remarks on approximation theorems on various Banach spaces, Jour. of Math. Anal. and Appl., Vol.77, (2), 1980, 567-576.

3. Z. DITZIAN, Lipschitz classes and convolution approximation processes, Math. Proc. Camb. Phil. Soc., 1981, (90), 51-61.

4. A.N. KOLMOGOROV, On inequalities between the upper bounds of the successive derivatives of an arbitrary function on an infinite interval, 1939, Amer. Math. Soc. transl. 4, 1949, 233-243.

5. J.J. SCHOENBERG and A. CAVARETTA, Solution of Landau's problem concerning higher derivatives on half line, Proceedings of the international conference on constructive function theory, Varna, May 19-25, 1970, 297-308.

6. E.M. STEIN, Functions of exponential type, Ann. of Math., 65, 1957, 582-592.

7. G.I. SUNOUCHI, Derivatives of trigonometric polynomials of best approximation, in "Abstract spaces and approximation", Proceedings of conference at Oberwolfach, 1968, (P.L. Butzer and B.Sz. Nagy Eds.), 233-241, Birkhäuser, Basel und Stuttgart, 1969.

Department of Mathematics
The University of Alberta
Edmonton, Canada T6G2G1

Functional Analysis, Holomorphy and
Approximation Theory II, G.I. Zapata (ed.)

A BIFURCATION SET ASSOCIATED TO THE COPY PHENOMENON IN THE SPACE OF GAUGE FIELDS

Francisco Antonio Doria

We show that gauge field copies are associated to a stratified bifurcation set in gauge field space. Such a set is noticed to be the locus of other bifurcation phenomena in gauge field theory besides the copy phenomenon.

1. INTRODUCTION

The phenomenon that we are going to discuss in the present paper has been discovered by two theoretical physicists, T.T. Wu and C.N. Yang, in search for differences between the so-called Abelian gauge theories and their non-Abelian counterparts [7]. Gauge field theories are physical interpretations for the usual theory of connections on a principal fiber bundle [5]. Physicists usually take spacetime manifold (a 4-dimensional real Hausdorff smooth manifold with a nowhere degenerate quadratic form, the "metric tensor") as base space for the bundle, while the fiber is identified with a finite-dimensional semi-simple Lie group. More general finite-dimensional differentiable manifolds are sometimes used as base space for bundles of physical interest, so that our results will not in general depend on the base manifold's dimension.

The above description is the mathematical setting for the so-called Wu-Yang ambiguity or gauge field copy phenomenon. Let us be given the expression for a curvature form on such a bundle in a local coordinate system:

$$F = (1/2)F_{\mu\nu}(x)dx^{\mu} \wedge dx^{\nu} \qquad (1.1)$$

Here the components $F_{\mu\nu}$ are Lie-algebra valued objects. F is the expression of the bundle's curvature form φ at a (local) identity cross-section $U \times \{1\}$, where U is an open trivializing domain in the base manifold M. Curvature and connection forms on the bundle $P(M,G)$ - where G, a semi-simple Lie group as described above, is the bundle's fiber, - are related by Cartan's structure equation:

$$\varphi = d\alpha + (1/2)[\alpha \wedge \alpha]. \qquad (1.2)$$

(In a local coordinate system, at the identity cross-section,

$$F_{\mu\nu} = \partial_{\mu}A_{\nu} - \partial_{\nu}A_{\mu} + [A_{\mu},A_{\nu}], \qquad (1.3)$$

$$F^{a}_{\mu\nu} = \partial_{\mu}A^{a}_{\nu} - \partial_{\nu}A^{a}_{\mu} + c^{a}_{bc}\, A^{b}_{\mu}\, A^{c}_{\nu}, \qquad (1.4)$$

where the c^{a}_{bc} are Lie-algebra structure constants.)

If we are given a curvature form φ, do we have a unique connection form α ? The answer is no, in general. If the group is the Abelian group $U(1)$, we can immediately check that uniqueness of φ can only be a local phenomenon provided that M is not simply connected. If the group is any non-Abelian semi-simple group it is easy to show that there exists a curvature form φ represented by a Lie-algebra valued 2-form F which can be obtained out of an infinite family of connection forms which are <u>not</u> related anywhere on spacetime by the so-called gauge transformations, that is by the natural action on the bundle induced by the right action of G on the fiber. The example is quite simple: let our spacetime M be the four-plane with Cartesian coordinates and let the fiber group be any non-Abelian Lie group (the other assumptions are not essential to the example.) If $L(G)$ is G's Lie algebra, we choose as components for the connection and curvature forms (at the local identity cross-section),

$$A: \begin{cases} A_1 = A_3 = A_4 = 0 \\ A_2 = (x^1)\theta, \quad \theta \in L(G), \end{cases} \tag{1.5}$$

$$F: \begin{cases} F_{12} = \theta, \\ F_{13} = F_{23} = F_{14} = F_{24} = F_{34} = 0. \end{cases} \tag{1.6}$$

It is now easy to check that

$$B: \begin{cases} B_1 = B_3 = B_4 = 0, \\ B_2 = A_2 + h(x^2)\theta', \quad \theta' \in L(G), \end{cases} \tag{1.7}$$

is also a connection form for F, whenever $h(x^2)$ is any C^1 function of the Cartesian coordinate (x^2). Now if $[\theta,\theta'] \neq 0$, A and B cannot be gauge-related, for if it were so, it is immediate that θ and θ' should commute [8]. The Wu and Yang argument has a rather more physical flavor: they formed the current $j^\mu = \partial_\nu F^{\mu\nu} + [A_\nu, F^{\mu\nu}]$. We then check that $j = 0$ for A and $j \neq 0$ for B. As there is no gauge transformation that can make the current vanish, A and B cannot be gauge-related.

The Wu-Yang ambiguity has remained a curiosity until now. However some recent work has opened the way for deeper understanding of the phenomenon along physical and mathematical lines [1].

2. MAIN RESULTS IN THE FIELD COPY PROBLEM

Since our goal is to describe the geometry of copied curvatures and connections in the space of all curvatures and connections, we review here the main characterizations for copied curvatures and connections.

Let M be a differentiable real n-manifold, G a semi-simple finite-dimensional Lie group, $\pi: P(M,G) \to M$ a principal G-bundle over M with projection π. Suppose that over $\pi^{-1}(U)$, $U \subset M$ a nonvoid open set, the curvature form φ can be derived from two different connection forms α^1 and α^2, $\alpha^2 = \alpha^1 + \theta$. We then conclude:

PROPOSITION 2.1. Under the above conditions θ satisfies:

$$d\theta + \frac{1}{2}[\theta \wedge \theta] + [\alpha^1 \wedge \theta] = 0. \tag{2.1}$$

PROOF. Calculate

$$\varphi = d\alpha^1 + (1/2)[\alpha^1 \wedge \alpha^1] =$$

$$= d(\alpha^1+\theta) + (1/2)[(\alpha^1+\theta) \wedge (\alpha^1+\theta)].$$

We now define the auxiliary connection form $\alpha^o = \alpha^1 + (1/2)\theta$ [4]. This implies:

COROLLARY 2.2. Condition (2.1) is equivalent to

$$d(\alpha^o)\theta \equiv_{\text{Def}} d\theta + [\alpha^o \wedge \theta] = 0, \tag{2.2}$$

where $d(\alpha^o)$ denotes the covariant exterior operator w.r.t. α^o.

PROOF. Substitute $\alpha^1 = \alpha^o - (1/2)\theta$ into (2.1).

Condition (2.1) implies also a well-known <u>necessary</u> condition for the existence of gauge copies:

COROLLARY 2.3.

$$d(d\theta+(1/2)[\theta \wedge \theta] + [\alpha^1 \wedge \theta]) = [\varphi \wedge \theta] = 0 \tag{2.3}$$

PROOF. One calculates the derivative and then substitutes in the result Cartan's structure equation and equation (2.1).

(2.3) was erroneously considered by the author to be also a

sufficient condition for the existence of connection ambiguities [8]. A counterexample is given elsewhere [12]. More on that below. Equation (2.2) can be solved if we suppose that $\theta = \mathcal{U} d\beta$, where β is a Lie-algebra-valued equivariant function on the bundle. Covariance considerations indicate that $\mathcal{U} = \mathrm{Ad}(u)$, the adjoint action of a (possibly local) gauge transformation u. If we then substitute $\theta = \mathcal{U} d\beta$ into (2.2) we get

$$[\tilde{\alpha}^{o} \wedge d\beta] = 0, \tag{2.4a}$$

$$[\tilde{\alpha}^{o},-] = \mathcal{U}^{-1}[\alpha^{o},-]\mathcal{U} + \mathcal{U}^{-1}d\mathcal{U}. \tag{2.4b}$$

Equation (2.4a) can be rewritten as

$$[\alpha^{1} \wedge d\beta] = -(1/2)[d\beta \wedge d\beta]. \tag{2.5}$$

If we delete the combined product symbols, we see that solving (2.1) or (2.2) is equivalent to solving the equation $(d\beta^{2}) = = (\alpha^{1})(d\beta)$. We have two possibilities: (i) the infinitesimal, continuous copies, given by

$$(d\beta)^{2} = 0 \quad \text{iff} \quad (\alpha^{1})(d\beta) = 0, \tag{2.5a}$$

and the (ii) discrete, paired copies,

$$\alpha^{1} = \bar{\alpha}^{1} - (1/2)d\beta, \quad (\bar{\alpha}^{1})d\beta = 0. \tag{2.5b}$$

Names are due to the following: if we put $\alpha^{2}-\alpha^{1} = \epsilon\rho$, $\epsilon > 0$, $\epsilon^{2} \cong 0$, we get

LEMMA 2.4. A connection form α^{1} is infinitesimally copied as above iff $[\alpha^{1} \wedge \rho] = 0$, $\rho = d\beta$.

PROOF. Equation (2.1) with $\theta = \epsilon\rho$ implies $d(\alpha^{1})\rho = 0$ (to the first order). And this last equation implies and is implied by $[\alpha^{1} \wedge \rho] = 0$, $\rho = d\beta$.

Solutions like (2.5b) are said to be discrete because $\theta = d\beta$ is a unique solution for (2.1) whenever $\alpha^1 = \bar{\alpha}^1 + d\beta$, $[\bar{\alpha}^1 \wedge d\beta] = 0$. We will soon see that "infinitesimal" copies form a boundary set in the space of all copied potentials.

We finally notice that combining (2.3) with (2.5) we see that copied curvatures should satisfy

$$[\varphi \wedge d\beta] = 0. \qquad (2.7)$$

That is, the (algebraic) operator $[\varphi \wedge -]$ should have a nontrivial, integrable nullspace. This property allows us to show that copied curvatures form a boundary set in the space of all curvatures that satisfy $[\varphi \wedge \theta] = 0$ for a nontrivial θ on an open set in the bundle.

A final result will be very useful in the next sections: we will need the fact that

PROPOSITION 2.5. φ has gauge-equivalent different potentials over $\pi^{-1}(U)$, U and open set in M, iff its Ambrose-Singer holonomy group $H(\varphi)$ has a nontrivial centralizer in G on $\pi^{-1}(U)$.

For the prof see [7]. This allows us to show that "true" copies are dense in the space of all copies, and that (locally at least) gauge-equivalent copies belong to a boundary set in the space of copied curvature and connection forms.

3. DEGENERACIES IN CONNECTION AND CURVATURE SPACE

We are going to describe some aspects of the geometry of connection and curvature spaces that have at least one of the degeneracies listed in the preceding section. In order to summarize these degeneracies for the benefit of our exposition, we notice that if θ is a (possibly local) 1-form and $d(\alpha)$ the exterior covariant derivative operator w.r.t. α, the condition $d^2(\alpha)\theta = 0$

is equivalent to condition (2.3), or $[\varphi \wedge \theta] = 0$, provided that θ be an ad-type tensorial form. We can thus list:

Covariant cohomology condition:

$$d^2(\alpha)\theta = 0 \text{ iff } [\varphi \wedge \theta] = 0, \tag{3.1}$$

Necessary condition for copies:

$$[\varphi \wedge d\beta] = 0, \tag{3.2}$$

Existence of copies:

$$d(\alpha^1 + (1/2)\theta)\theta = 0, \tag{3.3}$$

Discrete copies:

$$[\alpha^1 \wedge d\beta] = -(1/2)[d\beta \wedge d\beta], \tag{3.4}$$

Infinitesimal copies:

$$[\alpha^1 \wedge d\beta] = 0 = -(1/2)[d\beta \wedge d\beta], \tag{3.5}$$

False copies:

$$H(\varphi) \text{ with nontrivial centralizer.} \tag{3.6}$$

Our objects are connections and curvatures for principal fiber bundles $P(M,G)$ with a real n-dimensional smooth manifold as its base space and a fixed finite-dimensional semi-simple Lie group G as its fiber. The geometry of curvature and connection space is already a pretty well-known subject [5] and we will sketch here some of its main lines. Curvature forms can be identified with cross-sections of the bundle $\Omega^2(M,L(G))$ of Lie-algebra $L(G)$-valued 2-forms on M. Connection forms can be identified with <u>all</u> smooth cross-sections of the bundle of $L(G)$-valued 1-forms on M, $\Omega^1(M,L(G))$. Curvatures are ad-type objects; connection forms will be so provided that we fix and arbitrary connection (which can be the zero, or vacuum, connection 0) and identify $\alpha - 0$ with the cross-sections of $\Omega^1(M,L(G))$. Near any point $x_o \in M$, any $L(G)$-

valued 2-form f can be seen as the curvature of a particular connection form a via the construction

$$a_\mu(x) \cong (1/2)\ f_{\nu\mu}(x_o)(x^\nu - x_o^\nu) \tag{3.7}$$

(in a local coordinate system) [4]. Globally, while the map that sends a connection form over its curvature is pretty well-behaved, the inverse map is full of pathologies [13].

We will consider here connection $\mathcal{G} = C^\infty(\Omega^1(M,L(G)))$ and curvature $\mathfrak{F} \subset C^\infty(\Omega^2(M,L(G)))$ spaces to be endowed with a natural Fréchet structure if we endow these cross-section spaces with the C^∞ topology. This rather weak structure will be enough for our first series of results.

We will first restrict our remarks to the case when M is a spacetime, that is, a 4-dimensional real smooth manifold endowed with a nondegenerate metric tensor. We then define on M the Hodge $*$ operator w.r.t. the spacetime metric and then check that (3.1) becomes a spacetime-parametrized linear homogeneous system:

$$[\varphi \wedge \theta] = (\mathrm{Ad}{*}\varphi)\theta = 0. \tag{3.8}$$

This system will obviously have nontrivial solutions provided that $\det(\mathrm{Ad}{*}\varphi) = 0$ somewhere in spacetime. We now say that a curvature φ globally satisfies a property P iff P is verified by φ over the whole manifold M (or the whole bundle $P(M,G)$). With this definition in mind we assert:

PROPOSITION 3.1. Curvatures that globally satisfy property (3.1) form a closed and nowhere dense set in $\mathfrak{F}$ in the C^∞ topology.

PROOF. Let $\det \mathrm{Ad} * F$ be the determinant of the finite-dimensional function matrix $\mathrm{Ad} * F$, where F is the Lie-algebra valued 2-form over spacetime associated to a curvature φ. For curvatures that globally satisfy property (3.1) we have $\det \mathrm{Ad} * F = 0$ on the whole of M. And from the map

$$* \text{Det}: \mathfrak{F} \to C^{\infty}(M)$$
$$F \mapsto \det \text{Ad}{*}F$$

we see that $\text{Det}^{-1}(0) \subset \mathfrak{F}$ is closed and nowhere dense in $\mathfrak{F}$.

This immediately implies:

COROLLARY 3.2. Globally copied curvatures form a nowhere dense set in $\mathfrak{F}$ in the C^{∞} topology.

PROOF. The copy condition (3.3) implies (3.1), as shown in Corollary 2.3.

What about objects that satisfy one of the properties (3.1)-(3.6) only locally, that is, over a nonvoid open set in M ? (or in P(M,G)). For property (3.1) w.r.t. bundles over a spacetime we can settle that question with:

PROPOSITION 3.3. Curvatures that satisfy property (3.1) locally over spacetime form a closed and nowhere dense set in $\mathfrak{F}$ in the C^{∞} topology.

PROOF. We first form $*\text{Det}(\mathfrak{F}) = \mathfrak{D} \subset C^{\infty}(M)$ and endow $\mathfrak{D}$ with the topology induced by the map $*\text{Det}(-)$. For any $f \in \mathfrak{D}$, if f vanishes over an open set in M then $F \in \text{Det}^{-1}(f)$ satisfies property (3.1) locally over M. Let $\mathfrak{h}(f)$ be a neighborhood of f in $\mathfrak{D}$ in the induced topology. It is immediate that there exists a $g \in \mathfrak{h}(f)$ which is nowhere vanishing on M. Thus any $G \in \text{Det}^{-1}(g) \subset \text{Det}^{-1}(\mathfrak{h})$ will never satisfy property (3.1) on M. The conclusion follows immediately.

This result allows the solution of a question raised by M. Halpern [11]. Halpern suggested that ambiguous curvatures and connections should give extra contributions to the integrals in Feynmann quantization techniques. Despite the fact that we don't have a rigorous measure theoretical construction for a general Feynmann integral, we have several heuristic procedures for such calculations which try to characterize objects similar to Borel sets on connection and curvature spaces. As ambiguous curvatures and connections (which satisfy property (3.1)) are nowhere dense in $\mathfrak{F}$ and in $\mathcal{C}$ in a natural topology like the C^∞ topology, we have now reason to expect that they should be ignored during Feynmann integral calculations.

The next question is: are copied curvatures dense in the space of all curvatures that satisfy condition (3.1)? The answer is no:

PROPOSITION 3.4. Copied curvatures are nowhere dense in the space of all curvatures that satisfy (3.1).

PROOF. If φ is a copied curvature then it satisfies (3.2). Consider all tensorial $\xi = [\varphi \wedge \theta]$. If $C^\infty\ \Omega^1(M,L(G))$ is the space of all tensorial $L(G)$-valued 1-forms on the bundle, and $C^\infty\ \Omega^4(M,L(G))$ the space of all tensorial $L(G)$-valued 4-forms on the bundle, we can form the map

$$d\colon \mathfrak{F} \times \Omega^1(M,L(G)) \to C^\infty\ \Omega^4(M,L(G))$$

$$[\varphi \wedge \theta] \longmapsto d[\varphi\wedge\theta] = [\varphi\wedge d\theta].$$

where d is the standard exterior derivative. Due to the Bianchi differential identity we have that $d^{-1}(0) = \{[\varphi\wedge\theta],\ \theta \text{ a cocycle}\}$. It is immediate that $d^{-1}(0) \subset \mathfrak{F} \times C^\infty\Omega^1$ is a closed and nowhere dense set and so is $p_1(d^{-1}(0)) \subset \mathfrak{F}$, where p_1 denotes the projection $p_1\colon \mathfrak{F} \times C^\infty\Omega^1 \to \mathfrak{F}$. Condition $[\varphi \wedge \theta] = 0$ defines another closed and nowhere dense subset in $\mathfrak{F} \times C^\infty\Omega^1$, and so do the corresponding restrictions to $d^{-1}(0)$ and $p_1 d^{-1}(0)$. Thus globally copied curvatures are nowhere dense in the space of all curvatures that satisfy (3.1). For locally copied curvatures we must follow a reasoning similar to the one used in Proposition 3.3.

The same technique can be applied to prove:

PROPOSITION 3.5. Infinitesimally copied fields are nowhere dense in the space of all copied fields.

PROOF. Consider the map

$$f: \mathcal{G} \times C^{\infty}\Omega^{1} \to \mathcal{G} \times C^{\infty}\Omega^{1}$$

and apply the same reasoning as in Proposition 3.4. We notice that $f^{-1}(0)$ is the set of all infinitesimally copied curvatures.

What about "false" copies, that is, connection form ambiguities that can be (locally at least) eliminated modulo a gauge transformation? This question is settled by

PROPOSITION 3.6. Curvatures with false copies are nowhere dense in the space of all curvatures with potential ambiguities.

PROOF. Curvatures with false copies are stabilized by gauge transformations that take values in the centralizer of $H(\varphi)$, the Ambrose-Singer holonomy group generated by φ. We can thus apply an adequate slice theorem to get via this symmetry a stratification wherefrom one sees that these symmetric curvatures are nowhere dense in the space of all copied curvatures. We could also reproduce in $\mathfrak{F}$ the embedding of Lie algebras within $L(G)$ associated to falsely copied curvatures.

Propositions 3.4 - 3.6 are valid without any restriction on the dimension of M. But if we consider the case when $\dim M = 4$, we can apply the same technique as in Propositions 3.1 - 3.3 to get a stratification in the set of all curvatures that obey (3.1) induced by the embedding of ideals in the space of all matrices $Ad^{*}F$.

4. DIFFERENTIABLE VERSIONS OF OUR RESULTS

In the present section we suppose that all M-defined objects have complex-valued componentes that belong to a convenient Sobolev space. More precisely, if $U \subset M$ is an arbitrary open nonvoid subset, we will suppose that our objects have components in one of the Hilbert-Sobolev spaces $H^m(U) = H^{2,m}(U)$. We thus have a differentiable norm for our objects and a very simple differentiable structure in our function spaces, so that we can nov give a more refined version for our previous results.

When we consider objects that are in a Sobolev space we implicitly admit that our smooth (i.e. C^∞) objects are supposed to have compact supports. Such a supposition may introduce some problems when we deal with globally defined smooth objects on a noncompact spacetime; however we notice that physical calculations are always done in a particular coordinate domain, and that domain can be always restricted to a compact region, or to a neighborhood with compact closure. Another way of looking at this restriction is to suppose that "physical" objects become (approximately, at least) zero beyond a certain range (such a supposition is commonly encountered in the discussion of some results in classical field theory). Anyway the gauge field copy problem is an essentially local phenomenon.

With those remarks in mind, we can obtain the smooth versions of our previous results:

PROPOSITION 4.1. Let $U \subset M$ have compact closure and consider the class of all curvatures on the bundle that do <u>not</u> satisfy condition (3.1) anywhere on U. That set is an open submanifold in $\mathfrak{F}$.

PROOF. Consider the map

$$h: \mathfrak{F} \times \Omega^1 \longrightarrow \mathbb{R} ,$$
$$[\varphi\wedge\theta] \longmapsto \|[\varphi\wedge\theta]\|$$

and consider its inverse (the norm we use is any finite-dimensional norm composed with the Sobolev norm), $h^{-1}(\mathbb{R}^+ - \{0\}) \subset \mathfrak{F} \times \Omega^1$. Due to smoothness we have here an open submanifold in $\mathfrak{F}$.

Appropriate modifications can be made in the other results. A sample is:

PROPOSITION 4.2. Let U be as above and consider the class of all curvatures that have a discrete connection ambiguity all over U. They form an open, dense submanifold of $\mathfrak{F}^o$, the space of all copied curvatures.

PROOF. Consider the map

$$k: \mathfrak{a} \times \Omega^1 \longrightarrow \mathbb{R} ,$$
$$[\alpha\wedge d\beta] \longmapsto \|[\alpha\wedge d\beta]\|$$

and act as in the preceding proposition.

5. INTERPRETATION AND CONCLUSION

Stratified sets first appeared in the study of bifurcation problems in Geometry [14]. We have here a rather complex stratified system, which depends in part on symmetry properties of the systems (the embedding of Ambrose-Singer holonomy algebras.) Stratifications similar to this last one lead in General Relativity to the classification of spacetime geometries that are unstable in the linear approximation [3]. A similar phenomenon leads to the linearization instability of gauge fields uncoupled to any gravitational field [15].

Fields that satisfy condition (3.1) (our first stratum) can be shown to be associated to a nonvanishing torsion tensor that satisfies the same set of Bianchi identities. Such a degeneracy can be related to well-known "inconsistencies" in higher-spin field theory [2]. Fields that possess infinitesimal copies can be shown to generate a very interesting version of the Higgs mechanism [1], where the gauge field can be shown to generate a field that satisfies the standard electromagnetic wave equation [10]. Finally fields with false copies are shown to imply Nambu's condition for the existence of nontrivial topological effects such as magnetic monopoles and vortices. This class of fields exhibits also an inconsistency that appears when one tries to add a gauge-like interaction to spin-0 fields; here this inconsistency is shown to be a symmetry-breaking condition.

We do not have a clear interpretation for the coupled sets of nonequivalent potentials that form discrete copies systems, despite the fact that they were one of the first examples of copies to be found [6].

6. ACKNOWLEDGMENTS

The author wishes to thank Professor G. Zapata for his kind invitation to expose these ideas at the 1981 Holomorphy and Functional Analysis Symposium in Rio de Janeiro. He also thanks Professor Leopoldo Nachbin for his constant interest and encouragement.

REFERENCES

1. A.F. AMARAL, F.A. DORIA and M. GLEISER, Higgs fields as Bargmann-Wigner fields and classical symmetry breaking, J. Math. Phys. 24 (1983), 1888-1890.

2. A.F. AMARAL, The Teitler lagrangian and its interactions, D.Sc. Thesis, Rio de Janeiro (1983) (in Portuguese).

3. J.M. ARMS, Linearization instability of gauge fields, J. Math. Phys. 20 (1979), 443-453.

4. C.G. BOLLINI, J.J. GIAMBIAGI and J. TIOMNO, Gauge field copies, Phys. Lett. 83 B (1979), 185-187.

5. Y.M. CHO, Higher-dimensional unification of gravitational and gauge theories, J. Math. Phys. 16 (1975), 2029-2035.

6. S. DESER and F. WILCZEK, Non-uniqueness of gauge field potentials, Phys. Lett. 65 B (1976), 391-393.

7. F.A. DORIA, The geometry of gauge field copies, Commun. Math. Phys. 79 (1981), 435-456.

8. F.A. DORIA, Quasi-abelian and fully non-abelian gauge field copies: A classification, J. Math. Phys. 22 (1981), 2943-2951.

9. F.A. DORIA and A.F. AMARAL, Linearization instability implies gauge field copies, Preprint, Universidade Federal do Rio de Janeiro, 1983.

10. M. GLEISER, Gauge field copies and the Higgs mechanism, M.Sc. Thesis, Rio de Janeiro (1982) (in Portuguese).

11. M.B. HALPERN, Gauge field copies in the temporal gauge, Nucl. Phys. B 139 (1978), 477-489.

12. M.A. MOSTOW and S. SHNIDER, Counterexamples to some results on the existence of field copies, Preprint, Univ. North Carolina, 1982.

13. M.A. MOSTOW and S. SHNIDER, Does a generic connection depend continuously on its curvature? Preprint, Univ. North Carolina, 1982.

14. R. THOM, La estabilité topologique des applications polynomiales, L'enseignement mathématique Vol. 8 (1960), 24-33.

Interdisciplinary Graduate Research Program
Department of Theory of Communication
Universidade Federal do Rio de Janeiro
Av. Pasteur 250
22290 Rio de Janeiro, RJ, Brazil

Functional Analysis, Holomorphy and
Approximation Theory II, G.I. Zapata (ed.)

ON THE ANGLE OF DISSIPATIVITY OF ORDINARY AND PARTIAL DIFFERENTIAL OPERATORS*

H.O. Fattorini

1. INTRODUCTION

Let A be a densely defined, closed operator in a complex Banach space E. For each $u \in E$ denote by $\Theta(u)$ the <u>duality set</u> of u consisting of all u^* in the dual space E^* such that

$$\langle u^*, u\rangle = \|u\|^2 = \|u^*\|^2, \tag{1.1}$$

where $\langle u^*, u\rangle$ denotes the value of the functional u^* at u. The operator A is <u>dissipative</u> if

$$\operatorname{Re}\langle u^*, Au\rangle \leq 0 \quad (u \in D(A),\ u^* \in \Theta(u)). \tag{1.2}$$

If we have

$$(\lambda I - A)D(A) = E \tag{1.3}$$

for some $\lambda > 0$, then A is called m-<u>dissipative</u> and generates a strongly continuous contraction semigroup $\{S(t);\ t \geq 0\}$,

$$\|S(t)\| \leq 1 \quad (t \geq 0). \tag{1.4}$$

The converse is as well true. We note also that (1.2) need only be assumed for a single element u^* of $\Theta(u)$; equivalently, we may replace (1.2) by

$$\operatorname{Re}\langle \theta(u), Au\rangle \leq 0 \quad (u \in D(A)) \tag{1.5}$$

---

*This work was supported in part by the National Science Foundation, U.S.A. under grant MCS 79-03163.

where θ is a duality map, that is, an arbitrary map $\theta: E \to E^*$ with $\theta(u) \in \Theta(u)$ $(u \in D(A))$. We note also in passing that if E is a Hilbert space, (1.3) is equivalent to maximality of A in the class of dissipative operators. For all necessary facts on the theory see for instance [4].

In certain questions of control theory (related to the computation of the inverse $(I - \alpha S(t))^{-1}$) it is of importance to decide whether the semigroup $S(\cdot)$ can be extended to a sector $|\arg \zeta| \leq \varphi$ $(\varphi > 0)$ in the complex plane in such a way that (1.4) is preserved there. More generally, it is often enough to inquire whether there exists $\varphi > 0$ and $\omega = \omega(\varphi)$ such that

$$\|S(\zeta)\| \leq e^{\omega|\zeta|} \qquad (|\arg \zeta| \leq \varphi). \tag{1.6}$$

We denote by $\varphi(A)$, the angle of dissipativity of A, the supremum of all the $\varphi \geq 0$ such that (1.6) holds for some $\omega = \omega(\varphi) > 0$. Equivalently, the angle of dissipativity $\varphi(A)$ may be characterized as the supremum of all $\varphi \geq 0$ such that $e^{\pm i\varphi}(A - \omega I)$ satisfies (1.5) with respect to some duality map θ, with ω depending in general of φ. Some obvious manipulations show that this requirement translates to

$$\mathrm{Re}\langle \theta(u), Au\rangle \leq \pm\delta\,\mathrm{Im}\langle \theta(u), Au\rangle + \omega\|u\|^2 \qquad (u \in E) \tag{1.7}$$

for some duality map θ and $\delta = \mathrm{tg}\,\varphi$.

The object of the present paper is the computation of the angle of dissipativity of second order uniformly elliptic operators.

$$A = \sum_{j=1}^{m}\sum_{k=1}^{m} a_{jk}(x)D^jD^k + \sum_{j=1}^{m} b_j(x)D^j + c(x) \tag{1.8}$$

$(a_{jk}(x) = a_{kj}(x),\ x = (x_1,\dots,x_m),\ D^j = \partial/\partial x_j)$ in a bounded domain Ω of m-dimensional Euclidean space $\mathbb{R}^m$, or rather, of the restriction $A_p(\beta)$ in $L^p(\Omega)$ $(1 \leq p < \infty)$ or $A(\beta)$ in $C(\bar{\Omega})$ of A defined by a boundary condition β of one of the following types:

$$\text{(I)}\quad D^{\tilde{\nu}}u(x) = \gamma(x)u(x), \qquad \text{(II)}\quad u(x) = 0 \qquad (x \in \Gamma).$$

Here Γ is the boundary of Ω and $D^{\tilde{\nu}}$ indicates the derivative in the direction of the <u>conormal vector</u> $\tilde{\nu} = \{\Sigma_j\, a_{jk}(x)\nu_j\}$, $\nu = (\nu_1,\ldots,\nu_m)$ the outer normal vector on Γ; when the Dirichlet boundary condition (II) is used the space $C(\bar{\Omega})$ of all continuous functions in $\bar{\Omega}$ is replaced by the subspace $C_\Gamma(\bar{\Omega})$ consisting of all u that vanish on Γ.

The results are as follows. Under the standard smoothness assumptions on the coefficients of A, on γ and on Γ the angle of dissipativity turns out to be independent of the operator A and the boundary condition β and only depends on the space. In the space $E = L^p(\Omega)$ $(1 < p < \infty)$ the angle of dissipativity $\varphi(A_p(\beta))$ is

$$\varphi(A_p(\beta)) = \varphi_p = \text{arc tg}\{(\frac{p}{p-2})^2 - 1\}^{1/2} \tag{1.9}$$

(Theorem 2.1 and Section 5). Since $\varphi_p \to \infty$ when $p \to 1,\infty$ we may surmise that the angle of dissipativity is zero in the spaces $L^1(\Omega)$ and $C(\bar{\Omega})$. This is in fact true; moreover, if β is of type (I), $A(\beta) - \omega I$ will not be dissipative for any ω unless β satisfies additional assumptions ((6.1) for the space $L^1(\Omega)$ and (6.8) for the space $C(\bar{\Omega})$), although these restrictions can be bypassed through a renorming of the space (Section 6). These results are presented in detail in Sections 5 and 6.

We treat separately the case $m = 1$ in Sections 2 and 4; here the results are slightly more precise while many of the technical complications disappear. Finally, we include in Section 3 an application of the one-dimensional results to the estimation of the norm of certain multiplier operators.

2. ORDINARY DIFFERENTIAL OPERATORS IN L^p, $1 < p < \infty$.

Let A_0 be the formal differential operator

$$A_0u(x) = a(x)u''(x) + b(x)u'(x) + c(x)u(x). \tag{2.1}$$

Our standing assumptions on the (real) coefficients are: $a(\cdot)$ is twice continuously differentiable, $b(\cdot)$ is continuously differentiable, $c(\cdot)$ is continuous in $\bar{\Omega} = \{x;\ 0 \le x \le \ell\}$.

We denote by β_0 a boundary condition at $x = 0$ of one of the two types

$$\text{(I)} \quad u'(0) = \gamma_0 u(0) \qquad \text{(II)} \quad u(0) = 0 \tag{2.2}$$

and by β_ℓ a boundary condition at $x = \ell$ of one of the two types

$$\text{(I)} \quad u'(\ell) = \gamma_\ell u(\ell) \qquad \text{(II)} \quad u(\ell) = 0 \tag{2.3}$$

The coefficients γ_0, γ_ℓ are real.

The operator $A_p(\beta_0,\beta_\ell)$ $(1 \le p < \infty)$ is defined by $A_p(\beta_0,\beta_\ell)u = A_0u$ in the complex space $L^p(0,\ell)$, the domain of $A_p(\beta_0,\beta_\ell)$ consisting of all $u \in W^{2,p}(0,\ell)$ that satisfy the corresponding boundary condition at each endpoint. Here the space $W^{2,p}(0,\ell)$ consists of all $u \in L^p(0,\ell)$ such that the distributional derivatives u', u'' belong to $L^p(0,\ell)$ as well. We shall show in the rest of this section that $A_p(\beta_0,\beta_\ell)$ (or, rather, a translate) fits into the theory of Section 1. That (1.3) holds means in this case that there exists a sufficiently large $\lambda > 0$ such that, for each $f \in L^p$ there exists $u \in D(A_p(\beta_0,\beta_\ell))$ with

$$(\lambda I - A_p(\beta_0,\beta_\ell))u = f \tag{2.4}$$

This can be shown by elementary means: if $G(x,\xi)$ is the Green function of the boundary value problem $(\lambda I-A)u = f$ corresponding to the boundary conditions β_0, β_ℓ (which function will exist for sufficiently large λ) then

$$u(x) = \int_0^\ell G(x,\xi)f(\xi)d\xi \tag{2.5}$$

For details see [3].

We check now (1.7) for $1 < p < \infty$. To this end, we recall that the only duality map from L^p into $(L^p)^* = L^{p'}$ $(p'^{-1}+p^{-1} = 1)$ is

$$\theta(u) = \|u\|^{2-p}\ |u(x)|^{p-2}\ \bar{u}(x) \qquad (x\neq 0) \tag{2.6}$$

Accordingly, if $u = u_1 + iu_2 \neq 0$ is smooth (say, a Schwartz test function) and $p \geq 2$, $\theta(u)$ is continuously differentiable with

$$\begin{aligned}\|u\|^{p-2}\ \theta(u)'(x) &= (p-2)|u|^{p-4}\ (u_1u_1' + u_2u_2')\bar{u} + |u|^{p-2}\ \bar{u}' \\ &= (p-2)|u|^{p-4}\ \mathrm{Re}(\bar{u}u')\bar{u} + |u|^{p-2}\ \bar{u}'\end{aligned}$$

On the other hand, we have

$$(|u|^p)' = p|u|^{p-2}\ (u_1u_1' + u_2u_2') = p|u|^{p-2}\ \mathrm{Re}(\bar{u}u')$$

Assume both boundary conditions are of type (I). Then we have

$$\begin{aligned}&\|u\|^{p-2}(\mathrm{Re}\langle\theta(u),A_p(\beta_0,\beta_\ell)u\rangle \pm \delta\,\mathrm{Im}\langle\theta(u),A_p(\beta_0,\beta_\ell)u\rangle) = \\ &\quad = \mathrm{Re}\int_0^\ell ((au')' + (b-a')u' + cu)|u|^{p-2}\ \bar{u}dx \\ &\quad \pm \delta\,\mathrm{Im}\int_0^\ell ((au')' + (b-a')u' + cu)|u|^{p-2}\ \bar{u}dx = \\ &= \{\gamma_\ell a(\ell) - \tfrac{1}{p}(a'(\ell)-b(\ell))\}|u(\ell)|^p - \{\gamma_0 a(0) - \tfrac{1}{p}(a'(0)-b(0))\}|u(0)|^p - \\ &- (p-2)\int_0^\ell a|u|^{p-4}(\mathrm{Re}(\bar{u}u'))^2dx \mp \delta(p-2)\int_0^\ell a|u|^{p-4}\mathrm{Re}(\bar{u}u')\mathrm{Im}(\bar{u}u')dx - \\ &\quad - \int_0^\ell a|u|^{p-2}\ |u'|^2dx \pm \delta\int_0^\ell (b-a')|u|^{p-2}\ \mathrm{Im}(\bar{u}u')dx + \\ &\quad + \frac{1}{p}\int_0^\ell (a''-b'+pc)|u|^pdx\end{aligned} \tag{2.7}$$

We transform the sum of the first three integrals on the right-hand side using the following result: given a constant $\alpha > -1$,

$$|z|^2 + \alpha((\mathrm{Re}z)^2 \pm \delta(\mathrm{Re}z)(\mathrm{Im}z)) \geq 0 \tag{2.8}$$

for every $z \in C$ if and only if

$$1 + \delta^2 \leq (\frac{\alpha+2}{\alpha})^2 \tag{2.9}$$

To prove this we begin by observing that (2.7) is homogeneous in z, thus we may assume that $z = e^{i\varphi}$, reducing it to the trigonometric identity $1 + \alpha(\cos^2\varphi \pm \delta\cos\varphi\sin\varphi) \geq 0$ or, equivalently (setting $\psi = 2\varphi$), $2 + \alpha(1 + \cos\psi \pm \delta\sin\psi) \geq 0$. Since the minimum of the function $g(\psi) = \cos\psi \pm \delta\sin\psi$ equals $-(1+\delta^2)^{1/2}$, (2.8) will hold if $2 - \alpha((1+\delta^2)^{1/2} - 1) \geq 0$, which is (2.9). On the other hand, the maximum of g is $(1+\delta^2)^{1/2}$ so that if $-1 < \alpha < 0$, (2.8) will hold if $2 - |\alpha|((1+\delta^2)^{1/2} + 1) \geq 0$, which is again (2.9). In view of the homogeneity of (2.8) it is obvious that if (2.9) is strict there exists $\nu > 0$ (depending on δ) such that, for every $z \in C$,

$$|z|^2 + \alpha((\mathrm{Re}z)^2 \pm \delta(\mathrm{Re}z)(\mathrm{Im}z)) \geq \nu|z|^2. \tag{2.10}$$

We use (2.10) for $z = \bar{u}u'$:

$$-(p-2)\int_0^\ell a|u|^{p-4}(\mathrm{Re}(\bar{u}u'))^2dx \mp$$
$$\mp\delta(p-2)\int_0^\ell a|u|^{p-4}\mathrm{Re}(\bar{u}u')\mathrm{Im}(\bar{u}u')dx -$$
$$-\int_0^\ell a|u|^{p-2}|u'|^2dx = -\int_0^\ell a|u|^{p-4}f dx \tag{2.11}$$

where

$$f = |\bar{u}u'|^2 + (p-2)\{(\mathrm{Re}(\bar{u}u'))^2 \pm \delta\mathrm{Re}(\bar{u}u')\mathrm{Im}(\bar{u}u')\} \geq$$
$$\geq \nu|\bar{u}u'|^2 \tag{2.12}$$

for some $\nu > 0$ (depending on δ) if

$$0 \leq \delta < \{(\frac{p}{p-2})^2 - 1\}^{1/2} \tag{2.13}$$

We must now estimate the other terms on the right-hand side

of (2.7). To this end, consider a real valued continuously differentiable function ρ in $0 \leq x \leq \ell$. Since $(\rho|u|^p)' = \rho'|u|^p + p\rho|u|^{p-2}\operatorname{Re}(\bar{u}u')$ we obtain, for any $\varepsilon > 0$,

$$
\begin{aligned}
&|\rho(\ell)|u(\ell)|^p - \rho(0)|u(0)|^p| \leq \\
&\leq \int_0^\ell |\rho'||u|^p dx + p\int_0^\ell |\rho||u|^{p-2}|u||u'|dx \leq \\
&\leq \frac{\varepsilon^2 p}{2}\int_0^\ell |\rho||u|^{p-2}|u'|^2 dx + \int_0^\ell \left(|\rho'| + \frac{p\varepsilon^{-2}}{2}|\rho|\right)|u|^p dx
\end{aligned}
\tag{2.14}
$$

where we have applied the inequality

$$
2|u||u'| = 2(\varepsilon^{-1}|u|)(\varepsilon|u'|) \leq \varepsilon^2|u'|^2 + \varepsilon^{-2}|u|^2. \tag{2.15}
$$

We use (2.14) for any ρ such that

$$
\rho(\ell) = \gamma_\ell a(\ell) - \frac{1}{p}(a'(\ell)-b(\ell)), \qquad \rho(0) = \gamma_0 a(0) - \frac{1}{p}(a'(0)-b(0))
$$

(we may take ρ linear) to estimate the first two terms on the right-hand side of (2.7); for the fourth integral we use again (2.14), in both cases with ε sufficiently small (in function of the constant ν in (2.10)). We can then bound the right-hand side of (2.7) by an expression of the type

$$
-c\int_0^\ell |u|^{p-2}|u'|^2 dx + \omega\int_0^\ell |u|^p dx
$$

for some constants $\omega = \omega(\delta)$ and $c = c(\delta) > 0$. Upon dividing by $\|u\|^{p-2}$ we obtain

$$
\operatorname{Re}\langle\theta(u), A_p(\beta_0,\beta_\ell)u\rangle \leq \pm\delta\operatorname{Im}\langle\theta(u), A_p(\beta_0,\beta_\ell)u\rangle + \omega\|u\|^2 \tag{2.16}
$$

To extend (2.16) to any $u \in D(A_p(\beta_0,\beta_\ell))$ we use an obvious approximation argument. Inequality (2.16) is obtained in the same way when β_0, β_ℓ (or both) are of type (II).

In the case $1 < p < 2$ the function $\theta(u)$ may not be continuously differentiable; however a simple argument based on the Taylor formula shows that if u is a polynomial (or, more generally, an analytic function) then $\theta(u)$ is absolutely continuous and the computations can be justified in the same way. Details are omitted.

We have completed half of the proof of the following result:

THEOREM 2.1. Let $1 < p < \infty$. Then the angle of dissipativity of $A_p(\beta_0,\beta_\ell)$ in $L^p(0,\ell)$ is given by

$$\varphi(A_p(\beta_0,\beta_\ell)) = \varphi_p = \operatorname{arc\,tg}\{(\tfrac{p}{p-2})^2 - 1\}^{1/2} \tag{2.17}$$

If $S(\cdot)$ is the (analytic) semigroup generated by $A_p(\beta_0,\beta_\ell)$ then for every φ, $0 < \varphi < \varphi_p$ there exists $\omega = \omega(\varphi)$ such that (1.6) holds.

All we have shown so far is that $\varphi(A_p(\beta_0,\beta_\ell)) \geq \varphi_p$. To obtain the opposite inequality we must prove that (2.16) cannot hold if $\delta = \operatorname{tg}\varphi$ with $\varphi > \varphi_p$. We sketch the argument for boundary conditions of type (I). Assume that

$$\delta > \{(\tfrac{p}{p-1})^2 - 1\}^{1/2} \tag{2.18}$$

Then we can find a complex number z (say, of modulus 1) such that

$$|z|^2 + (p-2)((\mathrm{Re}z)^2 - \delta(\mathrm{Re}z)(\mathrm{Im}z)) = -\mu < 0.$$

Let η be a smooth real valued function in $0 \leq x \leq \ell$. Then the function

$$u(x) = e^{z\eta(x)} \tag{2.19}$$

belongs to $D(A_p(\beta_0,\beta_\ell))$ if η satisfies the boundary conditions

$$\eta'(0) = \gamma_0/z, \quad \eta'(\ell) = \gamma_\ell/z. \tag{2.20}$$

We have $\overline{u(x)}u'(x) = z\eta'(x)e^{2(\mathbf{Re}z)\eta(x)} = z\psi(x)$ with ψ real. Accordingly, if f is the function in (2.11),

$$f = -\mu\psi^2 = -\mu|u|^2|u'|^2. \tag{2.21}$$

Making use of this equality and estimating the rest of the terms in (2.7) in a way similar to that used in Theorem 2.1 we obtain an inequality of the form

$$\operatorname{Re}\langle\theta(u),A_p(\beta_0,\beta_\ell)u\rangle - \delta\operatorname{Im}\langle\theta(u),A_p(\beta_0,\beta_\ell)u\rangle$$
$$\geq c\|u\|^{2-p}\int_0^\ell |u|^{p-2}|u'|^2dx - C\|u\|^2. \tag{2.22}$$

Assume that (2.16) holds as well for the same value of δ. Then we obtain from (2.22) that

$$\int_0^\ell |u|^{p-2}|u'|^2dx \leq C'\int_0^\ell |u|^p dx \tag{2.23}$$

for all functions of the form (2.19) where η satisfies the boundary condition (2.20). But (2.23) is easily seen to be false, for instance taking η to be rapidly oscillating function. This completes the proof of Theorem 2.1.

REMARK 2.2. If A_0 is written in variational form,

$$A_0u(x) = (a(x)u'(x))' + b(x)u'(x) + c(x)u(x) \tag{2.24}$$

we only need to require that $a(\cdot)$, $b(\cdot)$ (resp. $c(\cdot)$) be continuously differentiable (resp. continuous) in $0 \leq x \leq \ell$. The same observation will apply in Section 4.

3. AN APPLICATION: COMPUTATION OF THE NORM OF CERTAIN MULTIPLIER OPERATORS

We limit ourselves to the following example. Consider the operator $A_0u = u''$ in the interval $0 \leq x \leq \pi$ with boundary conditions $u'(0) = u'(\pi) = 0$. Then the semigroup $S(\zeta)$ generated by $A_p(\beta_0,\beta_\ell)$ is the multiplier operator

$$S(\zeta)u = \sum_{n=0}^{\infty} e^{-n^2\zeta}a_n\cos nx \tag{3.1}$$

(for $u(x) \sim \Sigma\, a_n\cos nx$) in the space $L^p(0,\pi)$. Note that $S(\zeta)$ is defined for $\operatorname{Re}\zeta > 0$; the alternative formula

$$S(\zeta)u(x) = \frac{1}{2\sqrt{\pi\zeta}}\int_{-\infty}^{\infty} e^{-(x-\xi)^2/4\zeta}u(\xi)d\xi \tag{3.2}$$

can be used, where u is extended 2π-periodically to $-\infty < x < \infty$ in such a way that u is even about $x = 0$ and $x = \pi$. It follows from (3.1) that the norm of $S(\zeta)$ in $L^2(0,\pi)$ is $\|S(\zeta)\|_2 = \max\{e^{-n^2(\mathrm{Re}\zeta)};\ n \geq 0\} = 1$. On the other hand, the norm of $S(\zeta)$ in $C[0,\pi]$ can be estimated from (3.2):

$$\|S(\zeta)\|_C \leq \frac{1}{2\sqrt{\pi|\zeta|}}\int_{-\infty}^{\infty} e^{-x^2(\mathrm{Re}\zeta)/4|\zeta|^2}dx$$

$$= \sqrt{|\zeta|/\mathrm{Re}\zeta} \tag{3.3}$$

thus we obtain the following estimate for the norm of $S(\zeta)$ in $L^p(0,\pi)$, $2 \leq p < \infty$, using interpolation:

$$\|S(\zeta)\|_p \leq (|\zeta|/\mathrm{Re}\zeta)^{(p-2)/2p} > 1 \quad (\mathrm{Re}\zeta > 0) \tag{3.4}$$

A far more precise estimate can be obtained from Theorem 2.1 or, rather, from a close examination of (2.7). Noting that in this case $a = 1$, $b = c = \gamma_0 = \gamma_\ell = 0$ we see that the righ hand side of (2.7) is non-positive for $\delta \leq \mathrm{tg}\,\varphi_p$ (φ_p given by (2.17)). Accordingly,

$$\|S(\zeta)\|_p = 1 \tag{3.5}$$

in the sector $|\arg\zeta| \leq \varphi_p$ (that $\|S(\zeta)\|_p \geq 1$ is obvious). On the other hand, it follows from the necessity part of Theorem 2.1 that (3.5) does not extend to any sector $|\arg\zeta| \leq \varphi$ with $\varphi > \varphi_p$; in other words there exists ζ (in fact, a sequence ζ_n with $|\zeta_n| \to 0$) in the ray $\arg\zeta = \varphi$ ($|\varphi| > \varphi_p$) such that $\|S(\zeta)\|_p > 1$ ($\|S(\zeta_n)\|_p > 1$). The same results can be achieved in the range $1 < p \leq 2$ (for instance, by using duality).

4. ORDINARY DIFFERENTIAL OPERATORS IN L^1 AND C

We consider again the formal differential operator

$$A_0u(x) = a(x)u''(x) + b(x)u'(x) + c(x)u(x) \tag{4.1}$$

under the assumptions on the coefficients used in §2). The operator $A_1(\beta_0,\beta_\ell)$ in $L^1(0,\ell)$ has already been defined there for boundary conditions of any type. The definition of the operator $A(\beta_0,\beta_\ell)$ in the space $C[0,\ell]$ of continuous functions in $0 \le x \le \ell$ (endowed with its usual supremum norm) is $A(\beta_0,\beta_\ell)u = A_0u$ with domain $D(A)$ consisting of all functions u twice continuously differentiable satisfying the boundary condition at each end. Note, however, that if the boundary condition at zero is of type (II), $D(A)$ will not be dense in E; this is remedied replacing $E = C[0,\ell]$ by its subspace $C_0[0,\ell]$ consisting of all u with $u(0) = 0$. When the boundary condition at ℓ is of type (II) (resp. when both conditions are of type (II)) the corresponding subspace is $C_\ell[0,\ell]$ defined by $u(\ell) = 0$ (resp. $C_{0,\ell}[0,\ell]$ defined by $u(0) = u(\ell) = 0$).

The first difficulty we encounter here is that $A_1(\beta_0,\beta_\ell) - \omega I$ will not be dissipative for any ω unless the boundary conditions (if of type (I)) are adequately restricted. The same problem exists with the operator $A(\beta_0,\beta_\ell)$.

THEOREM 4.1. (a) Assume the boundary condition at 0 is of type (I). Then the inequality

$$\gamma_0 a(0) - a'(0) + b(0) \ge 0 \tag{4.2}$$

is necessary for dissipativity in $L^1(0,\ell)$ of any operator $A_1(\beta_0,\beta_\ell)$ using the boundary condition β_0 at $x = 0$. If the boundary condition β_ℓ at $x = \ell$ is of type (I) the corresponding inequality is

$$\gamma_\ell a(\ell) - a'(\ell) + b(\ell) \leq 0 \tag{4.3}$$

If (4.2) and (4.3) hold (or if the corresponding boundary conditions are of type (II)) then $A_1(\beta_0,\beta_\ell) - \omega I$ is m-dissipative in $L^1(0,\ell)$ for sufficiently large ω. (b) Assume the boundary condition at 0 is of type (I). Then the inequality

$$\gamma_0 \geq 0 \tag{4.4}$$

is necessary for dissipativity in $C[0,\ell]$ ($C_0[0,\ell]$, $C_\ell[0,\ell]$, $C_{0,\ell}[0,\ell]$) of any operator $A(\beta_0,\beta_\ell)$ using the boundary condition β_0 at $x = 0$. If the boundary condition β_ℓ at $x = \ell$ is of type (I) the corresponding inequality is

$$\gamma_\ell \leq 0 \tag{4.5}$$

If (4.4) and (4.5) hold (or if the corresponding boundary conditions are of type (II)) then $A(\beta_0,\beta_\ell) - \omega I$ is m-dissipative in $C[0,\ell]$ ($C_0[0,\ell]$, $C_{0,\ell}[0,\ell]$) for sufficiently large ω.

PROOF. Let $u \in L^1$, $u \neq 0$. Then the duality set $\Theta(u)$ of u in $(L^1)^* = L^\infty$ consists of all the functions of the form

$$u^*(x) = \|u\|_1 |u(x)|^{-1}\bar{u}(x) \tag{4.6}$$

where $u(x) = 0$; at those x where $u(x) = 0$ the definition of u^* is arbitrary save by the restriction that $|u^*(x)| \leq \|u\|_1$ (which will be irrelevant in what follows). To show that (4.2) is necessary for dissipativity of any $A_1(\beta_0,\beta_\ell)$ let $0 < \alpha < \ell$, u an element of $D(A_1(\beta_0,\beta_\ell)$ which is positive in the interval $0 \leq x < \alpha$ and zero for $x > \alpha$. Then any $u^* \in \Theta(u)$ equals $\|u\|_1$ in $0 \leq x < \alpha$ and we have

$$\|u\|_1^{-1}\langle u^*, A_1(\beta_0,\beta_\ell)u\rangle = \int_0^\alpha ((au')' + (b-a')u' + cu)dx$$
$$= -(\gamma_0 a(0) - a'(0) + b(0))u(0) +$$
$$+ \int_0^\alpha (a'' - b' + c)udx \qquad (4.7)$$

If $\gamma_0 a(0) - a'(0) + b(0) < 0$, the right hand side of (4.7) can be made positive taking $u(0) = 1$ and α sufficiently small. This shows the necessity of (4.2); the argument for (4.3) is identical.

We prove the necessity of (4.4). Recall that the dual space of $C[0,1]$ can be identified linearly and metrically with the space $\Sigma[0,1]$ of all finite Borel measures defined in $0 \le x \le 1$ endowed with the total variation norm, application of a functional $\mu \in \Sigma$ to an element $u \in C$ given by

$$\langle \mu, u\rangle = \int_0^\ell u d\mu \qquad (4.8)$$

If the boundary condition at 0 is $u(0) = 0$ then the relevant space is $C_0[0,\ell]$, whose dual can be identified through (4.8) to the subspace $\Sigma_0[0,\ell]$ of $\Sigma[0,\ell]$ consisting of all μ with $\mu(\{0\}) = 0$. Similar comments apply to the case where the boundary condition at ℓ is $u(\ell) = 0$ or where the two boundary conditions are $u(0) = u(\ell) = 0$. The duality set $\Theta(u) \in \Sigma$ of an element $u \in C$ consists of all measures supported by the set $m(u) = \{x; |u(x)| = \|u\|\}$ such that $u\mu$ is a positive measure and

$$\|\mu\| = \|u\|.$$

If $\gamma_0 < 0$ we can obviously construct a real element of $D(A(\beta_0,\beta_1))$ having a single positive maximum at $x = 0$ and such that $u(0) = 1$, $u''(0) = \alpha$ where α is arbitrary and will be fixed later. Then $\Theta(u) = \{\delta\}$, δ the Dirac delta and we have

$$\langle \delta, A(\beta_0,\beta_\ell)u\rangle = \alpha a(0) + \gamma_0 b(0) + c(0)$$

which can be made positive by judicious choice of α.

The statements concerning m-dissipativity of the operators $A_1(\beta_0,\beta_\ell) - \omega I$ and $A(\beta_0,\beta_\ell) - \omega I$ can be read off the following two more general results where we show that conditions (4.2), (4.3), (4.4), (4.5) can in fact be discarded if one is willing to perform a renorming of the spaces $L^1(0,\ell)$ and $C[0,\ell]$.

Let $1 \leq p < \infty$, ρ a continuous positive function in $0 \leq x \leq \ell$. Consider the norm

$$\|u\|_\rho = \left(\int_0^\ell |u(x)|^p \rho(x)^p dx\right)^{1/p} \tag{4.9}$$

in $L^p(0,\ell)$. Clearly $\|\cdot\|_\rho$ is equivalent to the original norm of L^p; we write $L^p(0,\ell)_\rho$ to indicate that L^p is equipped with $\|\cdot\|_\rho$ rather than with its original norm. The dual space $L^p(0,\ell)^*_\rho$ can be identified with $L^{p'}(0,\ell)$, $p'^{-1} + p^{-1} = 1$ <u>endowed with its usual norm</u>, an element $u^* \in L^{p'}(0,\ell)$ acting on $L^p(0,\ell)_\rho$ through the formula

$$\langle u^*,u\rangle_\rho = \int_0^\ell u^* u\rho dx \tag{4.10}$$

If $p > 1$ there exists only one duality map $\theta_\rho : L^p \to L^{p'}$ given by $\theta_\rho(u) = \theta(\rho u)$, θ the duality map corresponding to the case $\rho = 1$ (see (2.6)). For $p = 1$ the duality set of an element $u \in L^1(0,\ell)_\rho$ coincides with the duality set of $u\rho$ as an element of $L^1(0,\ell)$ (see (4.6)). We take now u smooth and perform the customary integrations by parts, assuming that ρ is twice continuously differentiable as well:

$$\|u\rho\|^{p-2}\mathrm{Re}\langle\theta_\rho(u),A_p(\beta_0,\beta_\ell)u\rangle_\rho =$$

$$= \|u\rho\|^{p-2}\mathrm{Re}\langle\rho\theta(\rho u),A_p(\beta_0,\beta_\ell)u\rangle$$

$$= \mathrm{Re}\int_0^\ell ((au')' + (b-a')u' + cu)|u|^{p-2}\bar{u}\rho^p dx$$

$$= \{\gamma_\ell a(\ell) - \frac{1}{p}(a'(\ell) - b(\ell))\}|u(\ell)|^p\rho(\ell)^p -$$

$$- \{\gamma_0 a(0) - \frac{1}{p}(a'(0) - b(0))\}|u(0)|^p \rho(0)^p -$$

$$- (p-2)\int_0^\ell a|u|^{p-4}(\mathrm{Re}(\bar{u}u'))^2 \rho^p dx -$$

$$- \int_0^\ell a|u|^{p-2}|u'|^2 \rho^p dx - p\int_0^\ell a|u|^{p-2}\mathrm{Re}(\bar{u}u')\rho^{p-1}\rho' dx +$$

$$+ \frac{1}{p}\int_0^\ell (a''-b'+pc)|u|^p \rho^p dx + \int_0^\ell (a'-b)|u|^p \rho^{p-1}\rho' dx =$$

$$= \{(\gamma_\ell - \frac{\rho'(\ell)}{\rho(\ell)})a(\ell) - \frac{1}{p}(a'(\ell) - b(\ell))\}|u(\ell)|^p \rho(\ell)^p -$$

$$- \{(\gamma_0 - \frac{\rho'(0)}{\rho(0)})\, a(0) - \frac{1}{p}(a'(0) - b(0))\}|u(0)|^p \rho(0)^p -$$

$$- (p-2)\int_0^\ell a|u|^{p-4}(\mathrm{Re}(\bar{u}u'))^2 \rho^p dx -$$

$$- \int_0^\ell a|u|^{p-2}|u'|^2 \rho^p dx +$$

$$+ \int_0^\ell \{(a\rho^{p-1}\rho')' + \frac{1}{p}(a''-b'+pc)\rho^p + (a'-b)\rho^{p-1}\rho'\}|u|^p dx. \qquad (4.11)$$

It is obvious that $\rho'(0)$ and $\rho'(\ell)$ can be chosen at will, hence we may do so in such a way that the quantities between curly brackets in the first two terms on the right-hand side of (4.11) are nonpositive, say, for $1 < p \leq 2$. Since the first two integrals together contribute a nonpositive amount, we can bound (4.11) by an expression of the form $\omega'\|u\|^p \leq \omega\|u\|_\rho^p$ <u>where</u> ω <u>does not depend on</u> p. Consider now the space $L^1(0,\ell)_\rho$. Again under the identification (4.10) the duality set $\Theta_\rho(u)$ of an element u consists of all $u^* \in L^\infty(0,\ell)$ with $u^*(x) = \|u\|_\rho |u(x)|^{-1}\bar{u}(x) = \|u\rho\|\,|u(x)|^{-1}\bar{u}(x)$ where $u(x) \neq 0$ and $|u^*(x)| \leq \|u\|_\rho$ elsewhere. We can then take limits in (4.11) as $p \to 1$ and obtain an inequality of the form

$$\mathrm{Re}\langle u^*, A_1(\beta_0,\beta_\ell)u\rangle_\rho \leq \omega\|u\|_\rho^2 \qquad (4.12)$$

in L^1. The inequality is extended to arbitrary $u \in D(A_1(\beta_0,\beta_\ell))$ by means of the usual approximation argument. Now that $A_1(\beta_0,\beta_\ell)-\omega I$ has been shown to be dissipative, m-dissipativity is established by using Green functions as in Section 2. The case where one (or both) of the boundary conditions are of type (II) is treated in an entirely similar way; naturally, the use of the weight function is unnecessary in the last case.

We have completed the proof of

THEOREM 4.2. Let ρ be a positive twice continuously differentiable function in $0 \leq x \leq \ell$ such that

$$\rho(0)(\gamma_0 a(0) - a'(0) + b(0)) - \rho'(0)a(0) \geq 0 \tag{4.13}$$

$$\rho(\ell)(\gamma_0 a(\ell) - a'(\ell) + b(0)) - \rho'(\ell)a(\ell) \leq 0 \tag{4.14}$$

Then the operator $A_1(\beta_0,\beta_\ell)$ generates a strongly continuous semigroup $S(\cdot)$ in $L^1(0,\ell)_\rho$ such that, for some $\omega > 0$,

$$\|S(t)\|_\rho \leq e^{\omega t} \qquad (t \geq 0)$$

Here $\|S(t)\|_\rho$ indicates the norm of $S(t)$ as an operator in $L^1(0,\ell)_\rho$. Assumption (4.13) (resp. (4.14)) does not apply if the boundary condition at 0 (resp. at ℓ) is of type (II).

To prove a similar result for the operator A we renorm the space C or the corresponding subspace by means of

$$\|u\|_\rho = \max_{0\leq x\leq \ell} |u(x)|\rho(x) \tag{4.15}$$

where ρ is a positive, twice continuously differentiable function in $0 \leq x \leq \ell$. The use of the weight function ρ is again unnecessary when both boundary conditions are of type (II): we treat below in detail the case where β_0 and β_ℓ are of type (I), the "mixed" case being essentially similar. Choose ρ in such a way that

$$\rho'(0) + \gamma_0\rho(0) \geq 0 \quad (\text{resp. } \rho'(\ell) + \gamma_\ell\rho(\ell) \leq 0) \tag{4.16}$$

if $\gamma_0 < 0$ (resp. $\gamma_\ell > 0$). The dual of $C[0,\ell]$ equipped with

$\|\cdot\|_\rho$ can again be identified with $\Sigma[0,\ell]$, an element $\mu \in \Sigma[0,\ell]$ acting on functions $u \in C[0,\ell]$ through the formula

$$\langle \mu, u\rangle = \int_0^\ell u(x)\rho(x)\mu(dx). \tag{4.17}$$

Accordingly, the norm of a measure $\mu \in \Sigma$ as an element of C^* is still $\|\mu\| = \int_0^\ell |\mu(dx)|$ and the identification of the duality sets $\Theta_\rho(u)$ is the same as before; $\Theta_\rho(u)$ consists of all $\mu \in \Sigma$ with support in $m_\rho(u) = \{x;\ |u(x)\rho(x)| = \|u\|_\rho\}$ and such that $u\rho\mu$ (or $u\mu$) is a positive measure in $m_\rho(u)$ with $\|\mu\| = \|u\|_\rho$. The same comments apply of course to the spaces C_0, C_ℓ, $C_{0,\ell}$ where the corresponding measures are required to vanish at 0, ℓ, 0 and ℓ.

We now show that $A(\beta_0,\beta_\ell) - \omega I$ is m-dissipative for ω large enough. Observe first that if $u'(0) = \gamma_0 u(\ell)$, $u'(\ell) = \gamma_\ell u(\ell)$ then $u\rho$ satisfies the boundary conditions

$$(u\rho)'(0) = \gamma_{0,\rho}(u\rho)(0), \qquad (u\rho)'(\ell) = \gamma_{\ell,\rho}(u\rho)(\ell) \tag{4.18}$$

where

$$\gamma_{0,\rho} = \gamma_0 + \rho'(0)\rho(0)^{-1} \geq 0, \qquad \gamma_{\ell,\rho} = \gamma_\ell + \rho'(\ell)\rho(\ell)^{-1} \leq 0. \tag{4.19}$$

Using elementary calculus we show that for any $u \in D(A(\beta_0,\beta_\ell))$, $u \neq 0$ the set $m_\rho(u)$ does not contain either endpoint if both $\gamma_{0,\rho}, \gamma_{\ell,\rho} > 0$, so that

$$(|u\rho|^2)'(x) = 0 \qquad (|u\rho|^2)''(x) \leq 0 \qquad (x \in m_\rho(u)) \tag{4.20}$$

On the other hand, if either $\gamma_{0,\rho}$ or $\gamma_{\ell,\rho}$ vanish, $m_\rho(u)$ may contain the corresponding endpoint but we can prove again that (4.20) holds. Writing $\eta = \rho^2$ we have $(|u\rho|^2)' = 2(u_1u_1'+u_2u_2')\eta + (u_1^2+u_2^2)\eta'$, $(|u\rho|^2)'' = 2(u_1u_1''+u_2u_2'')\eta + 2(u_1'^2+u_2'^2)\eta + 4(u_1u_1'+u_2u_2')\eta' + (u_1^2+u_2^2)\eta''$. Hence, it follows from (4.20) that

$$\operatorname{Re}(u^{-1}u') = \|u\|^{-2}(u_1u_1'+u_2u_2') =$$
$$= -\tfrac{1}{2}\|u\|^{-2}(u_1^2+u_2^2)\eta^{-1}\eta' = -\tfrac{1}{2}\eta^{-1}\eta', \tag{4.21}$$

$$\operatorname{Re}(u^{-1}u'') = \|u\|^{-2}(u_1u_1''+u_2u_2'') =$$
$$\leq -\|u\|^{-2}(u_1'^{\,2}+u_2'^{\,2}) + \|u\|^{-2}(u_1^2+u_2^2)\eta^{-2}\eta'^{\,2} -$$
$$-\tfrac{1}{2}\|u\|^{-2}(u_1^2+u_2^2)\eta^{-1}\eta'' = -\|u\|^{-2}|u'|^2+\eta^{-2}\eta'^{\,2} - \tfrac{1}{2}\eta^{-1}\eta''$$

Accordingly, if $\mu \in \Theta_\rho(u)$ we have

$$\operatorname{Re}\langle\mu,u\rangle = \int_{m_\rho(u)} \operatorname{Re}(u^{-1}A(\beta_0,\beta_\ell)u)\rho\, u d\mu =$$
$$= \int_{m_\rho(u)} (a\operatorname{Re}(u^{-1}u'') + b\operatorname{Re}(u^{-1}u') + c)\rho\, u d\mu \leq$$
$$\leq -\|u\|^{-2}\int_{m_\rho(u)} |u'|^2\rho\, u d\mu - \tfrac{1}{2}\int_{m_\rho(u)} \eta^{-1}(3\eta'+\eta'')\rho\, u d\mu \leq \omega\|u\|_\rho$$

for some constant ω, which shows that $A(\beta_0,\beta_\ell) - \omega I$ is dissipative. That $(\lambda I - A(\beta_0,\beta_\ell))u = v$ has a solution u for all v is once again shown by means of Green functions.

The following result, that settles completely the question of angles of dissipativity in L^1 and C is a simple consequence of the identification of angles of dissipativity in L^2 in Theorem 2.1 and of the theory of interpolation of operators between L^1 and L^2 and between L^2 and C.

THEOREM 4.3. (a) Let (4.2) and (4.3) be satisfied. Then $A_1(\beta_0,\beta_\ell) - \omega I$ is m-dissipative in L^1 for ω sufficiently large but $\varphi(A_1(\beta_0,\beta_\ell)) = 0$. (b) The same conclusion holds for $A(\beta_0,\beta_\ell)$ in C $(C_0,C_\ell,C_{0,\ell})$ if (4.4) and (4.5) hold.

5. ELLIPTIC PARTIAL DIFFERENTIAL OPERATORS IN L^p, $1 < p < \infty$

Since the conclusions and some of the arguments are the same as those for the one-dimensional case we only sketch the details. To simplify the notation we write A_0 in variational form,

$$A_0 = \sum_{j=1}^{m} \sum_{k=1}^{m} D^j(a_{jk}(x)D^k) + \sum_{j=1}^{m} b_j(x)D^j + c(x) \tag{5.1}$$

The domain Ω is bounded and of class $C^{(2)}$. We assume that the a_{jk} and the b_j are continuously differentiable in $\bar{\Omega}$ and that c is continuuus in $\bar{\Omega}$. If the boundary condition is of type (I) we suppose also that γ is continuously differentiable on the boundary Γ. Finally, we assume that A_0 is uniformly elliptic in Ω, that is, that there exists a constant $\varkappa > 0$ such that

$$\sum_{j=1}^{m} \sum_{k=1}^{m} a_{jk}(x)\xi_j\xi_k \geq \varkappa|\xi|^2 \qquad (\xi \in \mathbb{R}^m) \tag{5.2}$$

The domain of the operator $A_p(\beta)$ is the subspace of $W^{1,p}(\Omega)$ consisting of functions that satisfy the boundary condition β at Γ (in the sense of Sobolev's imbedding theorems); here $W^{1,p}(\Omega)$ is the Sobolev space of all functions with partial derivatives of order ≤ 2 in $L^p(\Omega)$. For a proof that the equation

$$(\lambda I - A_p(\beta))u = f \tag{5.3}$$

has a solution $u \in D(A_p(\beta))$ for every $f \in L^p(\Omega)$ see for instance [1]. The dissipativity computation (2.7) becomes

$$\|u\|^{p-2}(\mathrm{Re}\langle\theta(u), A_p(\beta)u\rangle \pm \delta\,\mathrm{Im}\langle\theta(u), A_p(\beta)u\rangle) =$$

$$= \mathrm{Re}\int_\Omega \{\Sigma\Sigma\, D^j(a_{jk}D^k u) + \Sigma b_j D^j u + cu\}|u|^{p-2}\bar{u}dx +$$

$$\pm\,\delta\,\mathrm{Im}\int_\Omega \{\Sigma\Sigma\, D^j(a_{jk}D^k u) + \Sigma b_j D^j u + cu\}|u|^{p-2}\bar{u}dx =$$

$$= \int_{\Gamma} (\gamma+\frac{1}{p}b)|u|^{p} d\sigma - (p-2)\int_{\Omega} |u|^{p-4}\{\Sigma\Sigma a_{jk}\mathrm{Re}(\bar{u}D^{j}u)\mathrm{Re}(\bar{u}D^{k}u)\}dx$$

$$\mp \delta(p-2)\int_{\Omega} |u|^{p-4}\{\Sigma\Sigma a_{jk}\mathrm{Re}(\bar{u}D^{j}u)\mathrm{Im}(\bar{u}D^{k}u)\}dx$$

$$- \int_{\Omega} |u|^{p-2}\{\Sigma\Sigma a_{jk}D^{j}uD^{k}\bar{u}\}dx \pm \delta \int_{\Omega} |u|^{p-2}\{\Sigma b_{j}\mathrm{Im}(\bar{u}D^{j}u)\}dx$$

$$+ \frac{1}{p}\int_{\Omega} \{pc - \Sigma D^{j}b_{j}\}|u|^{p}dx \tag{5.4}$$

We have

$$-(p-2)\int_{\Omega} |u|^{p-4}\{\Sigma\Sigma a_{jk}\mathrm{Re}(\bar{u}D^{j}u)\mathrm{Re}(\bar{u}D^{k}u)\}dx$$

$$\mp \delta(p-2)\int_{\Omega} |u|^{p-4}\{\Sigma\Sigma a_{jk}\mathrm{Re}(\bar{u}D^{j}u)\mathrm{Im}(\bar{u}D^{k}u)\}dx$$

$$- \int_{\Omega} |u|^{p-2}\{\Sigma\Sigma a_{jk}D^{j}uD^{k}\bar{u}\}dx$$

$$= - \int_{\Omega} |u|^{p-4}\{\Sigma\Sigma a_{jk}f_{jk}\}dx \tag{5.5}$$

where

$$f_{jk} = (\bar{u}D^{j}u)(uD^{k}\bar{u}) +$$

$$+ (p-2)\{\mathrm{Re}(\bar{u}D^{j}u)\mathrm{Re}(\bar{u}D^{k}u) \pm \delta\,\mathrm{Re}(\bar{u}D^{j}u)\mathrm{Im}(\bar{u}D^{k}u)\}. \tag{5.6}$$

Let $z_1,\dots,z_m$ be arbitrary complex numbers. Consider the matrix $Z = \{z_{jk}\}$ with elements

$$z_{jk} = z_j\bar{z}_k + \alpha\{(\mathrm{Re}z_j)(\mathrm{Re}z_k) \pm \delta(\mathrm{Re}z_j)(\mathrm{Im}z_k)\}$$

where α, δ are real constants with $\alpha > -1$. If $\zeta_1,\dots,\zeta_m$ are complex we have

$$\Sigma\Sigma z_{jk}\zeta_j\zeta_k = (\overline{\Sigma z_j\zeta_j})(\Sigma z_j\zeta_j) +$$

$$+ \alpha\{\mathrm{Re}(\Sigma z_j\zeta_j)\mathrm{Re}(\Sigma z_k\zeta_k) \pm \mathrm{Re}(\Sigma z_j\zeta_j)\mathrm{Im}(\Sigma z_k\zeta_k)\}$$

which is nonnegative if

$$1 + \delta^2 \le (\frac{\alpha+2}{\alpha})^2 \tag{5.7}$$

in view of (2.8) and following comments; it follows then that Z is positive definite. As in the one dimensional case it is easy to see that if (5.7) is strict then Z will satisfy $\Sigma\Sigma z_{jk}\zeta_j\zeta_k \geq \nu|z|^2|\zeta|^2$ for all complex $\zeta_1,\ldots,\zeta_m$, $z_1,\ldots,z_m$. This is seen to imply that

$$\Sigma\Sigma a_{jk}f_{jk} \geq \nu|u|^2\Sigma|D^ju|^2 \tag{5.8}$$

We estimate next the other terms in (5.4). Consider a real vector field $\rho = (\rho_1,\ldots,\rho_m)$ continuously differentiable in $\bar{\Omega}$ and such that $(\rho,\tilde{\nu}) = \gamma - p^{-1}b$ on Γ, where ν denotes the outer normal vector on Γ. Then $\mathrm{div}(|u|^p\rho) = |u|^p\mathrm{div}\,\rho + p|u|^{p-2}\Sigma\rho_j\mathrm{Re}(\bar{u}D^ju)$. By virtue of the divergence theorem,

$$\begin{aligned}\left|\int_\Gamma (\gamma+\tfrac{1}{p}b)|u|^p d\sigma\right| &= \left|\int_\Gamma (|u|^2\rho,\nu)d\sigma\right| \\ &\leq \int_\Omega |\mathrm{div}\,\rho||u|^p dx + p\int_\Omega |u|^{p-2}(\Sigma|\rho_j||\bar{u}D^ju|)dx \\ &\leq \frac{\varepsilon^2 p}{2}\int_\Omega |u|^{p-2}(\Sigma|\rho_j||D^ju|^2)dx \\ &\quad + \int_\Omega (|\mathrm{div}\,\rho + \frac{p\varepsilon^{-2}}{2}\Sigma|\rho_j|)|u|^p dx\end{aligned} \tag{5.9}$$

where we have used the inequality

$$2|u||D^ju| \leq \varepsilon^2|D^ju|^2 + \varepsilon^{-2}|u|^2 \tag{5.10}$$

This is also used in an obvious way to estimate the fourth volume integral in (5.4). Taking $\varepsilon > 0$ sufficiently small in the resulting inequality and in (5.9) and dividing by $\|u\|^{p-2}$ we obtain an inequality of the form

$$\mathrm{Re}\langle\theta(u),A_p(\beta)u\rangle \leq \pm\delta\,\mathrm{Im}\langle\theta(u),A_p(\beta)u\rangle + \omega\|u\|^2 \tag{5.11}$$

which is then extended to $D(A_p(\beta))$ in the way indicated before.

As in the one dimensional case, (5.10) shows that $\varphi(A_p(\beta)) \geq \varphi_p$, where φ_p is given by (2.17). The proof that $\varphi(A_p(\beta)) \leq \varphi_p$ is essentially similar to that for the case $m = 1$ and is therefore omitted.

6. ELLIPTIC DIFFERENTIAL OPERATORS IN L^1 AND C

The assumptions on the operator A and on the domain Ω are the same as in the previous section. We treat first the case $E = L^1(\Omega)$. Condition (4.2) becomes

$$\gamma(x) + \Sigma b_j(x)\nu_j \leq 0 \qquad (x \in \Gamma) \tag{6.1}$$

and (6.1) is necessary for dissipativity of $A_1(\beta)$ for any A_0. On the other hand, if (6.1) is satisfied then $A_1(\beta) - \omega I$ is m-dissipative for ω large enough. The domain $D(A_1(\beta))$ is defined as the set of all $u \in L^1(\Omega)$ such that there exists f $(= A_1(\beta)u)$ in $L^1(\Omega)$ such that

$$\int_\Omega f v dx = \int_\Omega u A_0' v dx \tag{6.2}$$

where A_0' is the formal adjoint $\Sigma\Sigma D^j(a_{jk}D^k) - \Sigma b_j D^j + c$, and v is an arbitrary smooth function in $\bar{\Omega}$ satisfying the adjoint boundary condition β' (if β is of type (II) then $\beta' = \beta$; if β is $D^{\tilde{\nu}}u = \gamma u$ then β' is $D^{\tilde{\nu}}v = \gamma' v$ with $\gamma' = \gamma + \Sigma\, b_j\nu_j$). For a proof that, for λ large enough,

$$(\lambda I - A_1(\beta))u = f \tag{6.3}$$

has a solution $u \in D(A_1(\beta))$ for any $f \in L^1(\Omega)$ see [2]. The proof that $A_1(\beta)$ is dissipative follows from the computation below (the higher dimensional counterpart of (4.11)) where we show in fact that condition (6.1) can be bypassed through a renorming of the space $L^1(\Omega)$. As in the case $m = 1$ we define

$$\|u\|_{\rho} = \left(\int_{\Omega} |u(x)|^p \rho(x)^p dx\right)^{1/p} \tag{6.4}$$

The space $L^p(\Omega)$ equipped with this norm will be denoted $L^p_\rho(\Omega)$. The identification of the dual space is the same as that in Section 4; we use the same notation for duality maps and for application of functionals to elements of $L^p_\rho(\Omega)$. Assuming that ρ is twice continuously differentiable we obtain

$$\|u\rho\|^{p-2}\mathrm{Re}\langle \theta_\rho(u), A_p(\beta)u\rangle_\rho = \|u\rho\|^{p-2}\mathrm{Re}\langle \rho\theta(\rho u), A_p(\beta)u\rangle$$

$$= \mathrm{Re}\int_{\Omega} (\Sigma\Sigma D^j(a_{jk}D^k u) + \Sigma b_j D^j u + cu)|u|^{p-2}\bar{u}\rho^p dx$$

$$= \int_{\Gamma} (\gamma + \tfrac{1}{p}b)|u|^p\rho^p d\sigma$$

$$- (p-2)\int_{\Omega} |u|^{p-4}\{\Sigma\Sigma a_{jk}\mathrm{Re}(\bar{u}D^j u)\mathrm{Re}(\bar{u}D^k u)\}\rho^p dx$$

$$- \int_{\Omega} |u|^{p-2}\{\Sigma\Sigma a_{jk}D^j u D^k \bar{u}\}\rho^p dx$$

$$- p\int_{\Omega} |u|^{p-2}\{\Sigma\Sigma a_{jk}\mathrm{Re}(\bar{u}D^j u)D^k\rho\}\rho^{p-1} dx$$

$$+ \frac{1}{p}\int_{\Omega} (pc - \Sigma D^j b_j)|u|^p\rho^p dx$$

$$- \int_{\Omega} (\Sigma b_j D^j\rho)|u|^p\rho^{p-1} dx \tag{6.5}$$

We transform now the third volume integral keeping in mind that $p|u|^{p-2}\mathrm{Re}(\bar{u}D^j u) = D^j(|u|^p)$ and using the divergence theorem for the vector $U = (U_j)$ of components

$$U_j = |u|^p\rho^{p-1}\,\Sigma a_{jk}D^k\rho \qquad (1 \le j \le m).$$

Once this is done, the right hand side of (6.5) can be written

$$\int_{\Gamma} \{\gamma - \frac{1}{\rho} D^{\tilde{\nu}}\rho + \frac{1}{p} b\} |u|^p \rho^p d\sigma$$

$$- (p-2)\int_{\Omega} |u|^{p-4}\{\Sigma\Sigma a_{jk} \mathrm{Re}(\bar{u}D^j u)\mathrm{Re}(\bar{u}D^k u)\}\rho^p dx$$

$$- \int_{\Omega} |u|^{p-2}\{\Sigma\Sigma a_{jk} D^j u D^k \bar{u}\}\rho^p dx$$

$$+ \int_{\Omega} \{\Sigma\Sigma D^j(a_{jk}\rho^{p-1}D^k\rho) + (c - \tfrac{1}{p}\Sigma D^j b_j)\rho^p - (\Sigma b_j D^j\rho)\rho^{p-1}\} |u|^p dx \qquad (6.6)$$

We can now choose $D^{\tilde{\nu}}\rho$ at the boundary in such a way that the quantity between curly brackets in the surface integral in (6.6) is nonpositive. Noting that the first two volume integrals combine to yield a nonpositive amount we can bound the right hand side of (6.5) by an expression of the form $\omega\|u\|_\rho$ where ω does not depend on p. Using a limiting argument as in the one-dimensional case we obtain

$$\mathrm{Re}\langle u^*, A_1(\beta)u\rangle_\rho \leq \omega\|u\|_\rho^2 \qquad (6.7)$$

where the expression between brackets indicates application of the functional $u^* \in \Theta(u) \subset L^1_\rho(\Omega)$ to $A_1(\beta)u \in L^1_\rho(\Omega)$ and $\|\cdot\|_\rho$ is the norm of $L^1_\rho(\Omega)$. The customary approximation argument shows that (6.7) holds in $D(A_1(\beta))$. Renorming of the space $L^1(\Omega)$ is of course unnecessary when β is the Dirichlet boundary condition.

The case $E = C(\bar{\Omega})$ ($C_\Gamma(\bar{\Omega})$ if the Dirichlet boundary condition is used) is handled in a similar way. Condition (4.5) is now

$$\gamma(x) \leq 0 \qquad (x \in \Gamma) \qquad (6.8)$$

which condition is necessary for dissipativity of $A(\beta)$ for any A_0. If (6.8) is satisfied then $A(\beta) - \omega I$ is m-dissipative in $C(\bar{\Omega})$ ($C_\Gamma(\bar{\Omega})$) if ω is large enough. The domain $D(A(\beta))$ is defined as the set of all u that belong to $W^{2,p}(\Omega)$ for every

$p \geq 1$, satisfy the boundary condition β on Γ and are such that $A_0 u \in C(\bar{\Omega})$ $(C_\Gamma(\bar{\Omega}))$. The fact that, for λ large enough,

$$(\lambda I - A(\beta))u = f \tag{6.9}$$

has a solution $u \in D(A(\beta))$ for arbitrary $f \in C(\bar{\Omega})$ $(C_\Gamma(\bar{\Omega}))$ is a particular case of results in [5]. The proof that $A(\beta)$ is dissipative if (6.8) holds corresponds to the particular case $\rho = 1$ in the computation below, which shows that even if (6.8) does not hold, $A(\beta) - \omega I$ will be dissipative for ω large enough after the replacement of the norm of $C(\bar{\Omega})$ by the equivalent norm

$$\|u\|_\rho = \max_{x \in \Omega} |u(x)|\rho(x) \tag{6.10}$$

where ρ is a continuous positive function in $\bar{\Omega}$; the identification of the duals is achieved along the lines of Section 4. If β is of type (I) and condition (6.8) is not satisfied we select ρ twice continuously differentiable and such that

$$D^{\tilde{\nu}}\rho(x) + \gamma(x)\rho(x) \leq 0 \qquad (x \in \Gamma). \tag{6.11}$$

Let u be a smooth function satisfying the boundary condition β on γ and $m_\rho(u) = \{x; |u(x)|\rho(x) = \|u\|_\rho\}$ so that any $\mu \in \Theta_\rho(u)$ is supported by $m_\rho(u)$ and is such that $u\rho\mu$ (or $u\mu$) is a positive measure in $m_\rho(u)$. Since $D^{\tilde{\nu}}u(x) = \gamma(x)u(x)$ at the boundary, $u\rho$ satisfies the boundary condition

$$D^{\tilde{\nu}}(u(x)\rho(x)) = \gamma_\rho(x)(u(x)\rho(x)) \qquad (x \in \Gamma) \tag{6.12}$$

where

$$\gamma_\rho(x) = \gamma(x) + \frac{1}{\rho(x)} D^{\tilde{\nu}}\rho(x) \leq 0 \qquad (x \in \Gamma) \tag{6.13}$$

The argument employed in Secion 4 shows that $m_\rho(u)$ does not meet the boundary Γ if $\gamma(x) < 0$ everywhere on Γ so that

$$D^j|u\rho|^2(x) = 0, \qquad -\mathfrak{H}(x;|u\rho|^2) \geq 0 \qquad (x \in m_\rho(u)) \tag{6.14}$$

where $\mathcal{H}(x;g)$ indicates the Hessian matrix of g at x. On the other hand, if $\gamma(x) = 0$ for some $x \in \Gamma$, $m_\rho(u)$ may contain points of Γ but we can prove in the same way that (6.14) holds. Writing $\eta = \rho^2$ and $u = u_1 + iu_2$ with u_1, u_2 real we obtain $D^j|u\rho|^2 = 2(u_1D^ju_1 + u_2D^ju_2)\eta + (u_1^2+u_2^2)D^j\eta$, $D^jD^k|u\rho|^2 =$
$= 2(u_1D^jD^ku_1 + u_2D^jD^ku_2)\eta + 2(D^ju_1D^ku_1 + D^ju_2D^ku_2)\eta +$
$+ 2(u_1D^ju_1 + u_2D^ju_2)D^k\eta + 2(u_1D^ku_1 + u_2D^ku_2)D^j\eta + (u_1^2 + u_2^2)D^jD^k\eta$.
Accordingly, it follows from (6.14) that if $x \in m_\rho(u)$,

$$\begin{aligned} \mathrm{Re}(u^{-1}D^ju) &= \|u\|^{-2}(u_1D^ju_1 + u_2D^ju_2) \\ &= -\tfrac{1}{2}\|u\|^{-2}|u|^2\eta^{-1}D^j\eta = -\tfrac{1}{2}\eta^{-1}D^j\eta. \end{aligned}$$

On the other hand, again with $x \in m_\rho(u)$, we have

$$\begin{aligned} &\mathrm{Re}(u^{-1}D^jD^ku) = \\ &= \|u\|^{-2}(u_1D^jD^ku_1 + u_2D^jD^ku_2) \\ &= \tfrac{1}{2}\|u\|^{-2}\eta^{-1}D^jD^k|u\rho|^2 - \|u\|^{-2}(D^ju_1D^ku_1+D^ju_2D^ku_2)+\eta^{-2}D^j\eta D^k\eta-\tfrac{1}{2}\eta^{-1}D^jD^k\eta. \end{aligned}$$

Accordingly, if $\mu \in \Theta_\rho(u)$,

$$\mathrm{Re}\langle\mu, A(\beta)\rangle_\rho = \int_{m_\rho(u)} \mathrm{Re}(u^{-1}A(\beta)u)\rho u \, d\mu \le \omega\|u\|_\rho \tag{6.15}$$

for some constant, showing (after the usual approximation argument) that $A(\beta) - \omega I$ is dissipative.

Theorem 4.3 has an obvious analogue here: when (6.1) holds, $A_1(\beta) - \omega I$ is m-dissipative for ω large enough but $\varphi(A_1(\beta)) = 0$. A similar observation holds for $A(\beta)$.

We point out finally that a "multiplicative" renorming of the spaces $L^p(\Omega)$ like that used in $L^1(\Omega)$ does not change the angle of dissipativity of $A_p(\beta)$.

REFERENCES

1. S. AGMON, On the eigenfunctions and the eigenvalues of general elliptic boundary value problems, Comm. Pure Appl. Math. 15 (1962), 119-142.

2. H. BRÉZIS and W.A. STRAUSS, Semi-linear second-order elliptic equations in L^1, J. Math. Soc. Japan 25 (1973), 565-590.

3. E.A. CODDINGTON and N. LEVINSON, Theory of Ordinary Differential Equations, McGraw-Hill, New York, 1955.

4. A. PAZY, Semi-groups of Linear Operators and Applications to Partial Differential Equations, Univ. of Maryland Lecture Notes #10, College Park, 1974.

5. B. STEWART, Generation of analytic semigroups by strongly elliptic operators, Trans. Amer. Math. Soc. 199 (1974), 141-162.

Departments of Mathematics and System Science
University of California
Los Angeles, California 90024

Functional Analysis, Holomorphy and
Approximation Theory II, G.I. Zapata (ed.)

TWO EQUIVALENT DEFINITIONS OF THE DENSITY NUMBERS FOR A PLURISUBHARMONIC FUNCTION IN A TOPOLOGICAL VECTOR SPACE

Pierre Lelong

1. INTRODUCTION

In the following E will be a vector space over the complex field $\mathbb{C}$; in a domain $G \subset E$, we denote by $P(G)$ the class of plurisubharmonic functions f. The topology is not supposed at the beginning to be locally convex; Φ denotes a basis of disked neighborhoods of the origin O in E (we recall that W is disked if and only if $\lambda W \subset W$ for $\lambda \in C$, $|\lambda| \leq 1$). We may suppose the topology of E to be Hausdorff. If it is not, we pass to $E_1 = E/N$ for $N = \bar{0}$. By the upper semicontinuity of $f \in P(G)$ there exists an open neighborhood W of the origin such $f(x') < f(x) + \varepsilon$, $\varepsilon > 0$, for $x' \in x + W$. Then by $W + N = W$, we have $f(x+N) < f(x) + \varepsilon$ and $f(x')$ is upper bounded for $x' \in x + N$; therefore $f(x') = f(x)$ for $x' - x \in N$. Using the projection $\pi: E \to E_1$, we write $f(x) = f_1 \circ \pi(x)$ and f_1 is a plurisubharmonic function defined in the domain $G_1 = \pi(G)$; the given function f reduces to $f_1 \in P(G_1)$ defined in the Hausdorff space E_1.

We denote by D_o the compact disk $|u| \leq 1$ in $\mathbb{C}$ and by $D_{x,y} = x + D_o y$ its linear image in E. A set $A \subset G$ is called <u>pluripolar in</u> G if there exists $f \in P(G)$ such

$$A \subset A' = [x \in G \ ; \ f(x) = -\infty]. \tag{1}$$

The set A is called a cone with vertex the origin if and only if

$\lambda A = A$ for all $\lambda \in C$, $\lambda \neq 0$. A cone is called pluripolar in E if there exists a neighborhood U of its vertex such $A \cap U$ is pluripolar in U (see [4,b]).

For a plurisubharmonic function f which is supposed to be defined in $x + W$, and $W \in \Phi$, we define in E the set

(2) $\quad g_x = [y \in E;$ there exists $r > 0$ such $f = -\infty$ on $D_{x,ry}]$.

Equivalently: g_x is the union of the complex lines through x which contain a disk $D_{x,y}$, $y \neq 0$ which is in the pluripolar set $\tilde{\eta} = [x \in G;\ f(x) = -\infty]$; if such lines do not exist, and if $f(x) = -\infty$, g_x is reduced to the origin of E; it is empty if $f(x) \neq -\infty$; we have by translation $[g_x \cap W] \subset (\eta - x) \cap W$; the set $\tilde{\eta}$, and its translated $\tilde{\eta} - x$ are pluripolar, then g_x is a pluripolar cone.

In C^n the density number $\nu(x,f)$ of a plurisubharmonic function f in x appears to be like a multiplicity and characterizes the concentration of positive laplacian measure Δf near the point x; $\nu(x,f)$, which is called the Lelong number (see [3] and [6]) is an invariant of the one to one holomorphic mappings (see [6]) and it has a geometrical (for complex analytic geometry) meaning. I recall here for the convenience of the reader three (equivalent) definitions of $\nu(x,f)$.

(I) $\nu(x,f)$ is the Lelong number of the closed and positive current $\frac{i}{\pi}\partial\bar{\partial} f = \frac{1}{2\pi} dd^c f$

(II) $\nu(x,f)$ is the regular density (in real dimension 2n-2) of the positive measure $\sigma = \frac{1}{2\pi}\Delta f$, defined by the quotient

$$\nu(x,f) = \lim_{r=0} [\tau_{2n-2}(r)]^{-1} \sigma(x,r)$$

$\sigma(x,r)$ is the mass of σ in the ball $B(x,r)$ and $\tau_m(r)$ is the volume of $B(0,r)$ in R^m.

(III) A direct calculation of $\nu(x,f)$ is possible using the mean value $\lambda(x,r,f)$ of f on the sphere $S(x,r)$, for $0 < r < 1$:

$$\nu(x,f) = \lim_{r=0} (\log r)^{-1} \lambda(x,r,f).$$

None of the definitions (I), (II), (III) is available if f is defined in an infinite dimensional space E. I succeed in [4,a] in giving a suitable definition using the property of the traces $f|_L$ of f on the finite dimensional affine subspaces L through x, specially the complex lines

$$(4) \qquad L = L_{x,y} = [z \in E;\ z = x+uy,\ u \in \mathbb{C}].$$

If $f|_L$ is not the constant $-\infty$ (or, equivalently $y \notin g_x$), then the restriction $f|_L$ is subharmonic and for $f|_L$ with exception of a "small" set for the direction y of L, we get a value independent of y given by $\inf_L \nu(x,f|_L)$. More precise we define <u>the tangential density number of</u> f <u>in</u> x <u>for the direction</u> y:

$$(5) \qquad \nu(x,y,f) = \lim_{r=0} (\log r)^{-1} \frac{1}{2\pi} \int_0^{2\pi} f(x+re^{i\theta}y)d\theta.$$

By definition, $\nu(x,y,f)$ is the density number of the subharmonic function $\varphi(u) = f(x+uy)$ in $u = 0$. Then the definition given in [4,a] was:

DEFINITION 1. Given a function f plurisubharmonic in a domain $G \subset E$, the density number in $x \in G$ is defined by:

$$(6) \qquad \nu(x,f) = \inf_y \nu(x,y,f) \quad \text{for } y \in E-\{0\}$$

and $\nu(x,y,f)$ has the value (5).

I give here a different definition. It aroses from a question of C. Kiselman (see [3] and [4,c]) on the limit in normed spaces of $(\log r)^{-1} M(x,r,f)$ for $r \searrow 0$ and

$$M(x,r,f) = \sup_{x'} f(x+x') \quad \text{for} \quad \|x'\| \leq r$$

for $f \in P(E)$ and E being a Banach space (see later Theorem 2).

DEFINITION 2. Given a domain G in a complex topological space E, and a disked neighborhood W of 0, and $f \in P(G)$; we suppose f is upper bounded in $x+W \subset G$. Then we define for $0 < r < 1$

(7) $$M_W(x,r,f) = \sup f(x+x') \quad \text{for} \quad x' \in rW$$

and

(8) $$\nu_W(x,f) = \lim_{r=0} (\log r)^{-1} M_W(x,r,f).$$

We prove in this paper

$$\nu(x,f) = \nu_W(x,f).$$

A consequence is: if we denote by $\Phi'_x \subset \Phi$ the family of the disked neighborhoods of 0 such f is upper bounded in $x + W$, then $\nu_W(x,f)$ does not depend of $W \in \Phi'_x$.

More precise results will be given: for a disked set $A \subset W$, and $W \in \Phi'_x$ we consider the cone $\gamma(A) = \bigcup_\lambda \lambda A$, and define for given A and $W \in \Phi'_x$:

$$M_A(x,r,f) = \sup f(x+uy) \quad \text{for} \quad |u| \leq r, \quad y \in \gamma(A) \cap W.$$

It is possible to use $M_A(x,r,f)$ instead $M_W(x,r,f)$ in (7) if $\gamma(A)$ is not "too small"? The definition of the "smallness" to obtain $\nu(x,f)$ will depend of the space E. If E is a Fréchet space, it is sufficient for $\gamma(A)$ to be a not pluripolar cone; then the restriction for $y \in \gamma(A)$ and $u \searrow 0$ of $f(x+uy)$ enables to calculate $\nu(x,y,f)$ for $y \in \gamma(A)$, and then we write $\nu(x,f) = \inf_y \nu(x,y,f)$ for $y \in \gamma(A)$, if $\gamma(A)$ is a cone which is not pluripolar in E.

2. THE TANGENTIAL DENSITY NUMBERS

Given f plurisubharmonic in a domain $G \subset E$ and a disk $D_{x,y} = x + D_o y \subset G$, we define

$$\ell(x,y,r,f) = \frac{1}{2\pi} \int_0^{2\pi} f(x+re^{i\theta}y)d\theta$$

$$m(x,y,r,f) = \sup f(x+uy) \quad \text{for} \quad |u| \leq r .$$

By a classical result, the functions $(x,y) \to \ell$ and $(x,y) \to m$ are defined locally and are plurisubharmonic functions of (x,y) if the compact disk $D_{x,y}$ is moving in G and r is fixed $0 \leq r < 1$. Then for $D_{x,y} \subset G$ we define two tangential density numbers

$$\text{(9)} \qquad \begin{aligned} \nu(x,y,f) &= \lim_{r=0} (\log r)^{-1} \ell(x,y,r,f) \\ \nu_1(x,y,f) &= \lim_{r=0} (\log r)^{-1} m(x,y,r,f). \end{aligned}$$

If $f(x) \neq -\infty$, we obtain by (9): $\nu(x,y,f) = \nu(x,y,f) = 0$ for all $y \in E$, as an obvious consequence of

$$f(x) \leq \ell(x,y,r,f) \leq m(x,y,r,f).$$

It is convenient to change the signs in order to calculate ν and ν_1 by increasing limits of plurisubharmonic and negative functions. For given $x \in G$, let us consider the family $\Phi'_x \subset \Phi_x$ of the disked neighborhoods W of 0 such $x+W \subset G$ and $f(x+x')$ is bounded above for $x' \in W$; $\Phi'_x \neq \emptyset$ by the upper semicontinuity of f. For $W \in \Phi'_x$ and $f(x') \leq a$ for $x' - x \in W$, we define

$$\text{(10)} \qquad \begin{aligned} -\nu(x,y,r,f) &= \lim_{r=0} \left(\log \frac{1}{r}\right)^{-1} \ell(x,y,r,f-a) = \lim_{r=0} \psi(x,y,r) \\ -\nu_1(x,y,r,f) &= \lim_{r=0} \left(\log \frac{1}{r}\right)^{-1} m(x,y,r,f-a) = \lim_{r=0} \psi_1(x,y,r). \end{aligned}$$

Then we obtain for $0 < r < 1$:

PROPOSITION 1. Given a domain $G \subset E$, $\xi \in G$, and f plurisubharmonic in G, there exist two disked neighborhoods W and W' of 0 such $-\nu$ and $-\nu_1$ are defined by (10) for $x - \xi \in W$, $y \in W'$ as limits of the increasing functions $\psi(x,y,r)$, $\psi_1(x,y,r)$ of r, for $r \searrow 0$. For fixed r, ψ and ψ_1 are plurisubharmonic functions of (x,y) for $x - \xi \in W$, $y \in W'$ and are negative. Moreover for fixed x, there exists a continuation of $\nu(x,y,f)$ for $y \in E$ which satisfies $\nu(x,\lambda y,f) = \nu(x,y,f)$ for $\lambda \in \mathbb{C}$, $\lambda \neq 0$. The same properties hold for $\nu_1(x,y,f)$.

For the proof, we take $W \in \Phi$, $W' \in \Phi$ such $W + W' \in \Phi'_\xi$ and denote by a the upper bound of f in $\xi + W + W'$. Then in (10), for $0 < r < 1$, we obtain ψ and ψ_1 as negative plurisubharmonic functions of (x,y) for $x - \xi \in W$, $y \in W'$.
The graph of the function

$$v = \log r \to \ell(x,y,r,f-a) = v'(v)$$

is increasing and convex for $-\infty < v \leq 0$. In $R^2(v,v')$ the set defined by $v' \geq \ell(v)$, $-\infty \leq v \leq 0$, is a convex set and contains the origin $v = v' = 0$. This has for consequence that $\psi(x,y,r)$ increases as v is decreasing and tends to $-\infty$. Then (10) defines $-\nu(x,y,f)$ by an increasing limit of negative plurisubharmonic functions $\psi(x,y,r)$ in $x - \xi \in W$, $y \in W'$. Moreover for $\lambda \neq 0$, $\lambda \in \mathbb{C}$,

$$\ell(x,\lambda y,r,f) = \ell(x,y,|\lambda|^{-1}r,f). \tag{11}$$

Writing $r' = |\lambda|^{-1}r$, we obtain by (11):

$$\begin{aligned} -\nu(x,\lambda y,f) &= \lim_{r=0} \left(\log \frac{1}{r}\right)^{-1} \ell(x,y,|\lambda|^{-1}r,f-a) = \\ &= \lim_{r'=0} \left(\log \frac{1}{r'}\right)^{-1} \ell(x,y,r',f-a) \end{aligned} \tag{12}$$

and $\nu(x,\lambda y,f) = \nu(x,y,f)$ is proved for $\lambda \in \mathbb{C}$, $\lambda \neq 0$ and $y \in W'$. Now for $m > 1$, $m \in \mathbb{N}$ the definition of $\nu(x,y,f)$ for $y \in mW'$, is given by $\nu(x,y,f) = \nu(x,m^{-1}y,f)$. It is obvious that (10) remains true for $y \in mW'$ and $0 < r < \frac{1}{m}$.

Then $\nu(x,y,f)$ is defined locally for $x \in G$, $y \in \bigcup_m mW' = E$. It is given as a limit of an increasing sequence of negative plurisubharmonic functions and $\nu(x,\lambda y,f) = \nu(x,y,f)$ for $\lambda \in \mathbb{C}$, $\lambda \neq 0$. The same holds for $\nu_1(x,y,f)$.

REMARK 1. For $y \in g_x$, we obtain $-\nu(x,y,f) = -\nu_1(x,y,f) = -\infty$. For $y = 0$, and $x \in \tilde{\eta} = [x \in G, f(x) = -\infty]$, we have $-\nu(x,0,f) = -\nu_1(x,0,f) = -\infty$. For $y = 0$, $x \notin \tilde{\eta}$, we obtain $-\nu(x,0,f) = -\nu_1(x,0,f) = 0$.

3. EQUALITY OF THE TWO TANGENTIAL NUMBERS

Now we prove

PROPOSITION 2. Given a plurisubharmonic function f in a domain $G \subset E$, for all $x \in G$, $y \in E$, the equality, for the definition, see [10]:

$$(14) \qquad \nu(x,y,f) = \nu_1(x,y,f)$$

holds. If $y \neq 0$, the value (14) will be called the tangential density number $\nu(x,y,f)$ of f in $x \in G$ for the direction y.

By the preceding remarks, (14) is proved for $y = 0$ and for $y \in g_x$. We have only to prove (14) for $x \in G$, $y \in W$, and $W \in \Phi'_x$ and $\varphi(u) = f(x+uy)$ being a subharmonic function of $u \in \mathbb{C}$, for $|u| < 1$. We put

$$(15) \qquad \ell(r) = \frac{1}{2\pi}\int_0^{2\pi} \varphi(re^{i\alpha})d\alpha; \quad m(r) = \sup_\alpha \varphi(re^{i\alpha}).$$

It is sufficient to prove that the quotient by $|\log r|$ of the difference $m(r) - \ell(r)$ tends to zero for $r \searrow 0$; such a result was given in $\mathbb{C}^n$ by V. Avanissian [1].

For $n = 1$ a direct calculation gives a more precise result.

LEMMA 1. (I) Given a function $\varphi(u)$, subharmonic for $|u| < 1$, $u \in \mathbb{C}$, and R, $0 < R < 1$, then for r in the interval $0 < r < R$, and $\ell(r)$ and $m(r)$ defined by (15):

$$0 \leq m(r) - \ell(r) \leq \mu(R) \log 2 + \frac{2Cr}{R-r} \tag{16}$$

$\mu(R)$ is the mass of $\mu = \frac{1}{2\pi}\Delta\varphi$ in the disk $|u| \leq R$ and $C = \frac{1}{2\pi}\int_0^{2\pi} |\varphi(Re^{i\alpha})|\,d\alpha$.

(II) Moreover for $r \searrow 0$:

$$\lim_{r=0} [m(r) - \ell(r)] = 0. \tag{17}$$

The proof of (16) uses the classical Riesz decomposition of φ in $|u| < R$:

$$\varphi(u) = H_R(u) + \int_{|a|<R} d\mu(a) \log |u-a|$$

$H_R(u)$ is harmonic for $|u| < R$; the mean value ℓ for H_R is equal to $H_R(0)$. Then for $0 < r < R$:

$$0 \leq m(r) - \ell(r) \leq \sup H_R(re^{i\theta}) - H_R(0) +$$

$$+ \int_0^R d\mu(t)\log(r+t) - \int_0^R d\mu(t)\sup(\log r, \log t).$$

For the first difference we have the bound

$$|H_R(re^{i\theta}) - H_R(0)| \leq \frac{2r}{R-r}\,\frac{1}{2\pi}\int_0^{2\pi} |\varphi(Re^{i\theta})|\,d\theta = \frac{2Cr}{R-r}\,.$$

Otherwise the two integrals in (17) are bounded by

$$\int_0^r d\mu(t) \log \frac{r+t}{2} + \int_r^R d\mu(t) \log \frac{r+t}{t}\,.$$

An upper bound of the integral in (18) is given by $\mu(R)\log(1+\frac{r}{r_o})$.
We obtain for given $\varepsilon > 0$ and $0 < r < r_o < R$

$$(19) \qquad 0 \leq m(r) - \ell(r) \leq \varepsilon \log 2 + r\left[\frac{2C}{R-r_o} + \frac{\mu(R)}{r_o}\right]$$

which proves (17)

Then by (10) and (16) the Proposition 2 is proved for $x \in G$, $y \in E$.

4. EQUIVALENCE OF THE DEFINITIONS 1 AND 2.

First we give properties of the function $y \in E \to \nu(x,y,f)$, it is of the tangential numbers ν and ν_1 for fixed $x \in G$, and y variable in E.

For a real valued function $h(y)$ we denote

$$h_x(y) = \liminf_{y' \to y} h(y') \quad \text{and} \quad h^*(y) = \limsup_{y' \to y} h(y')$$

the two regularizations of h.

THEOREM 1. (I) The lower regularization of $\nu(x,y,f)$ it is $\liminf_{y' \to y} \nu(x,y',f)$ has a finite positive value which does not depend of $y \in E$. Let us denote $\nu(x,f)$ its value. Then

$$(20) \qquad \nu(x,y,f) \geq \nu(x,f) \quad \text{for all } y \in E$$

and

$$(21) \qquad \nu(x,f) = \inf_{y \in \omega} \nu(x,y,f) \quad \text{if } \mathring{\omega} \neq \emptyset.$$

(II) For given x, the set $[y \in E,\ \nu(x,y,f) > \nu(x,f)]$ is a cone g'_x of vertex 0; $g'_x = \emptyset$ if and only if $f(x) \neq -\infty$; g'_x contains the cone g_x.
The cone g'_x is a pluripolar cone in E or $g'_x = E$.

(III) If E is a Fréchet space or if E is a Baire space and

f is a continuous plurisubharmonic function (it is e^f is continuous) then $g'_x \neq E$ is a pluripolar cone in E.

Let us write $h(y) = -\nu(x,y,f)$ for fixed x; by the Proposition 1, $h(\lambda y) = h(y)$ for $\lambda \in \mathbb{C}$, $\lambda \neq 0$ and $h(y) \leq 0$. Then $h^*(y)$ is a plurisubharmonic function in E because it is the upper regularization of the upper envelope in (10) of $\psi(x,y,r) \leq 0$, which is a plurisubharmonic functions of y. By $h^*(y) \leq 0$, h^* is upper bounded in E, and then, it is a constant h^* for $y \in E$ and fixed x. We have $h(y) \leq h^* = \nu(x,f)$, which proves (20); (21) is a consequence of the definition of the regularization. If $f(x) \neq -\infty$, we have $\nu(x,y,f) = 0$ for all y; and therefore $\nu(x,f) = 0$. The set $[x \in G;\ \nu(x,f) > 0]$ is contained in the set defined in G by $f(x) = -\infty$. For each $x \in G$, $\nu(x,f) \geq 0$ has a finite value; there exists $y \neq 0$ such $f(x+uy) \not\equiv -\infty$ for $|u| < 1$; then $\nu(x,y,f) < \infty$ and $\nu(x,f) = \inf_y \nu(x,y,f)$ has a finite value.

To prove (II): by $\nu(x,y,\lambda f) = \nu(x,y,f)$ for all $\lambda \in C$, $\lambda \neq 0$, $g'_x = [y \in E;\ \nu(x,y,f) > \nu(x,f)]$ is a cone of vertex 0. By the Remark 1, we have seen: $0 \in g'_x$ if $f(x) = -\infty$ and $g'_x = \emptyset$ if $f(x) \neq -\infty$. For $y \in g_x$, and $y \neq 0$, there exists a disk $D_{x,y}$ in which f has the constant value $-\infty$. Therefore $\nu(x,y,f) = +\infty$, and $\nu(x,y,f) > \nu(x,f)$; then $y \in g'_x$, and $g_x \subset g'_x$ is proved.

To end the proof of (II) and (III), we write for $r_n \to 0$:

$$s_n(y) = -\left(\log \frac{1}{r_n}\right)^{-1} \ell(x,y,r_n,f) + \nu(x,f).$$

By (I), we have $s_n(y) \in P(E)$ and $s_n(y) \leq 0$ and

$$s(y) = \sup_n s_n(y) \leq 0.$$

By the definition of $\nu(x,f)$, we have in E

$$s^*(y) \equiv 0.$$

Now we have two possible situations:

a) Suppose there exists y_o such $s(y_o) = \sup s_n(y_o) = 0$. Then by $s_n(y_o) \leq 0$, passing to a subsequence if necessary, we may suppose $\sum_n |s_n(y_o)| < \infty$. Then $V_p(y) = \sum_1^p s_n(y) < 0$ is a decreasing sequence of plurisubharmonic with finite limit in y_o. Then $V(y) = \sum_n s_n(y)$ is plurisubharmonic in G. For $y \in g'_x$, we have $s(y) < 0$, and $V(y) = -\infty$. As a consequence

$$g'_x = [y \in E;\ V(y) = -\infty]$$

is a pluripolar cone in E.

If y_o does not exist, then $s(y) < s^*(y)$ everywhere in E and $g'_x = E$, which makes the proof of (II) complete.

To obtain (III), first if E is a Fréchet space, we use a result of Coeuré [2]: in each neighborhood of $\xi \in G$, there exist compact disked and convex sets P (actually polycylinders of center ξ) and a probability measure μ of support in P such

$$s^*(\xi) \leq \int d\mu(a) s(\xi + a).$$

Then by $s^*(\xi) = 0$, and $s_n(\xi + a) \leq 0$, we obtain $s(y) = 0$ μ-everywhere on $\operatorname{supp} \mu \subset P$. As a consequence: <u>if</u> E <u>is a Fréchet space, and</u> s^* <u>is a constant, the set</u> $s(y) = s^*(y)$ <u>is dense in</u> E. Coming back to $\nu(x,y,f)$, we have $\nu(x,y,f) = \nu(x,f)$ in a dense set in E for y and, by (II), the cone g'_x is pluripolar in E.

Now, suppose E is a Baire space and f is continuous with values in $\underline{R}$. The sets

$$e_{n,q} = [y \in E;\ s_n(y) \leq -\tfrac{1}{q}]$$

are closed. The same property holds for $e_q = \bigcap_n e_{n,q}$. By (22), $\overset{\circ}{e}_q = \emptyset$ and as a consequence, $g'_x = \bigcup_q e_q$ is meager in E. Then $g'_x \neq E$; $y_o \in E$, $y_o \notin g'_x$ exists and g'_x is a meager and pluripolar cone in E. Now we prove:

THEOREM 2. In a complex topological vector space E the two density numbers given by the Definitions 1 and 2 coincide for a function f which is plurisubharmonic; it is for f upper bounded in a neighborhood $x+W$ of x:

$$\nu(x,f) = \nu_W(x,f).$$

For the proof, we write for $0 < r < 1$:

$$\nu_W(x,f) = \lim_{r=0} (\log r)^{-1} M_W(x,r,f)$$

$$M_W(x,r,f) = \sup_y m(x,y,r,f) \quad \text{for} \quad y \in W. \tag{22}$$

If a is an upper bound of f in $x+W$, by the convexity of the graph of $\log r \to m(x,y,r,f) = \sup_\theta f(x+re^{i\theta}y)$, we obtain:

$$m(x,y,r,f) \leq \nu(x,y) \log r + a.$$

By (22), and for $0 < r < 1$ (and $\log r < 0$):

$$\begin{aligned} M_W(x,r,f) &\leq \sup_y [\nu(x,y,f) \log r] + a \\ &\leq [\inf_y \nu(x,y,f)] \log r + a \end{aligned}$$

$$M_W(x,r,f) \leq \nu(x,f) \log r + a. \tag{23}$$

Then

$$\nu_W(x,f) = \lim_{r=0} (\log r)^{-1} M_W(x,r,f) \geq \nu(x,f). \tag{24}$$

Conversely, for $0 < r < 1$, $y \in W$, $W \in \Phi'_x$, (24) gives:

$$m(x,y,r,f) \leq M_W(x,r,f).$$

Dividing by $\log r < 0$, and taking the limit for $r \searrow 0$:

$$\nu_1(x,y,f) \geq \nu_W(x,f)$$

$$\nu(x,f) = \inf_y \nu(x,y,f) = \inf_y \nu_1(x,y,f) \geq \nu_W(x,f). \tag{25}$$

The comparison of (24) and (25) makes the proof complete.

5. CONSEQUENCES AND EXAMPLES.

First we give applications to the calculus of $\nu(x,f)$.

COROLLARY 1. If f is plurisubharmonic in a domain G of a complex topological vector space E, $\nu(x,f)$ can be calculated in $x \in G$ by

$$\nu(x,f) = \inf_{y \in \omega} \nu(x,y,f) \tag{26}$$

if the cone $\gamma_x = \bigcup_{\lambda} \lambda\omega$ of vertex 0 in E has a not empty interior.

As consequence: if W is a disked neighborhood of 0, and $x+W \subset G$, to calculate $\nu(x,f)$, we have to use only the values of $f(x+uy)$ for $y \in \gamma_x$, $|u| \searrow 0$, if $\mathring{\gamma}_x$ is supposed to be not empty; the condition is satisfied if γ is absorbing for an open set.

As a consequence of the Theorem 1 we state:

COROLLARY 2. If E is a Fréchet space or if E is a Baire space and $f \in P(G)$ is continuous, $\nu(x,f)$ is given by (26) if γ_x is supposed not to be a pluripolar cone in E.

COROLLARY 3. By the same hypothesis on E, there exists for each integer $p \geq 1$ an affine space L^p through x, of finite dimension p (for $p=1$ we obtain a complex line) such $\nu(x,f) = \nu(x,f|_{L^p})$.

PROOF. For $p = 1$ take the line L^1 not in the pluripolar cone g'_x. For $p > 1$, take L^p containing a line $L^1 \not\subset g'_x$ through x, then $\nu(x,f|_{L^1}) = \nu(x,f)$ and for $y_o \in L^1$, $y_o \neq 0$, we obtain $\nu(x,f) = \nu(x,f|_{L^1}) = \nu(x,y_o) \geq \nu(x,f|_{L^p}) \geq \nu(x,f)$ and $\nu(x,f|_{L^p}) = \nu(x,f)$.

COROLLARY 5. If E is a Banach space and f a plurisubharmonic function with finite upper bound $\|f\|_R$ in the ball $\|z\| < R$, then in the ball $\|z\| \leq r$, $r < R$, we have the bound:

$$\|f\|_r \leq \|f\|_R - \nu(0,f) \log \frac{R}{r}$$

$$\nu(0,f) = \inf_y \nu(0,y,f) \quad \text{for} \quad y \in \gamma \tag{27}$$

and (27) holds for all cones γ which are not pluripolar and meager in E, if f is supposed to be continuous; in (27) the number $\nu(x,y,f)$ can be calculated using $\ell(0,y,r,f)$ or $m(0,y,r,f)$ for $y \in \gamma$.

REMARK 2. If E is a Baire space, and $f = \log |F|$, for F holomorphic in $x+W$, $W \in \Phi$, $\nu(0,y,f)$ is the multiplicity of the zero $u = 0$ of $F(uy)$ for fixed y. The condition for γ to be not pluripolar and meager can be replaced by the condition that γ <u>is not contained in an algebraic cone defined by one equation it is</u> $[y \in E;\ P_\nu(y) = 0]$; P_ν is an homogeneous polynomial of y of degree $\nu = \nu(x,f)$; it is the first term of the Taylor series of F at the origin; a more precise result is available with a much more precise hypothesis.

In C^n, $\nu(x,f)$ is an upper semi-continuous function of x, and the sets $N(c,f) = [x \in G;\ \nu(x,f) \geq c]$, for $c > 0$, are closed (by a theorem of Y.T. Siu, cf. [6], $N(c,f)$ is proved to be an analytic set). We prove:

THEOREM 3. If E is a locally convex space, the density number $\nu(x,f)$ of a plurisubharmonic function f is an upper semi-continuous function of x.

We suppose the topology is given by a family $\{p_\alpha\}$ of semi-norms; B_α is the unit ball $p_\alpha(x) < 1$. We suppose f is plurisubharmonic and upper bounded in $x + B_\alpha$ and write for $0 < r < 1$:

$$M_\alpha(x,r,f) = \sup_y f(x+y) \quad \text{for} \quad p_\alpha(y) < r.$$

The function $\log r \to M_\alpha$ is increasing, convex; it is continuous. Writing $B_\alpha(x,r) = [x' \in E;\ p_\alpha(x'-x) < r]$, we have for $p_\alpha(x'-x) < \eta$, and $0 < \eta < r$:

$$B_\alpha(x,r-\eta) \subset B_\alpha(x',r) \subset B_\alpha(x,r+\eta)$$

(28) $$M_\alpha(x,r-\eta,f) \leq M_\alpha(x',r,f) \leq M_\alpha(x,r+\eta,f).$$

For $\varepsilon > 0$, there exists η such $0 \leq M_\alpha(x,r+\eta,f) - M_\alpha(x,r-\eta,f) \leq \varepsilon$.

Then for $p_\alpha(x'-x) < \eta$, and $|r'-r| < \eta$, we have proved

$$|M_\alpha(x',r',f) - M_\alpha(x,r,f)| < 2\varepsilon .$$

The function $(x,r) \to M_\alpha(x,r,f)$ is continuous for $r > r_o > 0$; M_α is a continuous function of x for given $r > 0$; then the upper semi-continuity of $x \to \nu(x,f)$ can be proved by the same way as in C^n using the limit (8). We state the following consequence.

COROLLARY 6. If E is a complex vector space with locally convex topology, the density sets

$$N(c,f) = [x \in G;\ \nu(x,f) \geq c], \quad c > 0$$

of a plurisubharmonic function f in a domain $G \subset E$ are pluripolar closed sets in G.

REMARK 3. By Corollary 6 a new problem arises: are the sets $N(c,f)$ analytic sets like in C^n for $c > 0$? We conjecture that the result of Y.T. Siu (cf. [6]) remains true in Banach spaces having the approximation property (see [7]). But such a result in an infinite dimensional space would not have the same precise geometrical consequences as in C^n, or in finite dimensional manifolds, if no further information is available on the codimension of $N(c,f)$ (see for example [5], p.33).

REMARK 4. Will Theorem 3 remain true if E is not supposed to be a locally convex space? In C^n the upper semi-continuity of $\nu(x,f)$ was a consequence of this elementary property: given a compact K and an open neighborhood ω of K, there exists a disked neighborhood W of 0 such $K+W \subset \omega$. If E is infinite dimensional, the basis Φ of the neighborhoods of 0 does not contain any relatively compact W; we have to take the sup of f on rW, $r > 0$, which is not relatively compact and to write $M_W(x,r,f) = \sup_y f(x+y)$ for $y \in rW$, $r > 0$. Indeed the proof of the Theorem 3 was obtained using the continuity of $M_W(x,r,f)$ for fixed x and variable $r > 0$ (such a property remains true if E is not locally convex) and a property (P) of the topology which is the following:

(P) - Given a disked neighborhood W of the origin, and $\eta > 0$, there exists a disked neighborhood W' of 0 such

$$W + W' \subset (1+\eta)W. \tag{29}$$

PROPOSITION 2. If the topology of the complex space E has the property (P), then the density number of $\nu(x,f)$ of the plurisubharmonic functions f in x is an upper semi-continuous function of x.

PROOF. By the convexity of $\log r \to M_W(x,r,f) = \sup_{|u| \le r < 1} f(x+uy)$, $y \in W$ and $f(x+y) \le a$, for $y \in W$, we obtain: for $r > 0$, the function $r \to M_W(x,r,f)$ is continuous. For given x, r, $0 < r < 1$ there exists η such $0 \le M_\alpha(x,r+\eta,f) - M_\alpha(x,r-\eta,f) < \varepsilon$. Then by (29) there exists W' such $W + W' \subset (1+\eta)W$; therefore for $x' - x \in W'$ we obtain for $(1+\eta)r < 1$:

$$x' + W \subset x + W' + W \subset x + (1+\eta)W$$

$$M_W(x',r,f) \le M_W[x,(1+\eta)r,f] \le M_W(x,r,f) + \varepsilon . \tag{30}$$

By (29), given η, $0 < \eta < 1$ and a disked neighborhood W of 0 there exists a disked neighborhood W'' of 0 such

$$(1-\eta)W + W'' \subset W$$

(31) $$M_W[x,(1-\eta)r,f] \leq M_W(x',r,f) \quad \text{if} \quad x'-x \in W''$$

or $$M_W(x,r-\eta,f) \leq M_W(x',r,f) \quad \text{if} \quad x'-x \in W''$$

(32) $$M_W(x',r,f) \geq M_W(x,r,f) - \varepsilon .$$

From (30) and (32) we obtain: if the topology of E has the property (P), then $M_W(x,r,f)$ is for $r > 0$ a continuous function of x.

From the continuity of M_W and (8) we then can deduce Proposition 2. If f a in $x + W$, we obtain $-\nu(x,f) = \lim(\log \frac{1}{r})^{-1} M_W(x,r,f-a)$. The quotient is an increasing limit of continuous functions, for $r \searrow 0$; the Proposition 2 is proved.

REMARK 5. Examples of non Baire spaces **E** with continuous plurisubharmonic functions f such in x the cone g'_x is not pluripolar and is all the space E, can be given. For instance (this example was given by C. Kiselman in an unpublished letter) let us consider the space

$$E = \oplus\, C^N$$

E is the space of the sequences $x = \{x_n\}$ such that $x_n = 0$ with the exception of a finite set for n. We define $\text{supp}\, x =$ $= [\, j \in N;\ x_j \neq 0]$.

A basis Φ of the neighborhoods of 0 will be the $W_\varepsilon = [x;\ |x_j| < \varepsilon_j]$, $\varepsilon_j > 0$, $j \in \mathbb{N}$. Then $f(x) = \sup c_j \log|x_j|$ is plurisubharmonic in E if we take $c_j > 0$. For $f_y(u) = f(yu)$, $u \in \mathbb{C}$, we obtain at the origin $x = 0$:

$$\nu(0,y,f) = \nu(0,f_y) = \inf c_j \quad \text{for} \quad j \in \text{supp}\, y$$

$$\nu(0,f) = \inf c_j, \quad j \in \mathbb{N}.$$

If $c_j = 2^{-j}$, then $\nu(0,y,f) = 2^{-s} > \nu(0,f) = 0$ for $y \in E$ and $s = [\sup j;\ j \in \operatorname{supp} y]$.

REMARK 6. Calculus of $\nu(x,y,f)$ and $\nu(x,f)$ if $E = \oplus E_n$.

More generally, let us consider a sequence E_n of complex locally convex complete spaces; in $M = \prod_n E_n$, we define E by

$$E = \oplus E_n \subset M$$

and take on E the locally convex direct sum topology T. We denote $p_n \colon E \to E_n$ the projections and $x_n = p_n(x)$; T is the topology of the mappings $j_n \colon E_n \to E$; $j_n(E_n)$ is the closed subspace of E defined by $p_m(x) = 0$, $m \neq n$ and $p_m(x) = x$ if $m=n$. We suppose on each E_n we have defined a continuous plurisubharmonic function $U_n(x_n)$, and we denote:

$$V_n(x) = U_n \circ p_n(x).$$

Let us define $\tilde{\eta}_n = [x_n \in E_n;\ U_n(x_n) = -\infty] \subset E_n$.

Given $x^o \in E$, we define a positive integer

$$s(x^o) = [\inf n,\ n \in \mathbb{N}, \text{ such } p_n(x^o) = 0 \text{ for } n > s(x^o)].$$

Now we suppose that the origin belongs to $\tilde{\eta}_n$ in E_n for each n. Given a complex line L in E:

$$x = x^o + uy, \qquad u \in \mathbb{C}$$

$$p_j(x) = p_j(x^o) + up_j(y).$$

For $x \in L$, the number $s(x)$ is bounded

$$s(x) \leq \sup[s(x^o), s(y)] = s(L).$$

Let us define like in the precedent example

$$f(x) = \sup_n U_n \circ p_n(x) = \sup_n V_n(x). \tag{33}$$

Then $p_n(x) = 0$ and $U_n \circ p_n(x) = -\infty$ for $n > s(L)$ and

$f|_L = \sup_j U_j \circ p_j(x^o + uy)$ for $1 \leq j \leq s(L)$ is a subharmonic function (or $\equiv -\infty$) defined by the formula

$$f|_L = \sup_j U_j[p_j(x^o) + up_j(y)] \quad \text{for} \quad 1 \leq j \leq s(L).$$

Now we suppose that the point x^o belongs to $\prod \tilde{\eta}_n$, it is $x_n^o = p_n(x^o) \in \tilde{\eta}_n$ for each n or $U_n(x_n^o) - \infty$ for each n. Then if L is a complex line through x^o, of direction $y = \{y_j\}$.

$$(33) \qquad \nu(x^o,y,f) = \nu(x^o,f|_L) = \inf_j \nu[x_j^o, p_j(y), U_j]$$

and we have to take in (33) the inf for given $y \in \oplus E_n$ and $1 \leq j \leq s(y)$. The density number $\nu(x^o,y,f)$ is

$$(34) \qquad \nu(x^o,f) = \inf_{y,f} \nu[x_j^o, p_j(y), U_j],$$

for $y \in E = \oplus E_n$, and $1 \leq j \leq s(y)$.

In the E_n are closed subspaces of E, the space E is not a Baire space; $f(x)$ is a continuous plurisubharmonic function. From (33) and (34) we get

$$(35) \qquad \nu(x^o,f) = \inf_y \nu(x^o,y,f) = \inf_j \nu[p_j(x^o), U_j].$$

The tangential density number (x^o,y,f), given by (33) is the inf of a finite set of positive numbers; we can choose in (35) the continuous plurisubharmonic functions $U_j \in P(E_j)$, in order to obtain $\nu(x^o,f) = c > 0$ and $\nu(x^o,y,f) > c$ for all $y \in E$.

BIBLIOGRAPHY

1. AVANISSIAN, V., Fonctions plurisousharmoniques et fonctions doublement sousharmoniques, Ann. E.N.S., t. 78, p.101-161.

2. COEURÉ, G., Fonctions plurisousharmoniques sur les espaces vectoriels topologiques, Ann. Inst. Fourier, 1970, p.361-432.

3. KISELMAN, Ch., Stabilité du nombre de Lelong par restriction à une sous-variété, Lecture Notes Springer nº 919, p. 324-337, (1980).

4. LELONG, P., a/ Plurisubharmonic functions in topological vector spaces. Polar sets and problems of measure. Lecture Notes, nº 364, 1973, p. 58-69.

b/ Fonctions plurisousharmoniques et ensembles polaires sur une aogèbre de fonctions holomorphes, Lecture Notes, nº 116, 1969, p. 1-20.

c/ Calcul du nombre densité $\nu(x,f)$ et lemme de Schwarz pour les fonctions plurisousharmoniques dans un espace vectoriel topologique, Lecture Notes Springer nº 919, p. 167-177, (1980).

d/ Intégration sur un ensemble analytique complexe, Bull. Soc. Math. de France, t. 85, p. 239-262, 1957.

5. RAMIS, J.-P., Sous-ensembles analytiques d'une variété banachique complexe, Ergebnisse der Math., t. 53, Springer, 1970.

6. SIU, Y.T., Analyticity of sets associated to Lelong numbers, Inv. Math., 6. 27, p. 53-156, 1974.

7. NOVERRAZ, Ph., Pseudo-convexité, convexité polynomiale et domaines d'holomorphie en dimension infinie, North Holland, Math. Studies, vol. 3 (1973).

Département de Mathématiques
Université de Paris VI
4 Place Jussieu
75230 Paris CEDEX 05

Functional Analysis, Holomorphy and Approximation Theory II, G.I. Zapata (ed.)
© Elsevier Science Publishers B.V. (North-Holland), 1984

CHEBYSHEV CENTERS OF COMPACT SETS WITH RESPECT TO STONE-WEIERSTRASS SUBSPACES

Jaroslav Mach

Let S be a compact Hausdorff space, X a Banach space, $C(S,X)$ the Banach space of all continuous functions on S with values in X equipped with the supremum norm. In this note two results concerning Chebyshev centers of compact subsets of $C(S,X)$ with respect to a Stone-Weierstrass subspace of $C(S,X)$ are established. In particular, a formula for the relative Chebyshev radius in terms of the Chebyshev radius of the corresponding set valued map is given. It is shown further that the proximinality of all Stone-Weierstrass subspaces implies the existence of relative Chebyshev centers for all compact subsets of $C(S,X)$.

The proximinality of Stone-Weierstrass subspaces has been studied by many authors. Mazur (unpublished, c.f., e.g., [6]) proved that any Stone-Weierstrass subspace is proximinal if X is the real line (a subspace G of a normed linear space Y is called proximinal if every $y \in Y$ possesses an element of best approximation x_o in G, i.e., if there is an $x_o \in G$ such that $\|y-x_o\| \leq \|y-x\|$ holds for every $x \in G$). The question for which Banach spaces X every Stone-Weierstrass subspace of $C(S,X)$ is proximinal is due to Pelczynski [4] and Olech [3]. Olech [3] and Blatter [1] showed that this is true if X is uniformly convex and an L_1-predual space, respectively. In [2] those Banach spaces X for which any Stone-Weierstrass subspace is proximinal were characterized.

We will employ the following notations and definitions. Let $x \in X$, $r > 0$. $B(x,r)$ will denote the closed ball in X with center x and radius r. $\mathcal{C}(X)$ will denote the class of all compact subsets of X. A subspace V of $C(S,X)$ is said to be a Stone-Weierstrass subspace of $C(S,X)$ if there is a compact Hausdorff space T and a continuous surjection $\varphi: S \to T$ such that V is the set of all functions f having the form $f = g \circ \varphi$ for some $g \in C(T,X)$. Let Φ be a set-valued mapping from S into 2^X. Φ is said to be upper Hausdorff semicontinuous (u.H.s.c.) (cf. [5] and [7]) if for every $s_o \in S$ and every $\varepsilon > 0$ there is a neighborhood U of s_o such that for every $s \in U$ we have

$$\sup_{x \in \Phi(s)} \operatorname{dis}(x, \Phi(s_o)) < \varepsilon .$$

Φ is lower semicontinuous if the set $\{s: \Phi(s) \cap G \neq \emptyset\}$ is open for any open set G. Φ is upper semicontinuous if the set $\{s: \Phi(s) \cap H \neq \emptyset\}$ is closed for any closed set H. A function $f \in C(S,X)$ is said to be a best approximation of Φ in $C(S,X)$ if the number

$$\operatorname{dist}(f,\Phi) = \sup_{s \in S} \sup_{x \in \Phi(s)} \|x - f(s)\|$$

is equal to $\inf \operatorname{dist}(g,\Phi)$ where the infimum is taken over all $g \in C(S,X)$. Let F be a bounded subset of X, G a subspace of X. The number

$$r_G(F) = \inf_{x \in G} \sup_{y \in F} \|x - y\|$$

is called the Chebyshev radius of F with respect to G. A point $x \in G$ is said to be a Chebyshev center of F with respect to G if $\|x-y\| \leq r_G(F)$ for all $y \in F$. The set of all such x will be denoted by $c_G(F)$. For a set-valued map $\Phi: S \to 2^X$, r_Φ will denote the number $\sup_{s \in S} r_X(\Phi(s))$.

THEOREM 1. Let V be a Stone-Weierstrass subspace of $C(S,X)$ defined by φ and T. Let F be a compact subset of $C(S,X)$. Then

$$r_V(F) = r_\Phi$$

where $\Phi: T \to \mathcal{C}(X)$ is the set-valued map

$$\Phi(t) = \{f(s): f \in F,\ s \in \varphi^{-1}(t)\}.$$

PROOF. To prove that Φ is u.s.c. we show that the set $\{t \in T: \Phi(t) \cap H \neq \emptyset\}$ is closed for any closed set H. Let $t_\alpha \in \{t: \Phi(t) \cap H \neq \emptyset\}$ be such that $t_\alpha \to t$. Then there are $s_\alpha \in \varphi^{-1}(t_\alpha)$ and $f_\alpha \in F$ such that $f_\alpha(s_\alpha) \in H$. Without loss of generality assume that $s_\alpha \to s$ and $f_\alpha \to f$. Then clearly $s \in \varphi^{-1}(t)$ and $f(s) \in H$, so $\Phi(t) \cap H \neq \emptyset$.

For any $g \in C(T,X)$ we have

$$\sup_{f\in F} \|f-g\| = \sup_{f\in F} \sup_{t\in T} \sup_{s\in\varphi^{-1}(t)} \|f(s)-g(t)\| =$$

$$\sup_{t\in T} \sup_{x\in\Phi(t)} \|x-g(t)\| = \operatorname{dist}(g,\Phi).$$

It was proved in [2] that

$$\inf_{g\in C(T,X)} \operatorname{dist}(g,\Phi) = r_\Phi.$$

It follows

$$r_V(F) = \inf_{g\in C(T,X)} \sup_{f\in F} \|f-g\circ\varphi\| = \inf_{g\in C(T,X)} \operatorname{dist}(g,\Phi) = r_\Phi.$$

THEOREM 2. Let X be such that every Stone-Weierstrass subspace of $C(S,X)$ is proximinal. Then $c_V(F) \neq \emptyset$ for every compact subset F of $C(S,X)$ and every Stone-Weierstrass subspace V.

PROOF. By Theorem 2 of [2], the proximinality of every Stone-Weierstrass subspace of $C(S,X)$ implies that for any compact Hausdorff space T, any u.H.s.c. map $\Phi: T \to \mathcal{C}(X)$ has a best approximation g in $C(T,X)$. Let

$$\Phi(t) = \{f(s): f \in F, s \in \varphi^{-1}(t)\}.$$

Then

$$\sup_{f\in F} \|f-g\circ\varphi\| = \mathrm{dist}(g,\Phi) = \inf_{h\in C(T,X)} (h,\Phi) =$$

$$\inf_{h\in C(T,X)} \sup_{f\in F} \|f-h\circ\varphi\| = r_V(F).$$

It follows $g\circ\varphi \in c_V(F)$.

The following corollary is a consequence of Blatter's result and Theorem 2.

COROLLARY 1. Let X be an L_1-predual space. Then $c_V(F) \neq \emptyset$ for any compact subset F of $C(S,X)$ and any Stone-Weierstrass subspace V.

In [8], a bounded subset of an L_1-predual space has been constructed whose set of Chebyshev centers is empty. This shows that Corollary 1 does not hold if compact subsets are replaced by bounded subsets.

It was shown in [2] that if X is a locally uniformly convex dual Banach space then every Stone-Weierstrass subspace of $C(S,X)$ is proximinal. The next corollary follows from this and Theorem 2.

COROLLARY 2. Let X be a dual locally uniformly convex Banach space. Then $c_V(F) \neq \emptyset$ for any compact subset F of $C(S,X)$ and any Stone-Weierstrass subspace V.

REFERENCES

1. J. BLATTER, Grothendieck spaces in approximation theory, Mem. Amer. Math. Soc. 120 (1972).

2. J. MACH, On the proximinality of Stone-Weierstrass subspaces, Pacific J. Math. 99 (1982), 97-104.

3. C. OLECH, Approximation of set-valued functions by continuous functions, Colloq. Math. 19 (1968), 285-293.

4. A. PELCZYNSKI, Linear extensions, linear averagings and their applications to linear topological classification of spaces of continuous functions, Dissert. Math. (Rozprawy Math.) 58, Warszawa 1968.

5. W. POLLUL, Topologien auf Mengen von Teilmengen und Stetigkeit von mengenwertigen metrischen Projektionen, Diplomarbeit, Bonn 1967.

6. Z. SEMADENI, Banach spaces of continuous functions, Monografje Matematyczne 55, Warszawa 1971.

7. I. SINGER, The theory of best approximation and functional analysis, Reg. conference ser. appl. math. 13, SIAM, Philadelphia 1974.

8. D. AMIR, J. MACH, K. SAATKAMP, Existence of Chebyshev centers, best n-nets and best compact approximants, Trans. Amer. Math. Soc. 271 (1982), 513-524.

Institut für Angewandte Mathematik der Universität Bonn
Wegelerstr. 6
5300 Bonn

This work was done while the author was visiting the Texas A&M University at College Station.

Functional Analysis, Holomorphy and
Approximation Theory II, G.I. Zapata (ed.)
© Elsevier Science Publishers B.V. (North-Holland), 1984

ON THE FOURIER-BOREL TRANSFORMATION AND SPACES OF ENTIRE FUNCTIONS IN A NORMED SPACE

Mário C. Matos

(Dedicated to the memory of Silvio Machado)

1. INTRODUCTION

We introduce here the spaces of entire functions in a normed space which are of order (respectively, nuclear order) k and type (respectively, nuclear type) strictly less than A.

Here $k \in [1,+\infty]$ and $A \in (0,+\infty]$. The corresponding spaces in which the type is allowed to be also equal to A are introduced when $k \in [1,+\infty]$ and $A \in [0,+\infty)$. These spaces have natural topologies and they are the infinite dimensional analogous of the spaces considered in Martineau [1]. In this paper we study the Fourier-Borel transformation in these spaces and we are able to show that these transformations identify algebraically and topologically the strong duals of the above spaces with other spaces of the same kind. In a second paper, to appear elsewhere, we prove existence and approximation theorems for convolution equations in these spaces.

The notations we used are those used by Nachbin [1] and Gupta [1]. Hence, if E is a complex normed space, $\mathcal{H}(E)$ is the vector space of all entire functions in E, $\mathcal{P}(^jE)$ the Banach space of all j-homogeneous continuous polynomials in E with the natural norm $\|\cdot\|$ and $\mathcal{P}_N(^jE)$ the Banach space of all j-homogeneous continuous polynomials of nuclear type with the nuclear norm $\|\cdot\|_N$ for all $j \in \mathbb{N}$.

2. SPACES OF ENTIRE FUNCTIONS IN NORMED SPACES

In this section E denotes a complex normed space.

2.1 DEFINITION. If $\rho > 0$ we denote $\mathcal{B}_\rho(E)$ the complex vector space of all $f \in \mathcal{H}(E)$ such that

$$\|f\|_\rho = \sum_{n=0}^{\infty} \rho^{-n} \|\hat{d}^n f(0)\| < +\infty \tag{1}$$

normed by $\|\cdot\|_\rho$. We denote $\mathcal{B}_{N,\rho}(E)$ the complex vector space of all $f \in \mathcal{H}(E)$ such that $\hat{d}^n f(0) \in \mathcal{P}_N({}^nE)$ for each $n \in \mathbb{N}$ and

$$\|f\|_{N,\rho} = \sum_{n=0}^{\infty} \rho^{-n} \|\hat{d}^n f(0)\|_N < +\infty \tag{2}$$

normed by $\|\cdot\|_{N,\rho}$.

2.2 PROPOSITION. For each $\rho > 0$, the normed spaces $\mathcal{B}_\rho(E)$ and $\mathcal{B}_{N,\rho}(E)$ are complete.

PROOF. If $(f_n)_{n=1}^{\infty}$ is a Cauchy sequence in $\mathcal{B}_\rho(E)$, for every $\varepsilon > 0$ there is $n_\varepsilon \in \mathbb{N}$, such that

$$\sum_{j=0}^{\infty} \rho^{-j} \|\hat{d}^j f_m(0) - \hat{d}^j f_n(0)\| < \varepsilon \tag{3}$$

for all $m \geq n_\varepsilon$ and $n \geq n_\varepsilon$. It follows that $(\hat{d}^j f_n(0))_{n=1}^{\infty}$ is a Cauchy sequence in the Banach space $\mathcal{P}({}^jE)$ and it converges to an element $P_j \in \mathcal{P}({}^jE)$. Hence, using (3), we have:

$$\sum_{j=0}^{\infty} \rho^{-j} \|\hat{d}^j f_m(0) - P_j\| \leq \varepsilon \tag{4}$$

for all $m \geq n_\varepsilon$. If we prove that

$$f(x) = \sum_{j=0}^{\infty} \frac{1}{j!} P_j(x) \qquad (x \in E)$$

defines an element of $\mathcal{H}(E)$, then we get:

$$\sum_{j=0}^{\infty} \rho^{-j}\|P_j\| \leq \sum_{j=0}^{\infty} \rho^{-j}\|\hat{d}^j f_{n_\varepsilon}(0) - P_j\| + \|f_{n_\varepsilon}\|_\rho < +\infty .$$

Therefore $f \in \mathcal{B}_\rho(E)$ and (4) implies the convergence of $(f_n)_{n=1}^{\infty}$ to f. In order to prove that $f \in \mathcal{H}(E)$ we note that

$$\rho^{-j}\|P_j\| \leq \rho^{-j}\|\hat{d}^j f_{n_\varepsilon}(0) - P_j\| + \rho^{-j}\|\hat{d}^j f_{n_\varepsilon}(0)\|$$

$$\leq \varepsilon + \|f_{n_\varepsilon}\|_\rho .$$

Hence

$$\limsup_{j\to\infty} \left\|\frac{1}{j!} P_j\right\|^{\frac{1}{j}} = 0$$

and $f \in \mathcal{H}(E)$. A similar proof may be used for $\mathcal{B}_{N,\rho}(E)$. □

2.3 DEFINITION. If $A \in (0,+\infty)$ we denote $\mathrm{Exp}^1_A(E)$ the complex vector space $\bigcup_{\rho<A} \mathcal{B}_\rho(E)$ with the locally convex inductive limit topology. The complex vector space $\mathrm{Exp}^1_{N,A}(E) = \bigcup_{\rho<A} \mathcal{B}_{N,\rho}(E)$ is equipped with the locally convex inductive limit topology. We consider the complex vector spaces $\mathrm{Exp}^1_{0,A}(E) = \bigcap_{\rho>A} \mathcal{B}_\rho(E)$ and $\mathrm{Exp}^1_{N,0,A}(E) = \bigcap_{\rho>A} \mathcal{B}_{N,\rho}(E)$ with the projective limit topologies. Finally, we consider $\mathrm{Exp}^1_\infty(E) = \bigcup_{\rho>0} \mathcal{B}_\rho(E)$ and $\mathrm{Exp}^1_{N,\infty}(E) = \bigcup_{\rho>0} \mathcal{B}_{N,\rho}(E)$ with the locally convex inductive limit topologies and set $\mathrm{Exp}^1_0(E) = \mathrm{Exp}^1_{0,0}(E) = \bigcap_{\rho>0} \mathcal{B}_\rho(E)$ and $\mathrm{Exp}^1_{N,0}(E) = \mathrm{Exp}^1_{N,0,0}(E) = \bigcap_{\rho>0} \mathcal{B}_{N,\rho}(E)$ with the projective limit topologies.

2.4 REMARK. It is possible to show:

$$\mathrm{Exp}^1_A(E) = \{f \in \mathcal{H}(E);\ \limsup_{j\to\infty} \|\hat{d}^j f(0)\|^{\frac{1}{j}} < A\}, \qquad \forall\, A \in (0,+\infty]$$

$$\mathrm{Exp}^1_{0,A}(E) = \{f \in \mathcal{H}(E);\ \limsup_{j\to\infty} \|\hat{d}^j f(0)\|^{\frac{1}{j}} \leq A\}, \qquad \forall\, A \in [0,+\infty).$$

On the other hand, if $f \in \mathcal{H}(E)$ is such that $\hat{d}^j f(0) \in \mathcal{P}_N(^jE)$ for all $j \in \mathbb{N}$, we have

(a) $f \in Exp^1_{N,A}(E)$ if, and only if, $\limsup_{j\to\infty} \|\hat{d}^j f(0)\|_N^{\frac{1}{j}} < A$ for all $A \in (0,+\infty]$

(b) $f \in Exp^1_{N,0,A}(E)$ if, and only if, $\limsup_{j\to\infty} \|\hat{d}^j f(0)\|_N^{\frac{1}{j}} \leq A$ for all $A \in [0,+\infty)$.

Therefore, it is natural to call the elements of $Exp^1_A(E)$ (respectively, $Exp^1_{N,A}(E)$) <u>entire functions of exponential type</u> (respectively, <u>of nuclear exponential type</u>) <u>strictly less</u> A. For $A = +\infty$ we drop out "strictly less than A ".

The elements of $Exp^1_{0,A}(E)$ (respectively, $Exp^1_{N,0,A}(E)$) are called <u>entire functions of exponential type</u> (respectively, <u>nuclear exponential type</u>) <u>less than or equal to</u> A.

2.5 PROPOSITION. (a) $Exp^1_A(E)$ and $Exp^1_{N,A}(E)$ are DF spaces for all $A \in (0,+\infty]$.

(b) $Exp^1_{0,A}(E)$ and $Exp^1_{N,0,A}(E)$ are Fréchet spaces for every $A \in [0,+\infty)$.

PROOF. If $A \in [0,+\infty)$ the proof is straightforward for part (b). The metrizability follows from the fact that we can get a sequence $(a_n)_{n=1}^{\infty}$ of positive real numbers, strictly decreasing and such that $\lim_{n\to\infty} a_n = A$. If $A \in (0,+\infty]$ part (a) follows from the fact that $Exp^1_A(E)$ and $Exp^1_{N,A}(E)$ are the inductive limit of a sequence of Banach spaces. It is enough to take a sequence $(b_n)_{n=1}^{\infty}$ of positive real numbers, strictly increasing, converging to A and then to consider the Banach spaces $\mathcal{B}_{b_n}(E)$ and $\mathcal{B}_{N,b_n}(E)$. □

Now we construct similar spaces for entire functions of finite order.

2.6 DEFINITION. If $\rho > 0$ and $k > 1$ we denote $\mathcal{B}^k_\rho(E)$ (respectively, $\mathcal{B}^k_{N,\rho}(E)$) the complex vector space of all $f \in \mathcal{H}(E)$ such that

$$\|f\|_{k,\rho} = \sum_{j=0}^{\infty} \rho^{-j} \left(\frac{j}{ke}\right)^{\frac{j}{k}} \left\|\frac{1}{j!}\hat{d}^j f(0)\right\| < +\infty \tag{5}$$

(respectively, such that $\hat{d}^j f(0) \in \mathcal{P}_N({}^jE)$ for all $j \in \mathbb{N}$ and

$$\|f\|_{N,k,\rho} = \sum_{j=0}^{\infty} \rho^{-j} \left(\frac{j}{ke}\right)^{\frac{j}{k}} \left\|\frac{1}{j!}\hat{d}^j f(0)\right\|_N < +\infty \tag{6}$$)

normed by $\|\cdot\|_{k,\rho}$ (respectively, by $\|\cdot\|_{N,k,\rho}$).

2.7 PROPOSITION. The normed spaces $\mathcal{B}^k_\rho(E)$ and $\mathcal{B}^k_{N,\rho}(E)$ are complete.

The proof of this result follows the pattern of the proof of 2.2.

2.8 DEFINITION. If $A \in (0,+\infty]$ and $k > 1$ we denote $\mathrm{Exp}^k_A(E)$ (respectively, $\mathrm{Exp}^k_{N,A}(E)$) the complex vector space $\bigcup_{\rho<A} \mathcal{B}^k_\rho(E)$ (respectively, $\bigcup_{\rho<A} \mathcal{B}^k_{N,\rho}(E)$) with the locally convex inductive limit topology. If $A \in [0,+\infty)$ and $k > 1$ the complex vector spaces $\mathrm{Exp}^k_{0,A}(E) = \bigcap_{\rho>A} \mathcal{B}^k_\rho(E)$ and $\mathrm{Exp}^k_{N,0,A}(E) = \bigcap_{\rho>A} \mathcal{B}^k_{N,\rho}(E)$ are endowed with the projective limit topologies. In order to simplify the notations, sometimes we write: $\mathrm{Exp}^k_\infty(E) = \mathrm{Exp}^k(E)$, $\mathrm{Exp}^k_{N,\infty}(E) = \mathrm{Exp}^k_N(E)$, $\mathrm{Exp}^k_{0,0}(E) = \mathrm{Exp}^k_0(E)$, $\mathrm{Exp}^k_{N,0,0}(E) = \mathrm{Exp}^k_{N,0}(E)$ (including the case $k = 1$).

2.9 REMARK. It is possible to show that

$$\mathrm{Exp}^k_A(E) = \{f \in \mathcal{H}(E);\ \limsup_{j\to\infty} \left(\frac{j}{ke}\right)^{\frac{1}{k}} \left\|\frac{1}{j!}\hat{d}^j f(0)\right\|^{\frac{1}{j}} < A\}$$

$$\mathrm{Exp}^k_{0,A}(E) = \{f \in \mathcal{H}(E);\ \limsup_{j\to\infty} \left(\frac{j}{ke}\right)^{\frac{1}{k}} \left\|\frac{1}{j!}\hat{d}^j f(0)\right\|^{\frac{1}{j}} \leq A\}$$

where $k > 1$ and $A \in (0,+\infty]$ in the first case and $A \in [0,+\infty)$ in the second case. On the other hand, if $f \in \mathcal{H}(E)$ is such that $\hat{d}^j f(0) \in \mathcal{P}_N({}^jE)$ for all $j \in \mathbb{N}$, we have:

(a) $f \in Exp^k_{N,A}(E)$ if, and only if, $\limsup_{j\to\infty} (\frac{j}{ke})^{\frac{1}{k}} \frac{1}{j!} \|\hat{d}^j f(0)\|_N^{\frac{1}{j}} < A$ for $k > 1$ and $A \in (0,+\infty]$.

(b) $f \in Exp^k_{N,0,A}(E)$ if, and only if, $\limsup_{j\to\infty} (\frac{j}{ke})^{\frac{1}{k}} \frac{1}{j!} \|\hat{d}^j f(0)\|_N^{\frac{1}{j}} \leq A$ for $k > 1$ and $A \in [0,+\infty)$.

Therefore, it is natural to call the elements of $Exp^k_A(A)$ (respectively, $Exp^k_{N,A}(E)$) <u>entire functions of order</u> k <u>and type strictly less than</u> A (respectively, <u>entire functions of nuclear order</u> k <u>and nuclear type strictly less than</u> A). When $A = +\infty$ we drop out "strictly less than A ". The elements of $Exp^k_{0,A}(E)$ (respectively, $Exp^k_{N,0,A}(E)$) are called <u>entire functions of order</u> k (respectively, <u>nuclear order</u> k) and <u>type less than or equal to</u> A (respectively, <u>nuclear type less than or equal to</u> A).

2.10 REMARK. We know that

$$\lim_{j\to\infty} \frac{j}{e} (\frac{1}{j!})^{\frac{1}{j}} = 1 .$$

Using this fact it is easy to see that, if we allow $k = 1$ in Definitions 2.6 and 2.8, the spaces we get coincide algebraically with the spaces of Definition 2.1 and 2.3 and are isomorphic topologically.

2.11 PROPOSITION. (a) If $k > 1$ and $A \in (0,+\infty]$, then $Exp^k_A(E)$ and $Exp^k_{N,A}(E)$ are DF-spaces.

(b) If $k > 1$ and $A \in [0,+\infty)$, then $Exp^k_{0,A}(E)$ and $Exp^k_{N,0,A}(E)$ are Fréchet spaces.

The proof follows as we have indicated in Proposition 2.5.

In order to consider similar spaces of functions of infinite order, we introduce some new definitions.

2.12 DEFINITION. If $A \in [0,+\infty)$ we denote $\mathcal{H}_b(B_{\frac{1}{A}}(0))$ the complex vector space of all $f \in \mathcal{H}(B_{\frac{1}{A}}(0))$ such that $\limsup_{j\to\infty} \|\frac{1}{j!}\hat{d}^j f(0)\|^{\frac{1}{j}} \leq A$ endowed with the locally convex topology generated by the family of seminorms $(p_\rho^\infty)_{\rho>A}$, where

$$p_\rho^\infty(f) = \sum_{j=0}^{\infty} \rho^{-j} \|\frac{1}{j!} \widehat{d^j f(0)}\|. \tag{7}$$

We denote $\mathcal{H}_{Nb}(B_{\frac{1}{A}}(0))$ the complex vector space of all $f \in \mathcal{H}(B_{\frac{1}{A}}(0))$ such that $\hat{d}^j f(0) \in \mathcal{P}_N({}^jE)$ for $j \in \mathbb{N}$ and $\limsup_{j\to\infty}\|\frac{1}{j!}\hat{d}^j f(0)\|^{\frac{1}{j}} \leq A$, endowed with the locally convex topology generated by the family of seminorms $(p_{N,\rho}^\infty)_{\rho>A}$, where

$$p_{N,\rho}^\infty(f) = \sum_{j=0}^{\infty} \rho^{-j} \|\frac{\hat{d}^j f(0)}{j!}\|_N . \tag{8}$$

These locally convex spaces are Fréchet spaces (see Gupta [1] and [2] and Matos [1]). These spaces are denoted respectively $\mathrm{Exp}_{0,A}^\infty(E)$ and $\mathrm{Exp}_{N,0,A}^\infty(E)$. We also write $\mathrm{Exp}_0^\infty(E) = \mathrm{Exp}_{0,0}^\infty(E)$ and $\mathrm{Exp}_{N,0}^\infty(E) = \mathrm{Exp}_{N,0,0}^\infty(E)$.

2.13 DEFINITION. If $A \in (0,+\infty]$ we denote $\mathcal{H}_b(\bar{B}_{\frac{1}{A}}(0))$ (respectively $\mathcal{H}_{Nb}(\bar{B}_{\frac{1}{A}}(0))$ the algebraic and locally convex inductive limit of Fréchet spaces $\mathcal{H}_b(B_{\frac{1}{\rho}}(0))$ for $\rho \in (0,A)$ (respectively, $\mathcal{H}_{Nb}(B_{\frac{1}{\rho}}(0))$ for $\rho \in (0,A)$). We remark that, except in the case $A = +\infty$, these spaces may not be the space of germs of holomorphic functions in a neighborhood of $\bar{B}_{\frac{1}{A}}(0)$ (for infinite dimensional normed spaces). An equivalent way of defining these spaces may be done as follows. We consider the Banach space $\mathcal{H}^\infty(B_{\frac{1}{\rho}}(0))$ (respectively, $\mathcal{H}_N^\infty(B_{\frac{1}{\rho}}(0))$ of all $f \in \mathcal{H}(B_{\frac{1}{\rho}}(0))$ (respectively, $f \in \mathcal{H}(B_{\frac{1}{\rho}}(0))$ such that $\hat{d}^j f(0) \in \mathcal{P}_N({}^jE)$ for every $j \in \mathbb{N}$) satisfying

$$\sum_{j=0}^{\infty} \frac{1}{\rho^j} \|\frac{\hat{d}^j f(0)}{j!}\| < +\infty$$

respectively,

$$\sum_{j=0}^{\infty} \frac{1}{\rho^j} \|\frac{\widehat{d^j f(0)}}{j!}\| < +\infty)$$

endowed with the norm p_ρ (respectively, $p_{N,\rho}$). (See (7) and (8) above). Hence we have the locally convex inductive limits

$$\underset{\rho\in(0,A)}{\text{ind.lim}}\ \mathcal{H}^{\infty}(B_{\frac{1}{\rho}}(0)) = \mathcal{H}_b(\bar{B}_{\frac{1}{A}}(0)) \tag{9}$$

$$\underset{\rho\in(0,A)}{\text{ind.lim}}\ \mathcal{H}_N^{\infty}(B_{\frac{1}{\rho}}(0)) = \mathcal{H}_{Nb}(\bar{B}_{\frac{1}{\rho}}(0)). \tag{10}$$

Now we use the following notations: $\text{Exp}_A^{\infty}(E) = \mathcal{H}_b(\bar{B}_{\frac{1}{A}}(0))$ and $\text{Exp}_{N,A}^{\infty}(E) = \mathcal{H}_{Nb}(\bar{B}_{\frac{1}{A}}(0))$ for each $A \in (0,+\infty]$. We also denote $\text{Exp}_{\infty}^{\infty}(E) = \text{Exp}^{\infty}(E) = \mathcal{H}_b(\{0\})$ and $\text{Exp}_{N,\infty}^{\infty}(E) = \text{Exp}_N^{\infty}(E) = \mathcal{H}_{Nb}(\{0\})$. In order to motivate these notations we note that

$$\lim_{k\to\infty} \left(\frac{j}{ek}\right)^{\frac{j}{k}} = 1.$$

Hence we may think p_ρ^{∞} as $\|\cdot\|_{\infty,\rho}$ and $p_{N,\rho}^{\infty} = \|\cdot\|_{N,\infty,\rho}$ and these notations will also be used.

2.14 PROPOSITION. (a) If $A \in (0,+\infty]$ the locally convex spaces $\text{Exp}_A^{\infty}(E)$ and $\text{Exp}_{N,A}^{\infty}(E)$ are DF.

(b) If $A \in [0,+\infty)$ the locally convex spaces $\text{Exp}_{0,A}^{\infty}(E)$ and $\text{Exp}_{N,0,A}^{\infty}(E)$ are Fréchet.

PROOF. As we remarked above (b) is proved in Gupta [1] and [2] and in Matos [1]. In order to prove (a) it is enough to see that these locally convex spaces are the inductive limits of <u>sequences</u> of Banach spaces. In order to get this it is enough to take an increasing sequence $(\rho_n)_{n=1}^{\infty}$ of positive numbers such that $\lim_{n\to\infty} \rho_n = A$. □

2.15 PROPOSITION. (a) If $k \in [1,+\infty]$ and $k \in (0,+\infty]$, then the Taylor series at 0 of each element of $\text{Exp}_A^k(E)$ (respectively, $\text{Exp}_{N,A}^k(E)$) converges to it in the topology of the space.

(b) If $k \in [1,+\infty]$ and $A \in [0,+\infty)$, then the Taylor series at

0 of each element of $Exp^k_{0,A}(E)$ (respectively, $Exp^k_{N,0,A}(E)$) converges to it in the topology of the space.

PROOF. It is enough to note that for each f in the correct space we have:

$$\|f - \sum_{j=0}^{n} \frac{1}{j!} \hat{d}^j f(0)\|_{k,\rho} = \sum_{j=n+1}^{\infty} \rho^{-j}\left(\frac{j}{ke}\right)^{\frac{j}{k}} \|\frac{1}{j!} \hat{d}^j f(0)\|, \quad k \in (1,+\infty).$$

$$\|f - \sum_{j=0}^{n} \frac{1}{j!} \hat{d}^j f(0)\|_{\rho} = \sum_{j=n+1}^{\infty} \rho^{-j} \|\hat{d}^j f(0)\|$$

$$\|f - \sum_{j=0}^{n} \frac{1}{j!} \hat{d}^j f(0)\|_{N,k,\rho} = \sum_{j=n+1}^{\infty} \rho^{-j}\left(\frac{j}{ke}\right)^{\frac{j}{k}} \|\frac{\hat{d}^j f(0)}{j!}\|_N, \quad k \in (1,+\infty)$$

$$\|f - \sum_{j=0}^{n} \frac{1}{j!} \hat{d}^j f(0)\|_{N,\rho} = \sum_{j=n+1}^{\infty} \rho^{-j} \|\hat{d}^j f(0)\|_N$$

$$p^{\infty}_{\rho}(f - \sum_{j=0}^{n} \frac{1}{j!} \hat{d}^j f(0)) = \sum_{j=n+1}^{\infty} \rho^{-j} \|\frac{\hat{d}^j f(0)}{j!}\|$$

$$p^{\infty}_{N,\rho}(f - \sum_{j=0}^{n} \frac{1}{j!} \hat{d}^j f(0)) = \sum_{j=n+1}^{\infty} \rho^{-j} \|\frac{\hat{d}^j f(0)}{j!}\|_N . \qquad \Box$$

2.16 PROPOSITION. (a) If $k \in (1,+\infty]$ and $A \in (0,+\infty]$, then e^{φ} belongs to $Exp^k_A(E)$ and to $Exp^k_{N,A}(E)$ for every $\varphi \in E'$.

(b) If $k \in (1,+\infty]$ and $A \in [0,+\infty)$, then e^{φ} belongs to $Exp^k_{0,A}(E)$ and to $Exp^k_{N,0,A}(E)$ for all $\varphi \in E'$.

(c) If $k = 1$ and $A \in (0,+\infty]$ then e^{φ} belongs to $Exp^1_A(E)$ and to $Exp^1_{N,A}(E)$ for all $\varphi \in E'$ such that $\|\varphi\| < A$.

(d) If $k = 1$ and $A \in [0,+\infty)$, then $e^{\varphi} \in Exp^1_{0,A}(E), Exp^1_{N,0,A}(E)$ for all $\varphi \in E'$ such that $\|\varphi\| \leq A$.

PROOF. It is enough to use the definitions of the spaces and to note that

$$\|\hat{d}^j(e^{\varphi})(0)\| = \|\varphi\|^j = \|\hat{d}^j(e^{\varphi})(0)\|_N . \qquad \Box$$

2.17 PROPOSITION. (1) The vector subspace generated by all e^{φ}, $\varphi \in E'$, is dense in

(a) $Exp^k_{N,A}(E)$ if $k \in (1,+\infty]$ and $A \in (0,+\infty]$.

(b) $Exp^k_{N,0,A}(E)$ if $k \in (1,+\infty]$ and $A \in [0,+\infty)$.

(c) $Exp^1_N(E)$.

(2) The vector subspace generated by e^{φ}, $\varphi \in E'$, $\|\varphi\| \leq A$, is dense in $Exp^1_{N,A}(E)$ if $A \in (0,+\infty)$.

(3) The vector subspace generated by e^{φ}, $\varphi \in E'$, $\|\varphi\| \leq A$, is dense in $Exp^1_{N,0,A}(E)$ if $A \in (0,+\infty)$.

PROOF. In all three cases we denote $\mathcal{S}$ the closure of the corresponding generated vector subspace. By 2.15 and by fact that $\mathcal{P}_f(^jE)$ is dense in $\mathcal{P}_N(^jE)$ for all $j \in \mathbb{N}$, it is enough to show that $\mathcal{P}_f(^jE) \in \mathcal{S}$ for all $j \in \mathbb{N}$. Therefore we only have to prove that $\varphi^n \in \mathcal{S}$ for all $n \in N$ and $\varphi \in E'$. We show this by induction in n. For $\lambda \in C$, $\lambda \neq 0$, $|\lambda|$ sufficiently small we have

$$e^{\lambda\varphi} = \sum_{j=0}^{\infty} \frac{1}{j!} \lambda^j \varphi^j$$

convergence in the sense of the topology of those spaces. Hence we get

$$\text{(i)} \quad \lim_{|\lambda|\to 0} \left\|\frac{e^{\lambda\varphi}-1}{\lambda} - \varphi\right\|_{N,k,\rho} = \lim_{|\lambda|\to 0} |\lambda| \left\|\sum_{j=2}^{\infty} \frac{\varphi^j}{j!}\right\|_{N,k,\rho} = 0$$

$$\text{(ii)} \quad \lim_{|\lambda|\to 0} \left\|\frac{e^{\lambda\varphi}-1}{\lambda} - \varphi\right\|_{N,\rho} = \lim_{|\lambda|\to 0} |\lambda| \left\|\sum_{j=2}^{\infty} \frac{\varphi^j}{j!}\right\|_{N,\rho} = 0$$

$$\text{(iii)} \quad \lim_{|\lambda|\to 0} p^{\infty}_{N,\rho}\left(\frac{e^{\lambda\varphi}-1}{\lambda} - \varphi\right) = \lim_{|\lambda|\to 0} |\lambda| p^{\infty}_{N,\rho}\left(\sum_{j=2}^{\infty} \frac{\varphi^j}{j!}\right) = 0.$$

From here we get $\varphi \in \mathcal{S}$ for all $\varphi \in E'$ in all possible cases. Now, if we suppose that $\varphi^j \in \mathcal{S}$ for $j \leq n-1$, we have

(i)
$$\lim_{|\lambda|\to 0} \left\| \frac{1}{\lambda^n}\left(e^{\lambda\varphi} - \sum_{j=0}^{n-1} \frac{1}{j!}\varphi^j\lambda^j\right) - \varphi^n \right\|_{N,K,\rho} =$$

$$= \lim_{|\lambda|\to 0} |\lambda| \left\| \sum_{j=n+1}^{\infty} \frac{1}{j!}\lambda^{j-n}\varphi^j \right\|_{N,k,\rho} = 0$$

(ii)
$$\lim_{|\lambda|\to 0} \left\| \frac{1}{\lambda^n}\left(e^{\lambda\varphi} - \sum_{j=0}^{n-1} \frac{1}{j!}\lambda^j\varphi^j\right) \right\|_{N,\rho} =$$

$$= \lim_{|\lambda|\to 0} |\lambda| \left\| \sum_{j=n+1}^{\infty} \frac{1}{j!}\lambda^{j-n}\varphi^j \right\|_{N,\rho} = 0$$

(iii)
$$\lim_{|\lambda|\to 0} p^{\infty}_{N,\rho}\left(\frac{1}{\lambda^n}\left(e^{\lambda\varphi} - \sum_{j=0}^{n-1} \frac{1}{j!}\lambda^j\varphi^j\right)\right) =$$

$$= \lim_{|\lambda|\to 0} |\lambda| p^{\infty}_{N,\rho}\left(\sum_{j=n+1}^{\infty} \frac{1}{j!}\lambda^{j-n}\varphi^j\right) = 0.$$

Hence $\varphi^n \in \mathcal{S}$ for all $\varphi \in E'$ in any possible case. □

3. BOUNDED SUBSETS

In this paragraph we give a characterization of the bounded subsets of $\mathrm{Exp}^k_{N,A}(E)$ and $\mathrm{Exp}^k_A(E)$ for $k \in [1,+\infty]$ and $A \in (0,+\infty]$, and of $\mathrm{Exp}^k_{N,0,A}(E)$ and $\mathrm{Exp}^k_{0,A}(E)$ for $k \in [1,+\infty]$ and $A \in [0,+\infty)$. These characterizations will be used in the proof that the Fourier-Borel transformations are topological isomorphisms.

The family of all sequences $\alpha = (\alpha_j)_{j=0}^{\infty}$ of real numbers $\alpha_j \geq 0$ such that $\overline{\lim_{j\to\infty}} [\alpha_j]^{\frac{1}{j}} \leqslant A$ will be denoted by $\mathcal{S}_A$.

3.1 PROPOSITION. For $k \in (1,+\infty)$, $A \in (0,+\infty]$, $\alpha \in \mathcal{S}_{\frac{1}{A}}$ the semi-norms $p_{N,k,\alpha}$ and $p_{k,\alpha}$ defined by

$$p_{N,k,\alpha}(f) = \sum_{j=0}^{\infty} \alpha_j \left(\frac{j}{ke}\right)^{\frac{j}{k}} \left\| \frac{\hat{d}^j f(0)}{j!} \right\|_N \tag{11}$$

$$p_{k,\alpha}(f) = \sum_{j=0}^{\infty} \alpha_j \left(\frac{j}{ke}\right)^{\frac{j}{k}} \left\| \frac{\hat{d}^j f(0)}{j!} \right\| \tag{12}$$

are continuous in $Exp^k_{N,A}(E)$ and in $Exp^k_A(E)$ respectively. For $k = 1$, $A \in (0,+\infty]$ and $\alpha \in \mathcal{S}_{\frac{1}{A}}$ the seminorms $p_{N,1,\alpha}$ and $p_{1,\alpha}$ defined by

$$p_{N,1,\alpha}(f) = \sum_{j=0}^{\infty} \alpha_j \|\hat{d}^j f(0)\|_N \tag{13}$$

$$p_{1,\alpha}(f) = \sum_{j=0}^{\infty} \alpha_j \|\hat{d}^j f(0)\|_N \tag{14}$$

are continuous in $Exp^1_{N,A}(E)$ and in $Exp^1_A(E)$ respectively. For $k = +\infty$, $A \in (0,+\infty]$ and $\alpha \in \mathcal{S}_{\frac{1}{A}}$ the seminorms $p_{N,\infty,\alpha}$ and $p_{\infty,\alpha}$ defined by

$$p_{N,\infty,\alpha}(f) = \sum_{j=0}^{\infty} \alpha_j \left\|\frac{\hat{d}^j f(0)}{j!}\right\|_N \tag{15}$$

$$p_{\infty,\alpha}(f) = \sum_{j=0}^{\infty} \alpha_j \left\|\frac{\hat{d}^j f(0)}{j!}\right\| \tag{16}$$

are continuous in $Exp^\infty_{N,A}(E)$ and in $Exp^\infty_A(E)$ respectively.

PROOF. If $\rho \in (0,A)$, then $\frac{1}{\rho} > \frac{1}{A}$. Since $\alpha \in \mathcal{S}_{\frac{1}{A}}$, there is $C(\rho)$ such that $\alpha_j \leq C(\rho)\frac{1}{\rho^j}$ for every $j \in \mathbb{N}$. Thus for $k \in (1,+\infty)$, $A \in (0,+\infty]$ we have:

$$p_{N,k,\alpha}(f) \leq C(\rho) \sum_{j=0}^{\infty} \rho^{-j}\left(\frac{j}{ke}\right) \left\|\frac{\hat{d}^j f(0)}{j!}\right\|_N$$

$$p_{k,\alpha}(f) \leq C(\rho) \sum_{j=0}^{\infty} \rho^{-j}\left(\frac{j}{ke}\right) \left\|\frac{\hat{d}^j f(0)}{j!}\right\| .$$

For $k = 1$, $A \in (0,+\infty]$ we get

$$p_{N,1,\alpha}(f) \leq C(\rho) \sum_{j=0}^{\infty} \rho^{-j}\|\hat{d}^j f(0)\|_N$$

$$p_{1,\alpha}(f) \leq C(\rho) \sum_{j=0}^{\infty} \rho^{-j}\|\hat{d}^j f(0)\|_N .$$

For $k = +\infty$ and $A \in (0,+\infty]$ we obtain:

$$p_{N,\infty,\alpha}(f) \leq C(\rho) \sum_{j=0}^{\infty} \rho^{-j} \left\| \frac{\hat{d}^j f(0)}{j!} \right\|_N$$

$$p_{\infty,\alpha}(f) \leq C(\rho) \sum_{j=0}^{\infty} \rho^{-j} \left\| \frac{\hat{d}^j f(0)}{j!} \right\|_N .$$

These last six inequalitites imply the results of our proposition.

□

3.2 PROPOSITION. For $k \in (1,+\infty)$, $A \in (0,+\infty]$ a subset $\mathcal{B}$ of $Exp^k_{N,A}(E)$ (respectively, $Exp^k_A(E)$) is bounded if and only if there is $\rho \in (0,A)$ such that

$$\overline{\lim_{j\to\infty}} \left(\frac{j}{ke}\right)^{\frac{1}{k}} \left[\sup_{f\in\mathcal{B}} \left\| \frac{\hat{d}^j f(0)}{j!} \right\|_N\right]^{\frac{1}{j}} \leq \rho \tag{17}$$

respectively,

$$\overline{\lim_{j\to\infty}} \left(\frac{j}{ke}\right)^{\frac{1}{k}} \left[\sup_{f\in\mathcal{B}} \left\| \frac{\hat{d}^j f(0)}{j!} \right\|\right]^{\frac{1}{j}} \leq \rho . \tag{18}$$

If $k = 1$, $A \in (0,+\infty]$, a subset $\mathcal{B}$ of $Exp^1_{N,A}(E)$, respectively, $Exp^1_A(E)$ is bounded if, and only if, there is $\rho \in (0,A)$ such that

$$\overline{\lim_{j\to\infty}} \left[\sup_{f\in\mathcal{B}} \|\hat{d}^j f(0)\|_N\right]^{\frac{1}{j}} \leq \rho \tag{19}$$

respectively

$$\overline{\lim_{j\to\infty}} \left[\sup_{f\in\mathcal{B}} \|\hat{d}^j f(0)\|\right]^{\frac{1}{j}} \leq \rho . \tag{20}$$

If $k = \infty$, $A \in (0,+\infty]$, a subset $\mathcal{B}$ of $Exp^\infty_{N,A}(E)$, respectively, $Exp^\infty_A(E)$, is bounded if, and only if, there is $\rho \in (0,A)$ such that

$$\overline{\lim_{j\to\infty}} \left[\sup_{f\in\mathcal{B}} \left\|\frac{1}{j!} \hat{d}^j f(0)\right\|_N\right]^{\frac{1}{j}} \leq \rho \tag{21}$$

respectively,

$$\overline{\lim_{j\to\infty}} \left[\sup_{f\in\mathcal{B}} \left\|\frac{1}{j!} \hat{d}^j f(0)\right\|\right]^{\frac{1}{j}} \leq \rho. \tag{22}$$

PROOF. (a) If one of the conditions (17), (18), (19), (20), (21), (22) holds it follows immediately that $\mathcal{B}$ is contained and bounded either in $\operatorname{Exp}^k_{N,A}(E)$ or in $\operatorname{Exp}^k_A(E)$, since either

$$\sup_{f\in\mathcal{B}} p(f) \leq C(\rho') \sup_{f\in\mathcal{B}} \|f\|_{N,k,\rho'} < +\infty$$

or

$$\sup_{f\in\mathcal{B}} p(f) \leq C(\rho') \sup_{f\in\mathcal{B}} \|f\|_{k,\rho'} < +\infty$$

for every $\rho' \in (\rho, A)$, when p is a continuous seminorm such that either

$$p(f) \leq C(\rho')\|f\|_{N,K,\rho'} \qquad (\forall\ f \in \operatorname{Exp}^k_{N,A}(E))$$

or

$$p(f) \leq C(\rho')\|f\|_{k,\rho'} \qquad (\forall\ f \in \operatorname{Exp}^k_A(E)).$$

(b) We suppose that $\mathcal{B}$ is bounded in $\operatorname{Exp}^k_{N,A}(E)$ (respectively, $\operatorname{Exp}^k_A(E)$). Since these spaces are inductive limits of a sequence of DF spaces of the type $\mathcal{B}^k_{N,\rho_n}(E)$ (respectively, $\mathcal{B}^k_{\rho_n}(E)$), it follows that $\mathcal{B}$ is contained in closure of a bounded subset of $\mathcal{B}^k_{N,\rho_n}(E)$ (respectively, $\mathcal{B}^k_{\rho_n}(E)$). In order to get our result it is enough to show that the closure (for the topology of $\operatorname{Exp}^k_{N,A}(E)$ [respectively, $\operatorname{Exp}^k_A(E)$]) of a closed ball of $\mathcal{B}^k_{N,\rho}(E)$ (respectively $\mathcal{B}^k_{\rho}(E)$) for $\rho \subset (0,A)$ is contained in a ball (of finite radius) of some $\mathcal{B}^k_{N,\rho_1}(E)$ (respectively, $\mathcal{B}^k_{\rho_1}(E)$) for some $\rho_1 \in (0,A)$. In fact, if this is true we have

$$\sup_{f\in\mathcal{B}} \|f\|_{N,k,\rho_1} \leq M \leq +\infty$$

(respectively,

$$\sup_{f\in\mathcal{B}} \|f\|_{k,\rho_1} \leq M < +\infty).$$

Then, for $k \in (1,+\infty)$, $A \in (0,+\infty]$

$$\rho_1^{-j} \left(\frac{j}{ke}\right)^{\frac{j}{k}} \sup_{f\in B} \left\|\frac{\hat{d}^j f(0)}{j!}\right\|_N \leq M \qquad (\forall\ j \in \mathbb{N})$$

(respectively,

$$\rho_1^{-j} \left(\frac{j}{ke}\right)^{\frac{j}{k}} \sup_{f\in\mathcal{B}} \left\|\frac{\hat{d}^j f(0)}{j!}\right\| \leq M \qquad (\forall\ j \in \mathbb{N})\ ).$$

Thus (17) (respectively, (18)) follows. The formulas (19), (20), (21) and (22) follow in analogous way.

Hence, let $k \in (1,+\infty)$, $A \in (0,+\infty]$, $\rho \in (0,A)$ and

$$\mathcal{B} = \{f \in \mathcal{B}^k_{N,\rho}(E);\ \|f\|_{N,k,\rho} \leq 1\}.$$

If g belongs to the closure $\bar{\mathcal{B}}$ of $\mathcal{B}$ in $\mathrm{Exp}^k_{N,A}(E)$, there is a net $(g_i)_{i\in I}$ in $\mathcal{B}$ such that $\lim_{i\in I} g_i = g$ in the sense of the topology of $\mathrm{Exp}^k_{N,A}(E)$. Hence

$$\rho^{-j} \left(\frac{j}{ke}\right)^{\frac{j}{k}} \sup_{i\in I} \left\|\frac{\hat{d}^j g_i(0)}{j!}\right\|_N \leq 1 \qquad (\forall\ j \in I) \tag{23}$$

By Proposition 3.1 $p_{N,k,\alpha}$ is a continuous seminorm in $\mathrm{Exp}^1_{N,A}(E)$, where

$$p_{N,k,\alpha}(f) = \rho^{-j} \left(\frac{j}{ke}\right)^{\frac{j}{k}} \left\|\frac{\hat{d}^j f(0)}{j!}\right\|_N$$

and $\alpha_j = \rho^{-j}$, $\alpha_k = 0$ se $k \neq j$. Hence, by (23), we have

$$\rho^{-j} \left(\frac{j}{ke}\right)^{\frac{j}{k}} \left\|\frac{\hat{d}^j g(0)}{j!}\right\|_N \leq 1 \qquad (\forall\ j \in \mathbb{N})$$

Thus we may write

$$\rho^{-j} \left(\frac{j}{ke}\right)^{\frac{j}{k}} \sup_{g\in\bar{\mathcal{B}}} \left\|\frac{\widehat{d^j g}(0)}{j!}\right\|_N \leq 1 \qquad (\forall\ j \in \mathbb{N})$$

Consequently

$$\lim_{j\to\infty} \left(\frac{j}{ke}\right)^{\frac{1}{k}} \sup_{g\in\bar{\mathcal{B}}} \left\|\frac{\hat{d}^j g(0)}{j!}\right\|_N^{\frac{1}{j}} \leq \rho\,.$$

Therefore, if $\rho_1 \in (\rho,A)$, there is $C(\rho_1) \geq 0$ such that

$$\left(\frac{j}{ke}\right)^{\frac{j}{k}} \sup_{g\in\bar{\mathcal{B}}} \left\|\frac{\hat{d}^j g(0)}{j!}\right\|_N \leq c(\rho_1)\rho_1^j \qquad (\forall\ j \in \mathbb{N})$$

Hence

$$\sup_{g\in \mathcal{B}} \|g\|_{N,k,\rho_2} \leq C(\rho_1) \sum_{j=0}^{\infty} \left(\frac{\rho_1}{\rho_2}\right)^j = C(\rho_1) \frac{1}{1-\frac{\rho_1}{\rho_2}} < +\infty$$

for $\rho_2 \in (\rho_1, A)$. Thus $\bar{\mathcal{B}}$ is contained in the closed ball of center 0 and radius $C(\rho_1)\frac{1}{1-\frac{\rho_1}{\rho_2}}$ in $\mathcal{B}^k_{N,\rho_2}(E)$. The other cases are proved analagously. □

3.3 COROLLARY. For $k \in [1,+\infty]$ and $A \in (0,+\infty]$, a subset $\mathcal{B}$ of $Exp^k_{N,A}(E)$ (respectively, $Exp^k_A(E)$) is bounded if, and only if, there is $\rho \in (0,A)$ such that $\mathcal{B}$ is contained and bounded in $\mathcal{B}^k_{N,\rho}(E)$ (respectively, $\mathcal{B}^k_\rho(E)$).

3.4 PROPOSITION. For $k \in (1,+\infty)$, $A \in [0,+\infty)$, a subset $\mathcal{B}$ of $Exp^k_{N,0,A}(E)$, respectively, $Exp^k_{0,A}(E)$, is bounded if, and only if,

$$\overline{\lim_{j\to\infty}} \left(\frac{j}{ke}\right)^{\frac{1}{k}} \left[\sup_{f\in\mathcal{B}} \left\|\frac{\hat{d}^j f(0)}{j!}\right\|_N\right]^{\frac{1}{j}} \leq A, \tag{24}$$

respectively,

$$\overline{\lim_{j\to\infty}} \left(\frac{j}{ke}\right)^{\frac{1}{k}} \left[\sup_{f\in\mathcal{B}} \left\|\frac{\hat{d}^j f(0)}{j!}\right\|\right]^{\frac{1}{j}} \leq A. \tag{25}$$

If $k = 1$ and $A \in [0,+\infty)$, a subset $\mathcal{B}$ of $Exp^1_{N,0,A}(E)$, respectively, $Exp^1_{0,A}(E)$, is bounded if, and only if,

$$\overline{\lim_{j\to\infty}} \left[\sup_{f\in\mathcal{B}} \|\hat{d}^j f(0)\|_N\right]^{\frac{1}{j}} \leq A, \tag{26}$$

respectively,

$$\overline{\lim_{j\to\infty}} \left[\sup_{f\in\mathcal{B}} \|\hat{d}^j f(0)\|\right]^{\frac{1}{j}} \leq A. \tag{27}$$

If $k = +\infty$ and $A \in [0,+\infty)$, a subset $\mathcal{B}$ of $Exp^\infty_{N,0,A}(E)$, respectively, $Exp^\infty_{0,A}(E)$, is bounded if, and only if,

$$\overline{\lim_{j\to\infty}} \left[\sup_{f\in\mathcal{B}} \left\|\frac{\hat{d}^j f(0)}{j!}\right\|_N\right]^{\frac{1}{j}} \leq A, \tag{28}$$

respectively,

$$\overline{\lim_{j\to\infty}} \left[\sup_{f\in\mathcal{B}} \left\| \frac{\hat{d}^j f(0)}{j!} \right\| \right]^{\frac{1}{j}} \leq A \ . \tag{29}$$

PROOF. Straightforward.

4. FOURIER-BOREL TRANSFORMATIONS

4.1 DEFINITION. If T is in any of the continuous duals of $\mathrm{Exp}_N(E)$, $\mathrm{Exp}^k_{N,A}(E)$ (for $k \in (1,+\infty]$ and $A \in (0,+\infty]$) and $\mathrm{Exp}^k_{N,0,A}(E)$ (for $k \in (1,+\infty]$ and $A \in [0,+\infty)$), the Fourier-Borel transform FT is defined by $FT(\varphi) = T(e^{\varphi})$ for all $\varphi \in E'$. If T is in the continuous dual of $\mathrm{Exp}_{N,A}(E)$ (for $A \in (0,+\infty)$) the Fourier-Borel transform FT of T is given by $FT(\varphi) = T(e^{\varphi})$ for all $\varphi \in E'$ such that $\|\varphi\| < A$.

4.2 NOTATIONS. As usual we set $A^{-1} = \frac{1}{A}$ for $A \in (0,+\infty)$. If $A = 0$ we denote $A^{-1} = +\infty$ and, if $A = +\infty$, we set $A^{-1} = 0$. For $k \in (1,+\infty)$ we define $k' \in (1,+\infty)$ as that number such that $\frac{1}{k} + \frac{1}{k'} = 1$. For $k = 1$, we set $k' = +\infty$ and for $k = +\infty$, we put $k' = 1$. We define $\lambda(k) = \dfrac{k}{(k-1)^{\frac{k-1}{k}}}$ for $k \in (1,+\infty)$. Since $\lim_{k\to 1} \lambda(k) = 1 = \lim_{k\to\infty} \lambda(k)$, we set $\lambda(1) = \lambda(+\infty) = 1$.

4.3 THEOREM. The Fourier-Borel transformation F is a vector-space isomorphism between:

(1) $[\mathrm{Exp}^k_{N,A}(E)]'$ and $\mathrm{Exp}^{k'}_{0,[\lambda(k)A]^{-1}}(E')$ for $k \in [1,+\infty]$ and $A \in (0,+\infty]$.

(2) $[\mathrm{Exp}^k_{N,0,A}(E)]'$ and $\mathrm{Exp}^{k'}_{(\lambda(k)A)^{-1}}(E')$ for $k \in (1,+\infty]$ and $A \in [0,+\infty)$.

PROOF. By the definition of F and by Proposition 2.17, it is clear that F is an injection. It is also quite clear that F is linear.

First we consider case (1) with $k \in (1,+\infty)$ and $A \in (0,+\infty]$. In this case, if $T \in [\mathrm{Exp}_{N,A}^{k}(E)]'$, then for all $\rho \in (0,A)$ there is $C(\rho) > 0$ such that

$$|T(f)| \leq C(\rho) \sum_{j=0}^{\infty} \frac{1}{\rho^j} \left(\frac{j}{ek}\right)^{\frac{j}{k}} \left\|\frac{\hat{d}^j f(0)}{j!}\right\|_N$$

for every $f \in \mathrm{Exp}_{N,A}^{k}(E)$. Hence, for $P \in \mathcal{P}_N({}^jE)$, we have

$$|T(P)| \leq C(\rho) \frac{1}{\rho^j} \left(\frac{j}{ek}\right)^{\frac{j}{k}} \|P\|_N .$$

If we set $T_j = T|\mathcal{P}_N({}^jE)$, we have, by a result of Gupta (see [1]) $\beta T_j \in \mathcal{P}({}^jE')$ with $\beta T_j(\varphi) = T_j(\varphi^j)$ for all $\varphi \in E'$ and

$$\|\beta T_j\| = \|T_j\| \leq C(\rho) \frac{1}{\rho^j} \left(\frac{j}{ek}\right)^{\frac{j}{k}} \tag{30}$$

for all $\rho < A$. Hence we may write

$$(FT)(\varphi) = \sum_{j=0}^{\infty} \frac{1}{j!} T(\varphi^j) = \sum_{j=0}^{\infty} \frac{1}{j!} \beta T_j(\varphi) \tag{31}$$

for all $\varphi \in E'$. By (30) we have

$$\limsup_{j\to\infty} \left[\frac{j}{ek'}\right]^{\frac{1}{k'}} \frac{1}{(j!)^{\frac{1}{j}}} \|\beta T_j\|^{\frac{1}{j}} \leq$$

$$\leq \limsup_{j\to\infty} [C(\rho)]^{\frac{1}{j}} \frac{1}{\rho} \left[\frac{j}{ek}\right]^{\frac{1}{k}} \left[\frac{j}{ek'}\right]^{\frac{1}{k'}} \frac{1}{(j!)^{\frac{1}{j}}} =$$

$$= \frac{1}{e\rho} \limsup_{j\to\infty} \frac{j}{(j!)^{\frac{1}{j}}} \left(\frac{1}{k}\right)^{\frac{1}{k}} \left(\frac{1}{k'}\right)^{\frac{1}{k'}} = \frac{1}{\rho\lambda(k)}$$

for all $\rho \in (0,A)$. Hence

$$\limsup_{j\to\infty} \left[\frac{j}{ek'}\right]^{\frac{1}{k'}} \frac{1}{(j!)^{\frac{1}{j}}} \|\beta T_j\|^{\frac{1}{j}} \leq \frac{1}{A\lambda(k)} . \qquad (32)$$

Since

$$\limsup_{j\to\infty} \left[\frac{j}{ek'}\right]^{\frac{1}{k'}} = +\infty$$

it follows from (32) that the radius of convergence of (31) is $+\infty$. Therefore $FT \in \mathrm{Exp}^{k'}_{0,(A\lambda(k))^{-1}}(E')$.

Now we consider $H \in \mathrm{Exp}^{k'}_{0,(A\lambda(k))^{-1}}(E')$. Hence, for each $\rho < A$, $\rho > 0$, there is $C(\rho) > 0$ such that

$$\left[\frac{j}{ek'}\right]^{\frac{j}{k'}} \frac{1}{j!} \|\hat{d}^j H(0)\| \leq C(\rho) \frac{1}{[\rho\lambda(k)]^j} .$$

Let $T_j \in [\mathcal{P}_N({}^jE)]'$ be such that $\beta T_j = \hat{d}^j H(0)$ and $\|T_j\| = \|\hat{d}^j H(0)\|$ (see Gupta [1]). Hence

$$\frac{1}{j!} \|T_j\| \leq C(\rho) \frac{1}{\rho^j} \frac{1}{[\lambda(k)]^j} \left[\frac{ek'}{j}\right]^{\frac{j}{k'}} . \qquad (33)$$

For $f \in \mathrm{Exp}^k_{N,A}(E)$ we define

$$T_H(f) = \sum_{j=0}^{\infty} T_j\left(\frac{1}{j!} \hat{d}^j f(0)\right).$$

Hence

$$|T_H(f)| \leq \sum_{j=0}^{\infty} \frac{1}{j!} \|T_j\| \, \|\hat{d}^j f(0)\|_N . \qquad (34)$$

By (33) we get

$$\frac{1}{j!} \|T_j\| \, \|\hat{d}^j f(0)\|_N \leq c(\rho) \frac{1}{\rho^j} \left[\frac{j}{ke}\right]^{\frac{j}{k}} \left[\frac{e}{j}\right]^j \|\hat{d}^j f(0)\|_N .$$

Since $\lim_{j\to\infty} \frac{e}{j} (j!)^{\frac{1}{j}} = 1$, for each $\varepsilon > 0$, there is $c(\varepsilon) > 0$ such that

$$\left[\frac{e}{j}\right]^j \leq c(\varepsilon)(1+\varepsilon)^j \frac{1}{j!} \qquad \forall\, j \in \mathbb{N}.$$

Hence

$$\frac{1}{j!}\,\|T_j\|\,\|\hat{d}^j f(0)\|_N \leq c(\rho)c(\varepsilon)\Big[\frac{1+\varepsilon}{\rho}\Big]^j\Big[\frac{j}{ek}\Big]^{\frac{j}{k}}\,\Big\|\frac{\hat{d}^j f(0)}{j!}\Big\|_N . \tag{35}$$

Therefore

$$|T_H(f)| \leq c(\rho)c(\varepsilon)\|f\|_{N,K,\frac{\rho}{1+\varepsilon}}$$

for all $f \in \mathrm{Exp}^k_{N,A}(E)$, all $\rho < A$, $\rho > 0$ and all $\varepsilon > 0$. This implies that $T_H \in [\mathrm{Exp}^k_{N,A}(E)]'$. It is very easy to show that $FT_H = H$.

Now we prove the case (1) with $k = 1$ and $A \in (0,+\infty]$. If $T \in [\mathrm{Exp}_{N,A}(E)]'$, for each $\rho \in (0,A)$, there is $c(\rho) > 0$ such that

$$|T(f)| \leq c(\rho) \sum_{j=0}^{\infty} \frac{1}{\rho^j}\,\|\hat{d}^j f(0)\|_N$$

for all $f \in \mathrm{Exp}_{N,A}(E)$. Hence, for $P \in \mathcal{P}_N({}^jE)$, we have

$$|T(P)| \leq c(\rho)\,\frac{j!}{\rho^j}\,\|P\|_N . \tag{36}$$

If $T_j = T|\mathcal{P}_N({}^jE)$, by a result of Gupta [1], we have $\beta T_j \in \mathcal{P}({}^jE')$ such that $\beta T_j(\varphi) = T_j(\varphi^j)$ for all $\varphi \in E'$ and $\|\beta T_j\| = \|T_j\|$. By (36)

$$\|\beta T_j\| \leq c(\rho)\,\frac{j!}{\rho^j}. \tag{37}$$

We have

$$FT(\varphi) = \sum_{j=0}^{\infty} \frac{1}{j!}\,T(\varphi^j) = \sum_{j=0}^{\infty} \frac{1}{j!}\,\beta T_j(\varphi) \tag{38}$$

for all $\varphi \in E'$, $\|\varphi\| \leq A$ (see Proposition 2.17). By (37) we have

$$\limsup_{j\to\infty} \Big\|\frac{1}{j!}\,\beta T_j\Big\|^{\frac{1}{j}} \leq \frac{1}{\rho}$$

for all $\rho \in (0,A)$. Hence the radius of convergence of (37) is $\geq\rho$ for all $\rho \in (0,A)$. Thus it is $\geq A$. It follows that

$$FT \in \mathcal{H}_b(B_A(0)) = \mathrm{Exp}^{\infty}_{0,\frac{1}{A}}(E').$$

Now, if $H \in \mathrm{Exp}^{\infty}_{0,\frac{1}{A}}(E') = \mathcal{H}_b(B_A(0))$, we have

$$\limsup_{j\to\infty} \left\|\frac{\hat{d}^j H(0)}{j!}\right\|^{\frac{1}{j}} \leq \frac{1}{A}. \tag{39}$$

By the result of Gupta mentioned earlier there is $T_j \in [\mathcal{P}_N(^jE)]'$ such that $\beta T_j = \hat{d}^j H(0)$ and $\|T_j\| = \|\hat{d}^j H(0)\|$. For all $f \in \mathrm{Exp}_{N,A}(E)$ we define

$$T_H(f) = \sum_{j=0}^{\infty} \frac{1}{j!} T_j(\hat{d}^j f(0)).$$

We have

$$\left|\frac{1}{j!} T_j(\hat{d}^j f(0)\right| \leq \frac{1}{j!} \|T_j\| \, \|\hat{d}^j f(0)\|_N =$$

$$= \frac{1}{j!} \|\hat{d}^j H(0)\| \, \|\hat{d}^j f(0)\|_N. \tag{40}$$

By (39), if $\rho \in (0,A)$ there is $c(\rho) > 0$ such that

$$\frac{1}{j!} \|\hat{d}^j H(0)\| \leq c(\rho) \frac{1}{\rho^j} \qquad \forall\, j \in \mathbb{N}. \tag{41}$$

Hence (40) and (41) imply

$$\sum_{j=0}^{\infty} \left|\frac{1}{j!} T_j(\hat{d}^j f(0)\right| \leq \sum_{j=0}^{\infty} c(\rho) \frac{1}{\rho^j} \|\hat{d}^j f(0)\|_N = c(\rho)\|f\|_{N,\rho}.$$

Hence $T_H \in [\mathrm{Exp}_{N.A}(E)]'$. As usual is easy to see that $FT_H = H$. Now, we prove case (1) with $k = +\infty$ and $A \in (0,+\infty]$.

Let $T \in [\mathrm{Exp}^{\infty}_{N,A}(E)]' = [\mathcal{H}_{Nb}(B_{\frac{1}{A}}(0))]'$. Thus, for all $\rho \in (0,A)$ there is $c(\rho) > 0$ such that

$$T(f) \leq c(\rho) \sum_{j=0}^{\infty} \frac{1}{\rho^j} \left\|\frac{\hat{d}^j f(0)}{j!}\right\|_N$$

for every $f \in \mathcal{H}_{Nb}(\bar{B}_{\frac{1}{A}}(0))$. Hence

$$|T(P)| \leq c(\rho) \frac{1}{\rho^j} \|P\|_N \tag{42}$$

for all $P \in \mathcal{P}_N({}^jE)$, $j \in \mathbb{N}$. If $T_j = T|\mathcal{P}_N({}^jE)$, by a result of Gupta [1] we have $\beta T_j \in \mathcal{P}({}^jE')$ such that $\|\beta T_j\| = \|T_j\|$ and $\beta T_j(\varphi) = T_j(\varphi^n)$ for all $\varphi \in E'$.

We have

$$(FT)(\varphi) = \sum_{j=0}^{\infty} \frac{1}{j!} \beta T_j(\varphi) \qquad \forall\ \varphi \in E'$$

and, by (23),

$$\limsup_{j\to\infty} \|\beta T_j\|^{\frac{1}{j}} \leq \frac{1}{\rho} \qquad \forall\ \rho \in (0,A).$$

Hence

$$\limsup_{j\to\infty} \|\beta T_j\|^{\frac{1}{j}} \leq \frac{1}{A}.$$

In clear that

$$\limsup_{j\to\infty} [\frac{1}{j!} \|\beta T_j\|]^{\frac{1}{j}} = 0.$$

Hence $FT \in \mathrm{Exp}_{0,\frac{1}{A}}(E')$.

Now, let $H \in \mathrm{Exp}_{0,\frac{1}{A}}(E')$. It follows that

$$\limsup_{j\to\infty} \|\hat{d}^jH(0)\|^{\frac{1}{j}} \leq \frac{1}{A}. \qquad (43)$$

Let $T_j \in [\mathcal{P}_N({}^jE)]'$ be such that $\beta T_j = \hat{d}^jH(0)$ and $\|T_j\| = \|\hat{d}^jH(0)\|$. This is possible by the result of Gupta mentioned earlier. We define

$$T_H(f) = \sum_{j=0}^{\infty} T_j(\frac{\hat{d}^jf(0)}{j!}) \qquad (44)$$

for all $f \in \mathrm{Exp}^{\infty}_{N,A}(E)$. Thus, for $\rho \in (0,A)$, $\exists\ c(\rho) > 0$ such that

$$|T_j(\frac{\hat{d}^jf(0)}{j!})| \leq \|T_j\| \frac{1}{j!} \|\hat{d}^jf(0)\|_N \leq c(\rho) \frac{1}{j!} \|\hat{d}^jf(0)\|_N \frac{1}{\rho^j}$$

for all $j \in \mathbb{N}$. This follows from (43). Now, from (44), we get

$$|T_H(f)| \leq c(\rho) \sum_{j=0}^{\infty} \frac{1}{\rho^j} \left\| \frac{\hat{d}^j f(0)}{j!} \right\|_N$$

for all $\rho \in (0,A)$. Thus $T_H \in (\mathrm{Exp}^{\infty}_{N,A}(E))'$. It is easy to prove that $FT_H = H$.

Now we prove case (2) with $k \in (1,+\infty)$ and $A \in [0,+\infty)$.

Let $T \in [\mathrm{Exp}_{N,0,A}(E)]'$ be given. Thus there are $\rho > A$ and $c(\rho) > 0$ such that

$$|T(f)| \leq c(\rho) \sum_{j=0}^{\infty} \frac{1}{j!} \frac{1}{\rho^j} \left(\frac{j}{ek}\right)^{\frac{j}{k}} \|\hat{d}^j f(0)\|_N$$

for all $f \in \mathrm{Exp}^k_{N,0,A}(E)$. For $T_j = T|\mathcal{P}_N({}^jE)$ we get

$$|T_j(P)| \leq c(\rho) \frac{1}{\rho^j} \left(\frac{j}{ek}\right)^{\frac{j}{k}} \|P\|_N \tag{45}$$

for all $P \in \mathcal{P}_N({}^jE)$, $j \in \mathbb{N}$. By Gupta's result, mentioned earlier, $\beta T_j \in \mathcal{P}({}^jE')$, $\beta T_j(\varphi) = T_j(\varphi^j)$ and $\|\beta T_j\| = \|T_j\|$. Thus, by (45),

$$\|\beta T_j\| \leq c(\rho) \frac{1}{\rho^j} \left(\frac{j}{ek}\right)^{\frac{j}{k}} \qquad \forall\ j \in \mathbb{N}$$

and

$$\limsup_{j\to\infty} \left\|\frac{\beta T_j}{j!}\right\|^{\frac{1}{j}} \left(\frac{j}{ek'}\right)^{\frac{1}{k'}} \leq \limsup_{j\to\infty} [c(\rho)]^{\frac{1}{j}} \left(\frac{1}{j!}\right)^{\frac{1}{j}} \frac{1}{\rho} \left(\frac{j}{ek}\right)^{\frac{1}{k}} \left(\frac{j}{ek'}\right)^{\frac{1}{k'}} =$$

$$= \frac{1}{\rho e} \limsup_{j\to\infty} \frac{j}{(j!)^{\frac{1}{j}}} \left(\frac{1}{k}\right)^{\frac{1}{k}} \left(\frac{1}{k'}\right)^{\frac{1}{k'}} = \frac{1}{\rho\lambda(k)} < \frac{1}{A\lambda(k)}.$$

Since $\limsup_{j\to\infty} \left(\frac{j}{ek'}\right)^{\frac{1}{k'}} = +\infty$ we get $\limsup_{j\to\infty} \left\|\frac{\beta T_j}{j!}\right\|^{\frac{1}{j}} = 0$.

Hence $FT \in \mathrm{Exp}^{k'}_{\frac{1}{A\lambda(k)}}(E')$.

Now let $H \in \mathrm{Exp}^{k'}_{\frac{1}{A\lambda(k)}}(E')$ be given. Hence there are $\rho \in A$ and $c(\rho) > 0$ such that

$$\left[\frac{j}{ek'}\right]^{\frac{j}{k'}} \left\|\frac{\hat{d}^j f(0)}{j!}\right\| \leq c(\rho)\, \frac{1}{\rho^j [\lambda(k)]^j}. \tag{46}$$

By a Gupta's result mentioned earlier there is $T_j \in [\mathcal{P}_N(^jE)]'$ such that $T_j = \beta^{-1}(\hat{d}^j H(0))$ and $\|T_j\| = \|\hat{d}^j H(0)\|$. Now, we define

$$T_H(f) = \sum_{j=0}^{\infty} T_j \left(\frac{\hat{d}^j f(0)}{j!}\right)$$

for every $f \in Exp^k_{N,0,A}(E)$. We use (46) and $\|T_j\| = \|\hat{d}^j H(0)\|$ and we obtain

$$\left|T_j \left(\frac{\hat{d}^j f(0)}{j!}\right)\right| \leq c(\rho)\, \frac{j!}{\rho^j} \left(\frac{1}{k}\right)^{\frac{j}{k}} \left(\frac{1}{k'}\right)^{\frac{j}{k'}} \left(\frac{ek'}{j}\right)^{\frac{j}{k'}} \left\|\frac{\hat{d}^j f(0)}{j!}\right\|_N =$$

$$= c(\rho)\, \frac{1}{\rho^j} \left(\frac{j}{ek}\right)^{\frac{j}{k}} \left(\frac{e}{j}\right)^j j!\, \left\|\frac{d^j f(0)}{j!}\right\|_N. \tag{47}$$

Since $\lim\limits_{j\to\infty} \frac{(j!)^{\frac{1}{j}}}{j} = e^{-1}$, for each $\varepsilon > 0$, there is $c(\varepsilon) > 0$ such that $\left(\frac{e}{j}\right)^j j! \leq c(\varepsilon)(1+\varepsilon)^j$ for all $j \in \mathbb{N}$.

This and (47) imply

$$\left|T_j \left(\frac{\hat{d}^j f(0)}{j!}\right)\right| \leq c(\rho) c(\varepsilon) \left(\frac{1+\varepsilon}{\rho}\right)^j \left(\frac{j}{ek}\right)^{\frac{1}{j}} \left\|\frac{\hat{d}^j f(0)}{j!}\right\|_N. \tag{48}$$

If $\varepsilon > 0$ is such that $\frac{\rho}{1+\varepsilon} > A$ we get

$$|T_H(f)| \leq c(\rho) c(\varepsilon)\, \|f\|_{N,k,\frac{\rho}{1+\varepsilon}}$$

for all $f \in Exp^k_{N,0,A}(E)$. Hence $T_H \in [Exp^k_{N,0,A}(E)]'$ and $FT_H = H$.

Now there is only the case (2) with $k = +\infty$ and $A \in [0,+\infty)$. In this case

$$Exp^{\infty}_{N,0,A}(E) = \mathcal{H}_{Nb}(B_{\frac{1}{A}}(0)).$$

The proofs were done in Gupta [1] and Matos [1] and [2].

Q.E.D.

4.4 REMARK. As we saw in Proposition 2.17, $e^{\varphi} \in \mathrm{Exp}_{N,0,A}(E)$ if $\varphi \in E'$ and $\|\varphi\| \leq A$. Hence if $T \in [\mathrm{Exp}_{N,0,A}(E)]'$, the natural definition for its Fourier-Borel transform $\mathcal{F}T$ would be $\mathcal{F}T(\varphi) = T(e^{\varphi})$ for all $\varphi \in E'$ with $\|\varphi\| \leq A$. However it can be proved that we can define $\mathcal{F}T$ for all $\varphi \in E'$ with $\|\varphi\| \leq \rho_T$, $\rho_T > A$, in such a very that it agrees with the previous definition for $\varphi \in E'$, $\|\varphi\| \leq A$.

4.5 DEFINITION. If $T \in [\mathrm{Exp}_{N,0,A}(E)]'$, we define its Fourier-Borel transform $\mathcal{F}T$ by

$$\mathcal{F}T(\varphi) = \sum_{j=0}^{\infty} \frac{1}{j!} \beta T_j(\varphi) \tag{49}$$

for all $\varphi \in E'$ such that (49) converges absolutely. Here $T_j = T|\mathcal{P}_N({}^jE)$ and $\beta T_j \in \mathcal{P}({}^jE')$ is given by $\beta T_j(\varphi) = T_j(\varphi^n)$. As we wrote previously $\|\beta T_j\| = \|\beta T_j\|$ by a Gupta's result.

4.6 PROPOSITION. If $A \in [0,+\infty)$ and $T \in [\mathrm{Exp}_{N,0,A}(E)]'$, there is $\rho > A$ such that $\mathcal{F}T \in \mathcal{H}_b(B_\rho(0))$, $B_\rho(0)$ the open ball in E'.

PROOF. If $T \in [\mathrm{Exp}_{N,0,A}(E)]'$ there are $\rho > 0$ and $c(\rho) > 0$ such that

$$|T(f)| \leq c(\rho) \sum_{j=0}^{\infty} \rho^{-j} \|\hat{d}^j f(0)\|_N$$

for all $f \in \mathrm{Exp}_{N,0,A}(E)$. Hence

$$|T(P)| \leq c(\rho)\, j!\, \rho^{-j} \|P\|_N$$

for all $P \in \mathcal{P}_N({}^jE)$. Using the notations of Definitions 4.5, we get

$$\|T_j\| = \|\beta T_j\| \leq c(\rho)\, j!\, \rho^{-j} \qquad \forall\ j \in \mathbb{N}.$$

Thus

$$\limsup_{j\to\infty} \left\|\frac{\beta T_j}{j!}\right\|^{\frac{1}{j}} \leq \frac{1}{\rho}. \tag{50}$$

It follows that $FT \in \mathcal{H}_b(B_\rho(0))$ since (50) says that the radius of convergence of FT is $\geq \rho$.

Q.E.D.

4.7 THEOREM. The Fourier-Borel transformation F is a vector space isomorphism between $[\mathrm{Exp}_{N,0,A}(E)]'$ and $\mathrm{Exp}^{\infty}_{\frac{1}{A}}(E')$ for $A \in [0,+\infty)$.

PROOF. By Proposition 4.6 and the definition of $\mathrm{Exp}^{\infty}_{\frac{1}{A}}(E')$ it is clear that $FT \in \mathrm{Exp}^{\infty}_{\frac{1}{A}}(E')$ for all $T \in [\mathrm{Exp}_{N,0,A}(E)]'$. If $T \in [\mathrm{Exp}_{N,0,A}(E)]'$ is such that $FT = 0$, then $(FT)(\varphi) = 0$ for all $\varphi \in E'$, $\|\varphi\| < \rho$ for some $\rho > A$ (see 4.6). Hence $0 = \|\beta T_j\| = \|T_j\|$ for all $j \in \mathbb{N}$. Thus $T(P) = 0$ for all $j \in N$ and $P \in \mathcal{P}_N({}^jE)$, it follows from Proposition 2.15 that $T = 0$. Hence F is an injection.

Now we consider $H \in \mathrm{Exp}^{\infty}_{\frac{1}{A}}(E') = \mathcal{H}_b(\overline{B_A(0)})$. Hence there is $\rho > A$ such that $H \in \mathcal{H}_b(B_\rho(0))$. Therefore

$$\limsup_{j\to\infty} \left\|\frac{\hat{d}^jH(0)}{j!}\right\|^{\frac{1}{j}} \leq \frac{1}{\rho}.$$

Therefore, for every $\varepsilon > 0$, there is $c(\varepsilon)$ such that

$$\frac{1}{j!}\|\hat{d}^jH(0)\| \leq c(\varepsilon)\left(\frac{1+\varepsilon}{\rho}\right)^j \tag{51}$$

for all $j \in N$. By a Gupta's result mentioned earlier there is $T_j \in [\mathcal{P}_N({}^jE)]'$ such that $\beta T_j = \hat{d}^jH(0)$ and $\|T_j\| = \|\hat{d}^jH(0)\|$. Thus, by (51), we have

$$\left|T_j\left(\frac{\hat{d}^jf(0)}{j!}\right)\right| \leq c(\varepsilon)\left(\frac{1+\varepsilon}{\rho}\right)^j \|\hat{d}^jf(0)\|_N \tag{52}$$

for all $j \in \mathbb{N}$ and all $f \in \mathrm{Exp}_{N,0,A}(E)$. Now we define

$$T_H(f) = \sum_{j=0}^{\infty} T_j\left(\frac{1}{j!}\hat{d}^jf(0)\right)$$

for all $f \in \mathrm{Exp}_{N,0,A}(E)$. We use (52) to obtain

$$|T_H(f)| \leq c(\varepsilon)\|f\|_{N,\frac{\rho}{1+\varepsilon}}$$

for all $f \in \mathrm{Exp}_{N,0,A}(E)$, all $\rho > A$ and all $\varepsilon > 0$ such that $\frac{\rho}{1+\varepsilon} > A$. Thus $T_H \in [\mathrm{Exp}_{N,0,A}(E)]'$. It is easy to see that $FT_H = H$.

Q.E.D.

4.8 THEOREM. The Fourier-Borel transformation F is a topological isomorphism between:

(a) $[\mathrm{Exp}^k_{N,A}(E)]'_\beta$ and $\mathrm{Exp}^{k'}_{0,[\lambda(k)A]^{-1}}(E')$ if $k \in [1,+\infty]$ and $A \in (0,+\infty]$.

(b) $[\mathrm{Exp}^k_{N,0,A}(E)]'_\beta$ and $\mathrm{Exp}^{k'}_{\frac{1}{\lambda(k)A}}(E')$ if $k \in [1,+\infty]$ and $A \in [0,+\infty)$.

Here $[\mathrm{Exp}^k_{N,A}(E)]'_\beta$ and $[\mathrm{Exp}^k_{N,0,A}(E)]'_\beta$ denote the duals $[\mathrm{Exp}^k_{N,A}(E)]'$ and $[\mathrm{Exp}^k_{N,0,A}(E)]'$ with the strong topologies.

PROOF. Since $\mathrm{Exp}^k_{N,A}(E)$ is a DF space, $\mathrm{Exp}^{k'}_{0,\frac{1}{\lambda(k)A}}(E')$ is a Fréchet space and F is an algebraic isomorphism, in order to prove (a) it is enough to show that F^{-1} is continuous. In fact, the Open Mapping Theorem implies that F is continuous.

Now we prove that F^{-1} is continuous for $k \in (1,+\infty)$ and $A \in (0,+\infty]$. Let $\mathcal{B}$ be a bounded subset of $\mathrm{Exp}^k_{N,A}(E)$.

By 3.2 there is $\rho \in (0,A)$ such that

$$\overline{\lim_{j\to\infty}} \left(\frac{j}{ke}\right)^{\frac{1}{k}} \left[\sup_{f\in\mathcal{B}} \left\|\frac{\hat{d}^j f(0)}{j!}\right\|_N\right]^{\frac{1}{j}} \leq \rho .$$

Hence, for every $\varepsilon > 0$ there is $c(\varepsilon) \geq 0$ such that

$$\left(\frac{j}{ke}\right)^{\frac{j}{k}} \sup_{f\in\mathcal{B}} \left\|\frac{\hat{d}^j f(0)}{j!}\right\|_N \leq c(\varepsilon)(\rho+\varepsilon)^j$$

for all $j \in \mathbb{N}$.

We have:

$$\sup_{f\in\mathcal{B}} |F^{-1}(H)(f)| = \sup_{f\in\mathcal{B}} |T_H(f)| \leq$$

$$\leq \sum_{j=0}^{\infty} \|\hat{d}^j H(0)\| \sup_{f\in\mathcal{B}} \frac{1}{j!} \|\hat{d}^j f(0)\|_N \leq$$

$$\leq c(\varepsilon) \sum_{j=0}^{\infty} (\rho+\varepsilon)^j \left(\frac{ke}{j}\right)^{\frac{j}{k}} j! \left\|\frac{\hat{d}^j H(0)}{j!}\right\| = \circledast .$$

Since

$$\left(\frac{ke}{j}\right)^{\frac{j}{k}} = \lambda(k)^j \left(\frac{j}{k'e}\right)^{\frac{j}{k'}} \frac{e^j}{j^j}$$

$$\circledast \leq c(\varepsilon) \sum_{j=0}^{\infty} \lambda(k)^j (\rho+\varepsilon)^j \frac{j!}{j^j} e^j \left(\frac{j}{k'e}\right)^{\frac{j}{k'}} \left\|\frac{\hat{d}^j H(0)}{j!}\right\| = \boxtimes .$$

We know that

$$\lim_{j\to\infty} \left(\frac{j!}{j^j}\right)^{\frac{1}{j}} e^{\frac{1}{j}} = 1.$$

Hence there is $d(\varepsilon) \geq 0$ such that

$$\frac{j!}{j^j} e^j \leq d(\varepsilon)(1+\varepsilon)^j \qquad \forall\ j \in \mathbb{N}.$$

Therefore

$$\boxtimes \leq d(\varepsilon)c(\varepsilon) \sum_{j=0}^{\infty} \lambda(k)^j (\rho+\varepsilon)^j (1+\varepsilon)^j \left(\frac{j}{k'e}\right)^{\frac{j}{k'}} \left\|\frac{\hat{d}^j H(0)}{j!}\right\|$$

$$= d(\varepsilon)c(\varepsilon) \|f\|_{k', \frac{1}{\lambda(k)(\rho+\varepsilon)(1+\varepsilon)}} \qquad (\forall\ \varepsilon > 0).$$

Now we choose $\varepsilon > 0$ such that $(\rho+\varepsilon)(1+\varepsilon) < A$. Hence $\frac{1}{\lambda(k)(\rho+\varepsilon)(1+\varepsilon)} > \frac{1}{\lambda(k)A}$ and F^{-1} is continuous. The proofs for all the other values of k for case (a) follow the same pattern.

Now we prove case (b) for $k \in (1,+\infty)$ and $A \in [0,+\infty)$.

Let p be a continuous seminorm in $\mathrm{Exp}^{k'}_{\frac{1}{\lambda(k)A}}(E')$. Hence, for every $\rho < A$, we have $\frac{1}{\lambda(k)} > \frac{1}{\lambda(k)A}$, and there is $c(\rho)$ such that

$$p(H) \leq c(\rho)\ \|H\|_{k',\frac{1}{\lambda(k)}} \qquad \forall\ H \in \mathrm{Exp}^{k'}_{\frac{1}{\lambda(k)A}}(E').$$

In particular

$$p(FT) \leq c(\rho) \sum_{j=0}^{\infty} (\lambda(k)\rho)^j \left(\frac{j}{k'e}\right)^{\frac{j}{k'}} \left\|\frac{\hat{d}^j(FT)(0)}{j!}\right\| =$$

$$= c(\rho) \sum_{j=0}^{\infty} (\lambda(k)\rho)^j \left(\frac{j}{k'e}\right)^{\frac{j}{k'}} \frac{1}{j!} \|T_j\| = (\#)$$

for every $T \in [\mathrm{Exp}^k_{N,0,A}(E)]'$.

We consider

$$\mathcal{B}_j = \{P \in \mathcal{P}_N(^jE);\ \left(\frac{j}{ke}\right)^{\frac{j}{k}} \|P\|_N \leq A^j\} \qquad (j \in \mathbb{N}).$$

If $\mathcal{B} = \bigcup_{j=0}^{\infty} \mathcal{B}_j$ we have

$$\left(\frac{j}{ke}\right)^{\frac{j}{k}} \sup_{f\in\mathcal{B}} \left\|\frac{\widehat{d^j f}(0)}{j!}\right\|_N = \left(\frac{j}{ke}\right)^{\frac{j}{k}} \sup_{P\in\mathcal{B}_j} \|P\|_N \leq A^j$$

for all $j \in \mathbb{N}$. Hence

$$\overline{\lim_{j\to\infty}} \left(\frac{j}{ke}\right)^{\frac{1}{k}} \left[\sup_{f\in\mathcal{B}} \left\|\frac{\hat{d}^j f(0)}{j!}\right\|_N\right]^{\frac{1}{j}} \leq A$$

and $\mathcal{B}$ is a bounded subset of $\mathrm{Exp}^k_{N,0,A}(E)$ by Proposition 3.4.

Now we write

$$(\#) \leq c(\rho) \sum_{j=0}^{\infty} [\rho\lambda(k)]^j \left(\frac{j}{k'e}\right)^{\frac{j}{k'}} \frac{1}{j!} A^{-j} \left(\frac{j}{ke}\right)^{\frac{j}{k}} =$$

$$= c(\rho) \sup_{f\in\mathcal{B}} |T(f)| \sum_{j=0}^{\infty} \left(\frac{\rho}{A}\right)^j \frac{j^j}{j!} e^{-j} = (\#\#).$$

Since

$$\lim_{j\to\infty} \frac{j}{(j!)^{\frac{1}{j}}} e^{-1} = 1$$

there is $c(\varepsilon) \geq 0$ such that

$$\frac{j^j}{j!} e^j \leq c(\varepsilon)(1+\varepsilon)^j \qquad (\S)$$

or all $j \in \mathbb{N}$. Hence

$$(\#\#) \leq c(\rho)c(\varepsilon) \sup_{f\in\mathcal{B}} |T(f)| \sum_{j=0}^{\infty} \left[\frac{\rho(1+\varepsilon)}{A}\right]^j \leq D_\varepsilon \sup_{f\in\mathcal{B}} |T(f)|$$

where

$$D_\varepsilon = c(\rho)c(\varepsilon)\frac{1}{1-\frac{(1+\varepsilon)}{A}}$$

for all $\varepsilon > 0$ such that $\rho(1+\varepsilon) < A$.

This proves the continuity of F.

In order to prove the continuity of F^{-1} we consider $\mathcal{B}$ bounded in $\mathrm{Exp}^k_{N,0,A}(E)$. Hence

$$\overline{\lim_{j\to\infty}} \left(\frac{j}{ke}\right)^{\frac{1}{k}} \left[\sup_{f\in\mathcal{B}} \left\|\frac{\hat{d}^j f(0)}{j!}\right\|_N\right]^{\frac{1}{j}} \leq A$$

by Proposition 3.4. Now, for every $\rho > A$, there is $c(\rho) \geq 0$ such that

$$\left(\frac{j}{ke}\right)^{\frac{j}{k}} \sup_{f\in\mathcal{B}} \left\|\frac{\hat{d}^j f(0)}{j!}\right\|_N \leq c(\rho)\rho^j$$

for all $j \in \mathbb{N}$. We have

$$\sup_{f\in\mathcal{B}} |F^{-1}(H)(f)| \leq \sum_{j=0}^{\infty} \|\hat{d}^j H(0)\| \sup_{f\in\mathcal{B}} \left\|\frac{\hat{d}^j f(0)}{j!}\right\|_N \leq$$

$$\leq c(\rho) \sum_{j=0}^{\infty} \rho^j \left(\frac{ke}{j}\right)^{\frac{j}{k}} j! \left\|\frac{\hat{d}^j H(0)}{j!}\right\| =$$

$$= c(\rho) \sum_{j=0}^{\infty} (\rho\lambda(k))^j \frac{j!}{j^j} e^j \left(\frac{j}{k'e}\right)^{\frac{j}{k'}} \left\|\frac{\hat{d}^j H(0)}{j!}\right\| \leq$$

$$\overset{(\text{see } \S)}{\leq} c(\varepsilon)c(\rho) \sum_{j=0}^{\infty} [\rho(1+\varepsilon)\lambda(k)]^j \left(\frac{j}{k'e}\right)^{\frac{j}{k'}} \left\|\frac{\hat{d}^j H(0)}{j!}\right\| =$$

$$= c(\varepsilon)c(\rho)\ \|H\|_{k',\frac{1}{\rho(1+\varepsilon)\lambda(k)}}$$

for all $\varepsilon > 0$ and $\rho > A$. Hence

$$\sup_{f\in\mathcal{B}} |F^{-1}(H)(f)| \leq D_r\|H\|_{k',r}$$

for all $r < \frac{1}{\lambda(k)A}$.

This proves that F^{-1} is continuous.

The proofs for the other values of k follow the same pattern. □

REFERENCES

1. C.P. GUPTA, Malgrange theorem for nuclearly entire functions of bounded type - Notas de Matemática 37, IMPA, Rio de Janeiro, 1968.

2. C.P. GUPTA, Convolution operators and holomorphic mappings on a Banach space, Seminaire d'Analyse Moderne, nº 2, Université de Sherbrooke, Sherbrooke, 1969.

1. L. NACHBIN, Topology on spaces of holomorphic mappings, Ergebnisse der Mathematik, 47 (1969), Springer-Verlag.

1. A. MARTINEAU, Équations différentielles d'ordre infini, Bull. Soc. Math. France, 95 (1967), p. 109-154.

1. M.C. MATOS, On Malgrange Theorem for nuclear holomorphic Functions in open balls of a Banach space, Math. Z. 162 (1978), 113-123.

2. M.C. MATOS, Correction to "On Malgrange Theorem Nuclear holomorphic Functions in open balls of a Banach Space", Math. Z., 171 (1980), 289-290.

Departamento de Matemática
IMECC - UNICAMP
Caixa Postal 6155
13100 - Campinas, S.P., Brasil

Functional Analysis, Holomorphy and Approximation Theory II, G.I. Zapata (ed.)

ON REPRESENTATIONS OF DISTANCE FUNCTIONS IN THE PLANE

John McGowan and Horacio Porta

INTRODUCTION

Suppose that $P(\theta) = r(\theta)(\cos\theta, \sin\theta)$ describes a centrally symmetric curve C in the plane (so that $r(\theta+\pi) = r(\theta)$). We assume that r is a continuous function and that $r(\theta) > 0$ on $[0,2\pi]$.

The <u>distance function</u> (sometimes called the <u>Minkowski functional</u>) of C is the real valued non-negative function L_C defined for $X = |X|(\cos\alpha, \sin\alpha)$ in $\mathbb{R}^2$ by $L_C(X) = |X|/r(\alpha)$. When C is convex, L_C is the norm associated to C, the unit ball for L_C consisting of the union of C and the region enclosed by C.

The objective of this paper is to find integral and differential representations for L_C. More specifically, for a class of curves (essentially those having finite angular variation and satisfying an interior cone condition) we obtain in §4 the formula

$$L_C(X) = \int_0^{\pi} |P(\theta)\times X| \, dM(\theta)$$

where $\times$ denotes cross-product of vectors and where dM is an appropriate Borel measure. This generalizes the Levy representation of plane norms (where $dM \geqq 0$; see 7.8 below). The measure dM is unique, and $dM \geqq 0$ on an interval (α,β) if and only if C is convex in the section $\alpha < \theta < \beta$.

As usual, this representation for L_C gives when C is convex the classical result that the normed space $(\mathbb{R}^2, L_C)$ can be embedded isometrically in $L^1[0,1]$ (as shown in §2 below using ultraproducts).

In view of the uniqueness result, it is reasonable to expect a simple expression for dM in terms of $P(\theta)$. We prove in §5 that

$$dM = \frac{1}{2} R\left(R + \frac{d^2R}{d\theta^2}\right) d\theta$$

where $R(\theta) = 1/r(\theta)$ and where the derivative $d^2/d\theta^2$ is to be interpreted in the sense of distributions. This formula can in turn be used for curves of class C^2 (convex or not) to find alternative expressions of dM in terms of geometric notions. Among other results, we prove in §6 that

$$dM = \frac{1}{2} \frac{\varkappa \quad d\theta}{r \sin^3\alpha}$$

where $\varkappa$ is the curvature of C and α is the angle formed by the ray through O and $P(\theta)$ and the tangent to C at $P(\theta)$.

Finally, in §7 we consider several notions related to dM, including its moments. Among several inequalities and identities we prove, for example, that $\int r^2 \, dM$ is invariant under inversion through the unit circle.

We want to thank R.P. Kaufman, H.P. Lotz and T. Morley for their valuable comments on parts of this work.

§1. POLYOGONAL EQUATIONS

Let us consider a polygon in the x, y plane, symmetric about the origin with vertices $P_1, P_2, \ldots, P_{2n}$ where $P_{n+i} = -P_i$ for $i = 1,2,\ldots,n$. Here and in the following we do not distinguish between points in the plane, and their position vectors with origin at $O = (0,0)$. Suppose that the $P_1, P_2, \ldots, P_{2n}$ are all non zero and that they form a radial sequence, i.e., they are totally ordered by their central angle (varying counterclockwise). We assume further that this order is strict so that no two distinct vertices lie on the same ray from the origin. For convenience we set $P_0 = -P_n$.

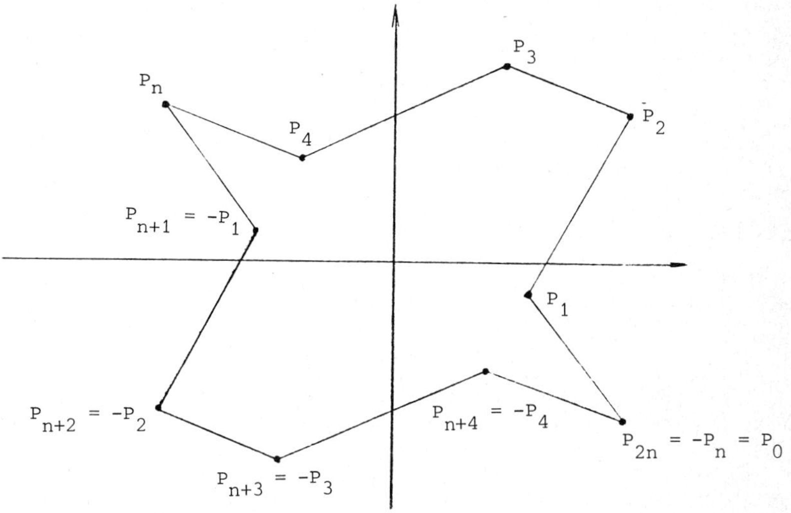

Let now α_{ij} denote the component of the cross product $P_i \times P_j$ in the positive z-direction, for $1 \leq i, \ j \leq n$ (here and in the following we consider the x,y-plane as the set in x,y,z-space characterized by $z = 0$).

1.1 PROPOSITION. The matrix with entries $|\alpha_{ij}|$ is invertible with inverse having entries β_{ij} given by

$$\beta_{1,n} = \beta_{n,1} = 1/(2\alpha_{1,n})$$

$$\beta_{ii} = \alpha_{i+1,i-1}/(2\alpha_{i+1,i}\alpha_{i,i-1})$$

$$\beta_{i,i+1} = \beta_{i+1,i} = 1/(2\alpha_{i,i+1})$$

for $1 \leq i \leq n-1$, and $\beta_{ij} = 0$ for all other i,j.

The proof of this proposition along with various observations on the matrices $(|\alpha_{ij}|)$ and (β_{ij}) will be found in the

Appendix, §8. We remark that by definition, the diagonal entries of (β_{ij}) are non-positive while the off diagonal entries are positive.

Consider now the expression

$$L(P) = \sum_{j=1}^{n} m_j |P_j \times P|$$

where $P = (x,y)$ is an arbitrary point in the plane and the m_i's are arbitrary real numbers not all zero. It is clear that for P between the rays from the origin through P_i and P_{i+1}, the equation $L(P) = 1$ is equivalent to

$$P \cdot \left(\sum_{j \leq i} m_j P_j \times k - \sum_{j > i} m_j P_j \times k \right) = 1$$

where $k = (0,0,1)$ is the unit vector in the z-direction; whence, it is the equation of a straight line.

Therefore, $L(P) = 1$ represents a polygonal line with vertices on the rays from the origin through the P_j's, and clearly symmetric about the origin

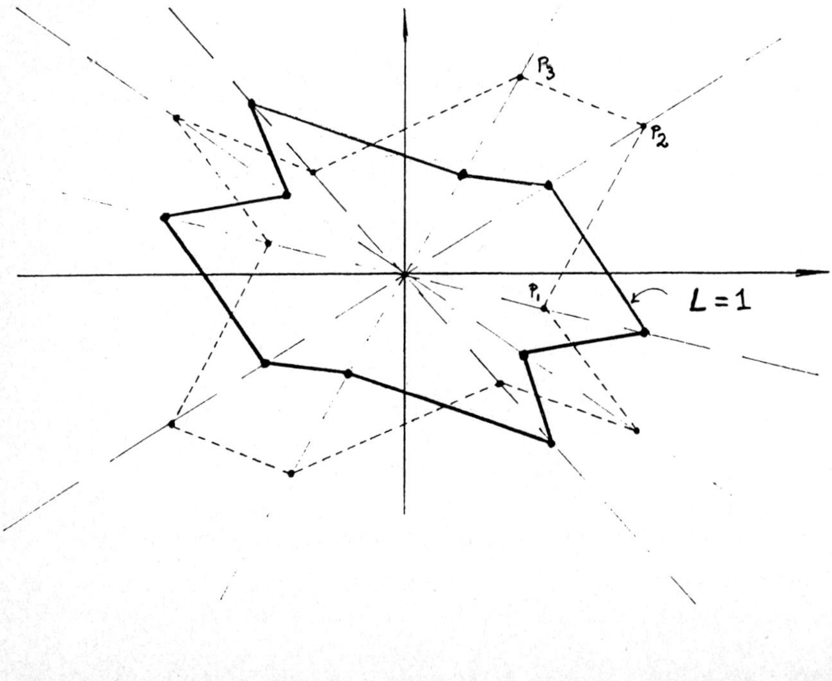

1.2 PROPOSITION. Given $P_1, P_2, \ldots, P_n, P_{n+1} = -P_1, \ldots, P_{2n} = -P_n$ in radial sequence, there exist unique $m_1, m_2, \ldots, m_n$ such that $L(P) = \Sigma\, m_i |P \times P_i| = 1$ is the equation of the polygon joining (in that order) $P_1, P_2, \ldots, P_n, -P_1, \ldots, -P_n, P_1$. Further, for each i, m_i is non-negative if and only if the quadrilateral $OP_{i-1}P_iP_{i+1}$ is convex (hence the polygon is convex if and only if all m_j are non-negative).

PROOF. $L(P) = 1$ represents the polygon $P_1, P_2, \ldots, P_n, -P_1, \ldots, -P_n, P_1$ if and only if $L(P_i) = 1$ for all $i = 1, 2, \ldots, 1, n$, i.e., $\sum_{j=1}^{n} m_j |P_i \times P_j| = 1$ for all $i = 1, 2, \ldots, n$, and the existence of a unique solution in the m_i follows from Proposition (1.1). Denote by A the matrix $A = (|\alpha_{ij}|)$, and let M and E be the vectors

$$M = \begin{pmatrix} m_1 \\ m_2 \\ \vdots \\ m_n \end{pmatrix} \qquad E = \begin{pmatrix} 1 \\ 1 \\ \vdots \\ 1 \end{pmatrix}$$

then $M = A^{-1}E$ is given by (use Proposition 1.1):

$$\begin{aligned} m_i &= \beta_{i,i-1} + \beta_{i,i} + \beta_{i,i+1} \\ &= (\alpha_{i-1,i} + \alpha_{i+1,i-1} + \alpha_{i,i+1})/(2\alpha_{i-1,i}\alpha_{i,i+1}). \end{aligned}$$

But since the area of the triangle $OP_{i-1}P_i$ is $\frac{1}{2}|P_{i-1} \times P_i| = \frac{1}{2}\alpha_{i-1,i}$, and similarly for all the other values of i, we get from the last equation

$$(1.2.i) \qquad 4\,\text{area}(OP_{i-1}P_i)\text{area}(OP_iP_{i+1})m_i = \\ = \text{area}(OP_{i-1}P_i) + \text{area}(OP_iP_{i+1}) - \text{area}(OP_{i-1}P_{i+1})$$

and therefore, $m_i \geqq 0$ if and only if

$$\text{area}(OP_{i-1}P_{i+1}) \leqq \text{area}(OP_{i-1}P_i) + \text{area}(OP_iP_{i+1}),$$

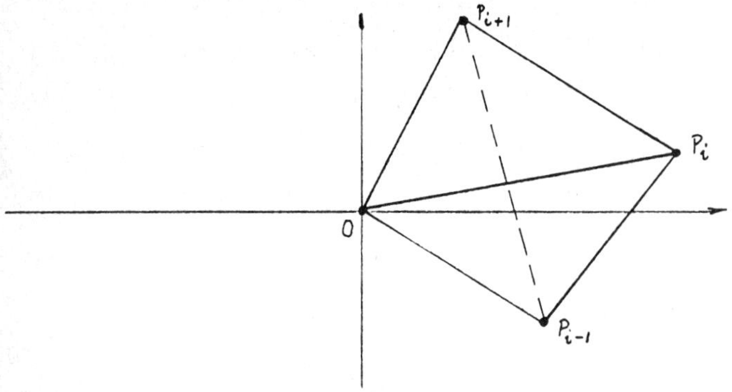

that is, if the quadrilateral $OP_{i-1}P_iP_{i+1}$ is convex.

§2. ISOMETRIC EMBEDDINGS

Consider a symmetric onvex polygon Γ with equation $1 = L_\Gamma(P) = \sum_{j=1}^{n} m_j|P_j \times P|$. Since $L_\Gamma = 1$ on Γ and $L_\Gamma(aP) = |a|L_\Gamma(P)$, it is clear that L_Γ is the norm on $\mathbb{R}^2$ associated to the convex body bounded by Γ. We shall denote by B_Γ the normed space $(\mathbb{R}^2, L_\Gamma)$.

Let $m = \sum_{j=1}^{n} m_j$, $\mu_j = m_j/m$ and decompose $[0,1)$ as the disjoint union of intervals $E(1) = [0,\mu_1)$, $E(2) = [\mu_1, \mu_1 + \mu_2)$, $E(3) = [\mu_1 + \mu_2, \mu_1 + \mu_2 + \mu_3)$, etc. with lengths $\lambda(E(j)) = \mu_j$, $1 \leq j \leq n$.

Suppose now that the coordinates of the P_j are $P_j = (a_j, b_j)$ and define functions f, g on $[0,1)$ by: $f = b_j$, $g = -a_j$ on $E(j)$, for $j = 1,2,\ldots,n$. Finally define $T\colon \mathbb{R}^2 \to L^1(0,1)$ by $TP = m(xf + yg)$ where $P = (x,y)$ in an arbitrary vector in $\mathbb{R}^2$. Then

$$\|TP\|_{L^1} = m \int_0^1 |xf(t) + yg(t)|\,dt =$$

$$= m \sum_{j=1}^{n} \int_{E(j)} |xb_j - ya_j|\,dt =$$

$$= m \sum_{j=1}^{n} (m_j/m)|xb_j - ya_j| =$$

$$= \sum_{j=1}^{n} m_j |P_j \times P| = L_\Gamma(P).$$

Thus T is an isometry from B_Γ into $L^1(0,1)$.

This is a particular case of the classical result that all 2-dimensional normed spaces can be isometrically embedded in $L^1 = L^1(0,1)$ (see for example [4]). The following argument shows that the general case follows from the case of polygons very simply.

Suppose that $\| \ \|$ is a norm on $\mathbb{R}^2$ and let C be the convex curve $\|P\| = 1$. We can approximate C by a sequence of polygons $\Gamma(n)$ inscribed in C in the sense that $\|P\| = \lim_n L_{\Gamma(n)}(P)$. Let now $T_n\colon B_{\Gamma(n)} \to L^1$ be isometries. The family (T_n) defines a linear operator $T\colon \mathbb{R}^2 \to L^1/\mathcal{U}$ into the ultraproduct of countably many copies of L^1 (see [4], p. 121ff), which is itself an L^1-space ([4], Proposition II 2.10). Since

$$\lim_n \|T_n P\|_{L^1} = \lim_n L_{\Gamma(n)}(P) = \|P\|$$

we have $\|TP\|_{L^1/\mathcal{U}} = \|P\|$ so that T is an isometry. But $T(\mathbb{R}^2)$ is a separable subspace of $L^1/\mathcal{U}$, and thus contained in a separable sublattice of $L^1/\mathcal{U}$, which, by the classical Kakutani representation theorem can be identified with $L^1(0,1)$. It follows that all 2-dimensional normed spaces can be isometrically embedded in $L^1(0,1)$.

§3. SOME ESTIMATES

Consider again the vertices $P_1, P_2, \ldots, P_n, P_{n+1} = -P_1, \ldots, P_{2n} = -P_n$, (ordered counterclockwise in radial sequence) of a symmetric polygon Γ. Denote by γ_i the angle from $P_{i+1} - P_i$ to $P_{i-1} - P_i$, chosen so that $0 < \gamma_i < 2\pi$.

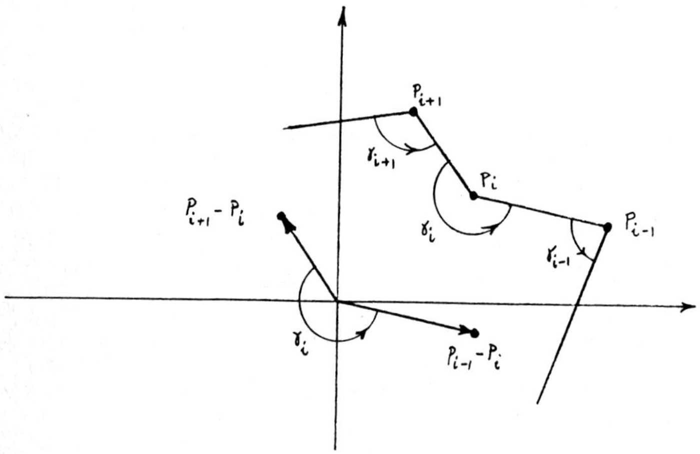

and let $\sum_{i=1}^{n} m_i |P_i \times P| = 1$ be the equation of Γ.

We know that (see formula (1.2.i) in the proof of Proposition (1.2)):

$$m_i = \frac{\text{area}(OP_{i-1}P_i) - \text{area}(OP_{i-1}P_{i+1}) + \text{area}(OP_iP_{i+1})}{4\ \text{area}(OP_{i-1}P_i)\text{area}(OP_iP_{i+1})}$$

$$= \frac{\pm\ \text{area}(P_{i-1}P_iP_{i+1})}{4\ \text{area}(OP_{i-1}P_i)\text{area}(OP_iP_{i+1})}$$

and the sign is + or - depending on whether the polygon is or is not convex "at P_i".

Since $\text{area}(P_{i-1}P_iP_{i+1}) = \frac{1}{2} |P_{i-1} - P_i||P_{i+1} - P_i||\sin \gamma_i|$ and $\text{area}(OP_jP_k) = \frac{1}{2} |P_j - P_k| a_{j,k}$ where $a_{j,k}$ denotes the distance from the origin to the straight line through P_j and P_k, we get

$$m_i = \frac{\sin \gamma_i}{2a_{i-1,i}a_{i,i+1}} . \tag{3.i}$$

Notice that the signs match automatically.

Assume now that $a_{k,k+1} \geqq a > 0$ for all k. This means that the distance from the line through P_k, P_{k+1} (which surely misses the origin) to the origin is bounded below by a. Then

$$\sum_{i=1}^{n} |m_i| \leqq \frac{1}{2a^2} \sum_{i=1}^{2} |\sin \gamma_i| .$$

Denote now $\delta_i = \pi - \gamma_i$; δ_i is the (signed) angle formed by $P_i - P_{i-1}$ and $P_{i+1} - P_i$, so that $-\pi < \delta_i < \pi$.

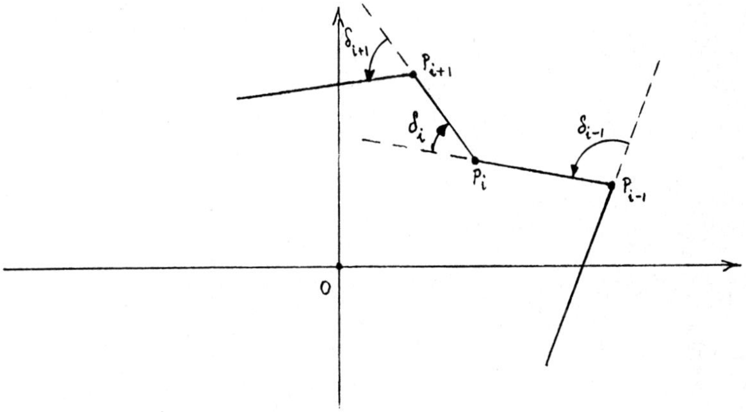

Since $|\sin \gamma_i| = |\sin \delta_i| \leqq |\delta_i|$ we have

$$\sum_{i=1}^{n} |m_i| \leqq \frac{1}{2a^2} \sum_{i=1}^{n} |\delta_i| . \tag{3.ii}$$

We remark that, since $|\sin \gamma_i| = |\sin \delta_i|$ is also majorated by $\eta_i = \min(|\delta_i|, \pi - |\delta_i|)$, we can improve (3.ii) somewhat:

$$\sum_{i=1}^{n} |m_i| \leqq \frac{1}{2a^2} \sum_{i=1}^{n} \eta_i . \tag{3.iii}$$

If the polygon is convex, then $\delta_i \geqq 0$ for all i, and

$\sum_{i=1}^{n} \delta_i = \pi$, whence (recall that $m_i \geqq 0$ by Proposition (1.2)):

$$\sum_{i=1}^{n} m_i \leqq \frac{\pi}{2a^2}. \tag{3.iv}$$

§4. INTEGRAL REPRESENTATION

Suppose now that C is a symmetric curve described in polar coordinates by $P(\theta) = (x(\theta), y(\theta))$, $0 \leq \theta \leq 2\pi$ where θ is the central angle measured as usual counterclockwise from the positive x-axis. Symmetry of C means that $P(\theta+\pi) = -P(\theta)$. We also assume that $P(\theta)$ is actually defined for all values of θ, periodic with period 2π, and that C does not pass through the origin. We will also denote by Ω the interior of C ($\Omega = \{(r\cos\theta, r\sin\theta);\ r < |P(\theta)|\}$). For each partition

$$\Pi = \{0 = \theta_1 < \theta_2 < \ldots < \theta_{n+1} = \pi\}$$

we can consider the polygon $\Gamma(\Pi)$ whose vertices are $P_j = P(\theta_j)$ and $P_{n+j} = P(\pi+\theta_j) = -P(\theta_j) = -P_j$, for $j = 1,2,\ldots,n$. The sum $\sum_{j=1}^{n} |\delta_j|$ is the <u>total angular variation</u> of $\Gamma(\Pi)$ on $0 \leq \theta \leq \pi$ (δ_j denotes the angle described in §3); the (possibly infinite) quantity

$$T(C) = 2 \operatorname{Sup}_{\Pi} \sum_{j=1}^{n} |\delta_j|$$

is the <u>total angular variation</u> of C. The reason for the factor 2 is that we want to assign to $T(C)$ the value that corresponds to the whole curve C, and this is double the angular variation on $[0,\pi]$. Observe that adding a new partition point to Π results in a larger value of $\Sigma|\delta_j|$, and therefore, we can calculate $T(C)$ as the supremum taken only over all partitions finer than a given partition Π_0.

Recall [8] that the <u>kernel</u> $\ker\Omega$ of a set Ω is the set of all P_0 such that the segment P_0P is contained in Ω for each P in Ω. For example, $\ker\Omega = \Omega$ if Ω is convex and $\ker\Omega = \{0\}$ if Ω is the interior of the astroid $|x|^{2/3} + |y|^{2/3} = 1$. Since $\ker\Omega$ is always convex (see [8] for the general case; [5] Problem III.111 for the plane) and clearly $\ker\Omega$ is centrally symmetric if Ω is, it follows in this case that $\ker\Omega$ has non-empty interior if and only if it contains a disc around the origin.

4.1 PROPOSITION. Suppose that $\ker\Omega$ contains the disc of radius a around the origin. Then for any $\varepsilon > 0$ there is a partition Π_ε such that for all Π finer than Π_ε, the coefficients of the polygon $\Gamma(\Pi)$ satisfy

$$\sum_{j=1}^{n} |m_j| \leq \frac{T(C)}{4(a-\varepsilon)^2}.$$

The proof follows easily from (3.ii).

Consider now, for each partition Π, the measure dM^Π on $[0,\pi)$ defined by $\Sigma\, m_i\, \delta_{\theta_i}$ where $0 = \theta_1 < \theta_2 < \ldots < \theta_n < \pi$ are the partition points and m_i, $i = 1,\ldots,n$, the coefficients of the equation of the polygon $\Gamma(\Pi)$. The measure δ_θ denotes, as usual, the unit point mass concentrated at θ.

Clearly the total mass of dM^Π is

$$\int_0^\Pi |dM^\Pi(\theta)| = \Sigma |m_i|$$

and according to the proposition above, these numbers are all bounded by $T(C)/4(a-\varepsilon)^2$. But then the net $\{dM^\Pi\}_{\Pi \geqq \Pi_\varepsilon}$ of measures is bounded: hence a subset $\{dM^{\Pi'}\}_{\Pi'}$ converges in the weak* topology of measures (= vague topology for continuous integrands): denote by $dM = \lim dM^{\Pi'}$ the limit, which clearly satisfies

$$\int_0^\Pi |dM(\theta)| \leqq T(C)/4a^2.$$

For $0 \neq X \in \mathbb{R}^2$, denote by $L_C(X)$ the number λ defined by the property that X/λ is on the curve C, and set also $L_C(0) = 0$. Then, for each X, $L_C(X)$ is the limit of the numbers $L_{\Gamma(\Pi')}(X)$ since they actually agree as soon as the argument of X is a partition point of Π', and this will happen eventually as $\{\Pi'\}$ is cofinal in all the partitions.

On the other hand,

$$L_{\Gamma(\Pi)}(X) = \Sigma\, m_i |P(\theta_i) \times X| = \int_0^{\pi} |P(\theta) \times X| \, dM^{\Pi}(\theta)$$

and so, taking limits along $\{\Pi'\}$ we conclude that

$$L_C(X) = \int_0^{\pi} |P(\theta) \times X| \, dM(\theta).$$

These remarks prove the first half of the following result:

4.2 THEOREM. Let $\Omega \subset \mathbb{R}^2$ be a centrally symmetric bounded domain with boundary the curve C. Let $P(\theta) = (x(\theta), y(\theta))$, $0 \leq \theta \leq 2\pi$, and $L_C(X)$ denote the equation and the distance function of C. Suppose that

(4.2.i) $\ker \Omega$ has non-empty interior;

(4.2.ii) C has bounded angular variation.

Then there exists a unique finite measure dM on $[0,\pi]$ such that for all $X \in \mathbb{R}^2$:

$$\text{(4.2.iii)} \qquad L_C(X) = \int_0^{\pi} |P(\theta) \times X| \, dM(\theta)$$

and it satisfies

$$\text{(4.2.iv)} \qquad \int_0^{\pi} |dM| \leq T(C)/4a^2$$

where a is the radius of the largest circle contained in $\ker \Omega$.

Further, if $dM \geqq 0$ then Ω is convex and if Ω is convex, then (i) and (ii) are satisfied automatically and $dM \geqq 0$. In this case

$$\int_0^\pi |dM| \leqq \pi/2a^2.$$

PROOF. It only remains to establish the uniqueness of dM. Let $|X|$ denote the euclidean length of the vector X, and abbreviate $r(\theta) = |P(\theta)|$. Then

$$|P(\theta)xX| = r(\theta)|X||\sin(\theta-t)|$$

where $X = (|X|\cos t, |X|\sin t)$. Thus, for all $X \in \mathbb{R}^2$:

$$L_C(X) = |X| \int_0^\pi |\sin(\theta-t)|r(\theta)dM(\theta). \tag{4.3}$$

Suppose now that dM_1 and dM_2 satisfy (4.3). Then the difference $dN = dM_1 - dM_2$ satisfies

$$\int_0^\pi |\sin(\theta-t)|r(\theta)dN(\theta) = 0, \tag{4.4}$$

for all t. This, however, implies that $dN = 0$. In fact, if we consider the group $G = [0,\pi)$ (with addition mod π as operation and $\frac{1}{\pi}d\theta$ as invariant measure). Then (4.4) reads

$$|\sin| * rdN = 0 \tag{4.5}$$

where $*$ denotes convolution and $|\sin|(\theta) = |\sin\theta|$, $\theta \in G$.

The characters of G are $e^{2in\theta}$, $n = 0,\pm1,\pm2,\ldots$ and taking the Fourier transforms:

$$|\sin|(\theta) \sim -\frac{2}{\pi}\sum_{-\infty}^{+\infty} \frac{1}{4n^2-1} e^{2ni\theta},$$

$$rdN \sim \sum_{-\infty}^{\infty} (rdN)^\wedge(n)e^{2ni\theta},$$

we conclude from (4.5) that

$$- \frac{2}{\pi} \frac{1}{4n^2-1} (rdN)^{\wedge}(n) = 0$$

and therefore, $(rdN)^{\wedge}(n) = 0$ for all n. Thus $rdN = 0$ and uniqueness follows.

§5. DIFFERENTIAL REPRESENTATION

Let us apply formula (4.3) of the last section to $X = P(t)$. Since X is on the curve C, we have $L_C(X) = 1$ and therefore, (4.3) reads

$$(5.1) \qquad 1 = r(t) \int_0^{\pi} |\sin(\theta - t)| r(\theta) dM(\theta)$$

or

$$1/r = |\sin| * rdM.$$

Define now a distribution T on the compact group $G = [0,\pi)$ by its Fourier transform

$$T \sim - \frac{\pi}{2} \sum_{-\infty}^{\infty} (4n^2-1) e^{2ni\theta}.$$

Clearly $T * |\sin| = \delta$, the unit point mass at $0 \in G$. But then

$$T * (1/r) = T * |\sin| * rdM = rdM$$

whence we get for dM the value

$$dM = (1/r)(T * 1/r).$$

We can give a more explicit form to this expression as follows. First, observe that from

$$\delta \sim \sum_{-\infty}^{\infty} e^{2ni\theta}$$

we get

$$\delta'' \sim \sum_{-\infty}^{\infty} - 4n^2 e^{2ni\theta}$$

and therefore,

$$T = \frac{\pi}{2}(\delta + \delta'').$$

But then

5.2 PROPOSITION. Under the hypothesis of 4.2, if $R(\theta) = 1/r(\theta) = 1/|P(\theta)|$, then

$$\text{(5.2.i)} \qquad dM = \frac{\pi}{2} R\left(R + \frac{d^2R}{d\theta^2}\right)\frac{d\theta}{\pi}$$

where $d^2R/d\theta^2$ in taken in the sense of distributions. In particular, if C is a smooth curve (of class C^2), then dM has the continuous density $(\pi/2)R(R + R'')$ with respect to $d\theta/\pi$.

We can illustrate the formula (5.2.i) in the case of the p-norms for $2 \leqq p < +\infty$, i.e., when C is the curve $|x|^p + |y|^p = 1$. We need

5.3. Let $p \geqq 2$ and $E^{1/p}(\theta) = (C^p(\theta) + S^p(\theta))^{1/p}$ where

$$C(\theta) = |\cos\theta|, \quad S(\theta) = |\sin\theta|$$

then

$$\left(\frac{d^2}{d\theta^2} + 1\right)E^{1/p} = (p-1)C^{p-2}S^{p-2}E^{-2+1/p}.$$

Observe now that for the p-norm, C can be described by $P(\theta) = E^{-1/p}(\theta)(\cos\theta, \sin\theta)$ so that $R(\theta) = E^{1/p}(\theta)$ and therefore, according to 5.2 and 5.3:

5.4 COROLLARY. For the p-norm,

$$dM = \frac{1}{2}(p-1)C^{p-2}S^{p-2}E^{-2+2/p}\, d\theta.$$

In particular $dM = \frac{1}{2}\,d\theta$ for the euclidean norm corresponding to the case where C is the unit circle and $p = 2$.

Another application of 5.2 is the following. Suppose that $\mathbb{R}^2$ with the norm $\|\ \|$ is isometric to a two dimensional subspace of ℓ^1. That is, there are vectors $Y_n \in \mathbb{R}^2$ such that for all $X \in \mathbb{R}^2$:

$$\|X\| = \sum_{n=1}^{\infty} |X \cdot Y_n| .$$

We may assume that $Y_n \neq 0$ for each n. Write

$$Y_n = w_n P(\theta_n + \tfrac{\pi}{2})$$

where $w_n > 0$ and where $P(\theta)$ is the equation of $\|X\| = 1$. Then

$$\|X\| = \Sigma\ |X \cdot Y_n| = \Sigma\ w_n |X \times P(\theta_n)|$$

$$= \int_0^{\pi} |X \times P(\theta)|\, dN$$

where $dN = \Sigma\ w_n\, \delta_{\theta_n}$. By uniqueness, this measure must coincide with dM. Thus:

5.5 COROLLARY. The normed space $(\mathbb{R}^2, \|\ \|)$ is isometric to a subspace of ℓ^1 if and only if dM is purely atomic. In particular, $\mathbb{R}^2$ with the p-norm, $2 \leqq p < +\infty$ is not isometric to a subspace of ℓ^1.

This result appears also in [3], first paragraph on p. 494 and last two paragraphs on p. 498.

Observe that if C coincides with a straight line between θ_1 and θ_2, then $r = a \sec(\theta - \theta_0)$ for appropriate a and θ_0, and therefore, $R + R'' = 0$ there. In particular, if C is a polygon, dM has support in the set $\{\theta_1, \theta_2, \ldots, \theta_n\}$ of values of θ for which $P(\theta)$ is a vertex. This gives back the expression $dM = \Sigma\ m_j \delta_{\theta_j}$ used in §4.

§6. GEOMETRIC INTERPRETATION

Suppose that C is smooth (of class C^2, say) and denote by α the angle formed by $P(\theta)$ and the tangent vector $dP(\theta)/d\theta$.

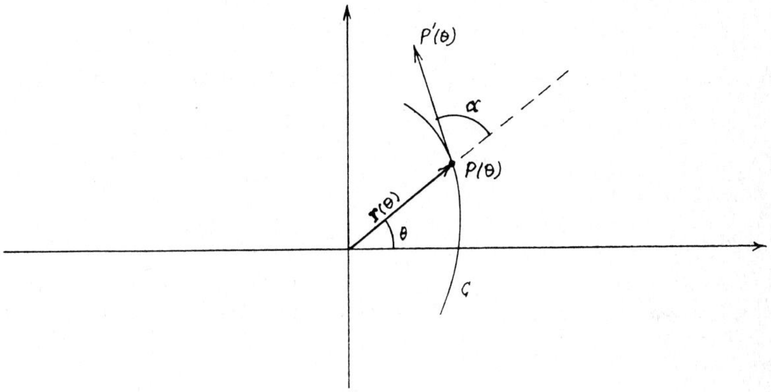

6.1 THEOREM. Let C be a smooth centrally symmetric curve. Then

(6.1.i) $$dM = \frac{1}{2}\,\frac{\varkappa}{r\sin^3\alpha}\,d\theta$$

where $\varkappa$ is the curvature of C at $P(\theta)$.

PROOF. Since $r = r(\theta)$ is the equation of C in polar coordinates, we have (primes denote derivatives with respect to θ):

$$\varkappa = \frac{2(r')^2 - rr'' + r^2}{((r')^2 + r^2)^{3/2}}\,.$$

But then (use $R = 1/r$):

$$(1 + (r'/r)^2)^{3/2}\varkappa = R + R''$$

and since $\cot\alpha = r'/r$ we get also

$$\varkappa \sin^{-3}\alpha = R + R''$$

so that, from (5.2.i),

$$dM = \frac{1}{2} R(R + R'')d\theta = \frac{1}{2} \frac{\varkappa}{r \sin^3\alpha} d\theta$$

as claimed.

Minor modifications allow us to obtain an expression similar to (6.1.i) for arbitrary smooth parametrizations of C.

Let then $P(t) = (x(t),y(t))$, $a \leqq t \leqq b$ describe the upper half of C (corresponding to $0 \leqq \theta \leqq \pi$) and assume that $P(t)$ is C^2, $P' = dP/dt \neq 0$, and that θ is an increasing function of t. For simplicity, let us use the rotation

$$\rho = \begin{pmatrix} 0 & -1 \\ 1 & 0 \end{pmatrix}$$

to express cross products as dot products by means of $(\rho X \cdot Y)k = X \times Y$ where $k = (0,0,1)$.

Using primes to indicate d/dt we have

$$\varkappa = (\rho P' \cdot P'')/(P' \cdot P')^{3/2}$$

and

$$d\theta/dt = (\rho P' \cdot P)/(P \cdot P).$$

Use now

$$\rho P' \cdot P = (P' \times P) \cdot k = |P'|\ |P| \sin\alpha$$

to get

$$\sin^3\alpha = (\rho P' \cdot P)^3/(P' \cdot P')^{3/2}(P \cdot P)^{3/2}.$$

Hence

$$\frac{\varkappa}{r \sin^3\alpha} = \frac{(\rho P' \cdot P'')/(P' \cdot P')^{3/2}}{[(P \cdot P)^{1/2}(\rho P' \cdot P)^3]/[(P' \cdot P')^{3/2}(P \cdot P)^{3/2}]}$$

$$= (\rho P' \cdot P'')(P \cdot P)/(\rho P' \cdot P)^3,$$

and finally using $d\theta/dt = (\rho P' \cdot P)/(P \cdot P)$, we get

(6.1.ii) $$dM = \frac{1}{2}\,\frac{\rho P'\cdot P''}{(\rho P'\cdot P)^2}\,dt.$$

Of special interest is, of course, the parametrization of C by arclength. Using again $\rho P'\cdot P = |\rho P'|\ |P|\sin\alpha$ and since $|P'| = 1$ we get $\rho P'\cdot P = r\sin\alpha$. Also $\varkappa = \rho P'\cdot P''$ so that

(6.1.iii) $$dM = \frac{1}{2}\,\frac{\varkappa}{r^2\sin^2\alpha}\,ds$$

(which also follows from (6.1.i) using $\frac{1}{r}\,ds = d\theta/\sin\alpha$).

Observe now that letting $r\sin\alpha = h$, it is easy to see that $h(\theta)$ is the distance from the origin to the tangent at $P(\theta)$. Clearly we can write the last formula as

(6.1.iv) $$dM = \frac{1}{2}\,\frac{\varkappa}{h^2}\,ds.$$

Each of (6.1.i) through (6.1.iv) provides a formula for the distance function of C, according to (4.2). For example, using (6.1.iv):

$$L_C(X) = \frac{1}{4}\int_C |P(\theta)xX|\,\frac{\varkappa}{h^2}\,ds.$$

§7. MISCELLANEOUS REMARKS AND PROPERTIES

7.1 Suppose that C is a centrally symmetric curve with equation $P(\theta)$ for which a finite Borel measure dM exists making (4.2.iii) valid. Then, using (5.1) we get

$$r(u) - r(t) = r(u)r(t)\left(\frac{1}{r(t)} - \frac{1}{r(u)}\right) =$$

$$= r(u)r(t)\int_0^{\pi} r(\theta)(|\sin(\theta-t)|-|\sin(\theta-u)|)\,dM(\theta),$$

so

$$|r(u)-r(t)| \leq \{r(u)r(t) \int_0^{\pi} r(\theta)\ |dM(\theta)|\}|u-t|,$$

and therefore, $r(\theta)$ in a Lipschitz function with constant not larger than

$$K = (\max_{0\leq\theta\leq\pi} t^3(\theta)) \int_0^{\pi} |dM|.$$

We can show by elementary calculations that whenever $r(\theta)$ is Lipschitz then $\ker \Omega$ contains a disc of radius $m(1+(1+\frac{\pi K}{2m})^2)^{-1/2}$ where $m = \inf r(\theta)$, $0 \leq \theta \leq \pi$. Here are the details.

Assume that

$$|r(u)-r(v)| \leq K|u-v|$$

holds for all u, v. Then

$$r(\theta) \geq r(\theta_0) - K|\theta-\theta_0|$$

for each fixed θ_0 and all θ. Denote by m the infimum of $r(\theta)$ and by α the angle indicated in the figure:

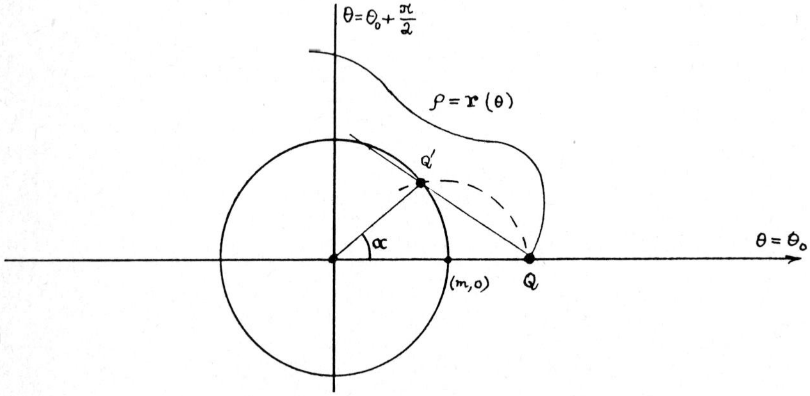

where the dotted curve has the equation $\rho = r(\theta_0) - K|\theta-\theta_0|$ (only the upper half is shown in the figure, corresponding to $\theta \geq \theta_0$). Clearly $m = r(\theta_0) - K\alpha$. Also, using the Lipschitz condition and the symmetry of C, we get

$$r(\theta_0) \leqq m + \frac{\pi}{2} K$$

which implies that $\alpha = (r(\theta_0) - m)/K$ satisfies $0 \leqq \alpha \leqq \pi/2$. Denote by S the absolute value of the slope of the line through the points $Q = (r(\theta_0),0)$ and $Q' = (m \cos \alpha, m \sin \alpha)$, i.e.,

$$S = \frac{m \sin \alpha}{r(\theta_0) - m \cos \alpha} .$$

We will show below that $S \geqq m/(m + \frac{\pi}{2} K)$. This implies that the distance a_0 from the origin to the line through Q and Q' is larger than the distance a_1 from the origin to the line through Q with slope $-m/(m + \frac{\pi}{2} K)$, that is

$$a_0 \geqq a_1 = (1 + (1 + \frac{\pi K}{2m})^2)^{-1/2} r(\theta_0).$$

But then $\ker \Omega$ contains the disc of radius $(1 + (1 + \frac{K}{2m})^2)^{-1/2}$ as claimed.

In order to show the inequality $S \geqq m/(m + \frac{\pi}{2} K)$ we define $\sigma(t) = \mu(t)/\nu(t)$ where

$$\mu(t) = m \sin t$$

$$\nu(t) = m + Kt - m \cos t.$$

Thus, $S = \sigma(\alpha)$. We will show that $\sigma(t) \geqq m/(m + \frac{\pi}{2} K)$ for all $0 \leqq t \leqq \pi/2$, and this will include the estimate for S.

Observe that $d\sigma/dt = \beta(t)/\nu^2(t)$ where $\beta(t) = \mu'\nu - \mu\nu' = m^2 \cos t + m K t \cos t - m K \sin t - m^2$ now $d\beta/dt = -m(m+kt)\sin t < 0$ for $0 < t \leqq \pi/2$, and so β is decreasing, which shows that $\beta(t) < \beta(0) = 0$ for those values of t. Hence, $d\sigma/dt = \beta/\nu^2 < 0$ and σ is also decreasing. But then $S = \sigma(\alpha) > \alpha(\pi/2) = m/(m + \frac{\pi}{2} K)$ and the proof is complete.

This implies that the fact that $\ker \Omega$ has non-empty interior is necessary for the existence of a finite measure dM.

Observe that the K obtained above is too large: the best K in terms of the interior of $\ker \Omega$ and the extrema of $r(\theta)$ was determined by Toranzos [8].

7.2. Concerning the hypotheses of (4.2) we observe that assuming that $\ker \Omega$ has non-empty interior does not guarantee that $T(C)$ is finite: it suffices to consider the curve defined by $r(\theta) = 2 + |\theta^2 \sin(\pi/\theta)|$ for $-\frac{\pi}{2} \leq \theta \leq \frac{\pi}{2}$ and by symmetry on the other half. Since $r(\theta)$ is Lipschitz, it follows from 7.1 above that $\ker \Omega$ has interior, but the usual argument shows that $T(C) = \infty$.

7.3. If Ω is given and $\gamma(t)$, $a \leq t \leq b$ is a curve in $\mathbb{R}^2$ of class C^1 we can define the length of γ according to the distance function of Ω by

$$\ell_\Omega(\gamma) = \int_a^b \int_0^\pi \left|\frac{d\gamma}{dt}(t) \times P(\theta)\right| dt\, dM(\theta)$$

$$= \int_0^\pi \Phi(\theta)\, dM(\theta)$$

where

$$\Phi(\theta) = \int_a^b \left|\frac{d\gamma}{dt}(t) \times P(\theta)\right| dt =$$

$$= \int_a^b \left|\frac{dg}{dt}(t)\right| dt$$

for $g(t) = \gamma(t) \cdot \rho P(\theta)$.

In particular the length of C with respect to its own distance is given by

$$\ell_\Omega(C) = 4 \int_0^\pi \{\sup_{0 \leq t \leq \pi} |P(t) \times P(\theta)|\}\, dM(\theta)$$

(which makes sense regardless of any smoothness assumptions); by

$$\ell_\Omega(C) = 2\int_0^\pi\int_0^\pi |P'(t)\times P(\theta)|\,dt\,dM(\theta)$$

for C^1 curves C; or finally by

$$\ell_\Omega(C) = \int_0^\pi\int_0^\pi |P'(t)\times P(\theta)|\,|P'(\theta)\times P''(\theta)|\,|P'(\theta)\times P(\theta)|^{-2}dt\,d\theta$$

for C^2 curves C.

It is an old result of Gołąb (see [6]) that $6 \leqq \ell_\Omega(C) \leqq 8$ when Ω is convex, but we don't see now how to obtain these estimates from the formulas above, or what happens when convexity is not assumed.

7.4. The moments of dM are related by equations easily derived from (5.2.i), which can be written as

$$2r\,dM = (R + R'')d\theta .$$

For any n, multiply the above times r^n and integrate to get

$$2\int_0^\pi r^{n+1}dM = \int_0^\pi r^{n-1}d\theta + \int_0 r^n R''\,d\theta .$$

Since r and $R = 1/r$ are periodic, integration by parts gives

$$\int r^n R''\,d\theta = \int (r^n)'R'\,d\theta =$$

$$= -n\int r^{n-1}r'(-r'/r^2)d\theta =$$

$$= n\int r^{n-3}(r')^2 d\theta,$$

thus, for all n:

$$2\int_0^\pi r^{n+1}dM = \int_0^\pi r^{n-1}d\theta + n\int_0^\pi r^{n-3}(r')^2 d\theta . \tag{7.4.i}$$

Some special case are of interest:

$$(7.4.\text{ii}) \qquad 2\int_0^{\pi} r\,dM = \int_0^{\pi} \frac{d\theta}{r}$$

$$(7.4.\text{iii}) \qquad 2\int_0^{\pi} r^2 dM = \pi - \int_0^{\pi} r'\left(\frac{1}{r}\right)' d\theta$$

$$(7.4.\text{iv}) \qquad 2\int_0^{\pi} r^2 dM = \pi + \int_0^{\pi} \cot^2\alpha\, d\theta$$

(these formulas correspond to $n = 0$ and $n = 1$ in (7.4.i); observe that (7.4.iii) and (7.4.iv) are actually the same formula with the last term written in two alternative ways).

Using now the formula

$$A = \int_0^{\pi} r^2 d\theta$$

for the area of Ω and (7.4.i) with $n = 3$ we get

$$(7.4.\text{v}) \qquad 2\int_0^{\pi} r^4 dM = A + 3\int_0^{\pi} (r')^2 d\theta = A - 3\int_0^{\pi} rr''\, d\theta,$$

as $\int_0^{\pi} (r')^2 d\theta = -\int_0^{\pi} rr''\, d\theta$ by integrating by parts.

Observe that (7.4.iv) implies that

$$\int_0^{\pi} r^2 dM \geqq \frac{\pi}{2}$$

with equality only for the circle. Since by (6.1.iii) $2dM = (\varkappa/r^2\sin^2\alpha)ds$, the last inequality reads:

$$\int_C \frac{\varkappa}{\sin^2\alpha}\, ds \geqq 2\pi$$

(with equality only for the circle), an easy fact to establish independently of dM.

7.5. Consider two curves C_1 and C_2 (with corresponding r_1, r_2, dM_1, dM_2 etc.) inverse of each other with respect to the unit circle, i.e., $r_1(\theta)r_2(\theta) = 1$ for each θ. This means that $R_2 = r_1$ and $R_1 = r_2$. Hence,

$$2dM_1 = R_1(R_1 + R_1'')d\theta = r_2(r_2 + r_2'')d\theta$$

and therefore,

$$(7.5.i) \qquad 2\int_0^\pi dM_1 = A_2 + \int_0^\pi r_2 r_2'' d\theta.$$

Now (7.4.v) gives

$$2\int_0^\pi r_2^4 dM_2 = A_2 + 3\int_0^\pi (r_2')^2 d\theta = A_2 - 3\int_0^\pi r_2 r_2'' d\theta,$$

and combining this with (7.5.1) we get

$$3\int_0^\pi dM_1 + \int_0^\pi r_2^4 dM_2 = 2A_2 .$$

Combining again these two formulas in a different way gives also

$$\int_0^\pi r_2^4 dM_2 - \int_0^\pi dM_1 = 2\int_0^\pi (r_2')^2 d\theta$$

and therefore,

$$\int_0^\pi r_2^4 dM_2 \geqq \int_0^\pi dM_1$$

with equality only for circles.

Finally, according to (7.4.iii) we have

$$\int r_1^2 dM_1 = \int r_2^2 dM_2$$

and so $\int r^2 \, dM$ is invariant under inversion.

7.6. Denote now by $y(\theta) = R(\theta)$ and $dM(\theta) = f(\theta)d\theta$. Then (5.2.i) says that y is a solution of

$$y(y'' + y) = f,$$

subject to:

$$y(0) = y(\pi),$$

$$y'(0) = y'(\pi).$$

It would be interesting to investigate further this boundary value problem. In particular, we do not know any necessary conditions on the periodic function f for the existence of solutions y. Obviously, if $f = c$, a positive constant, then $y = c^{1/2}$ is the solution. Robert Kaufman has observed that existence also can be obtained for f near a constant. In fact, if $D(y) = y(y'' + y)$, the linearized operator near $y = 1$ is $D'(h) = h'' + 2h$ and since the period is π $(\neq \sqrt{2})$, D' is surjective. By the implicit function theorem (as in, for example, [7]) it follows that D covers a neighborhood of $D(1) = 1$, as claimed.

7.8. We can rewrite (4.2.iii) in the form

$$L_C(X) = \int_0^{\pi} |E(\theta)\cdot X| \, dP(\theta)$$

where $E(\theta) = (\cos\theta, \sin\theta)$ is a variable point in the unit circle and where $dP(\theta) = r(\theta + \frac{\pi}{2})dM(\theta)$. Such formula for the norm L_C associated to a convex curve C is called the Levy representation of L_C, and the existence of such dP is obtained from the fact that in $\mathbb{R}^2$ all norms are negative definite functions. This is Ferguson's approach in [2] where he uses for convex C a technique similar to the one used in §4 above. We also remark that Dor's treatment of higher dimensional cases in [1] uses a potential theoretic approach related to the harmonic analysis approach of our §4.

Finally, it should be mentioned that Hamel in 1903 had already used expressions similar to our formulas for $\ell_\Omega(C)$ in §7 in his study of the geometries for which the straight line segments are the shortest curves (see volume 57, p. 251, of Mathematische Annalen).

§8. APPENDIX

Let $P_j = (a_j, b_j)$, $j = 1, 2, \ldots, n$ be points in R^2 such that $P_1, P_2, \ldots, P_n, -P_1, \ldots, -P_n$ is a radial sequence ordered counter-clockwise. For convenience we set $P_{n+1} = -P_1$ and $P_0 = -P_n$. Define $\alpha_{ij} = a_i b_j - a_j b_i$ and let $\underset{\sim}{k} = (0,0,1)$ denote the unit vector in the z-direction. Clearly the cross product $P_i \times P_j$ satisfies $P_i \times P_j = \alpha_{ij}\underset{\sim}{k}$. Further, by the assumed radial order, $\alpha_{ij} > 0$ if $i < j$ and, of course, $\alpha_{ij} = -\alpha_{ji}$ for all i, j. Denote by A the $n \times n$ matrix $A = (|\alpha_{ij}|)$, and consider the system

(1) $$Ax = y$$

where

$$x = \begin{pmatrix} x_1 \\ \vdots \\ x_n \end{pmatrix}$$

$$y = \begin{pmatrix} y_1 \\ \vdots \\ y_n \end{pmatrix}.$$

Clearly (1) is equivalent to

$$-\sum_{i \le j} x_i P_j \times P_i + \sum_{i > j} x_i P_j \times P_i = y_j \underset{\sim}{k}, \qquad 1 \le j \le n.$$

Define now $T_j = \sum_{i \le j} x_i P_i - \sum_{i > j} x_i P_i$, so that the last equations also reads

(2) $$P_j \times T_j = -y_i \underset{\sim}{k}, \qquad 1 \le j \le n.$$

Since each pair p_j, p_{j+1} is linearly independent we can write

$$T_j = \xi_j P_j + \eta_j P_{j+1}, \qquad 1 \le j \le n$$

(recall that P_{n+1} denotes $-P_1$). Using (2) we get

(3)
$$\eta_j \alpha_{j,j+1} = -y_j, \qquad 1 \le j \le n$$

and from $T_{j+1} = T_j + 2x_{j+1}P_{j+1}$ and (2) again applied to $P_{j+1} \times T_{j+1}$ we get also

(4)
$$\xi_j \alpha_{j,j+1} = y_{j+1}, \qquad 1 \le j \le n$$

(where $\alpha_{n,n+1} = -\alpha_{n,1}$ and $y_{n+1} = y_1$).

Calculate now

$$2x_j P_j = T_j - T_{j-1} = \xi_j P_j + \eta_j P_{j+1} - \xi_{j-1} P_{j-1} - \eta_{j-1} P_j$$

or

$$(2x_j - \xi_j + \eta_{j-1}) P_1 = \eta_j P_{j+1} - \xi_{j-1} P_{j-1} ;$$

taking cross products with P_{j-1} we get

$$(2x_j - \xi_j + \eta_{j-1}) \alpha_{j,j-1} = \eta_j \alpha_{j+1,j-1},$$

whence

$$2x_j = -\eta_{j-1} + \eta_j (\alpha_{j-1,j+1} / \alpha_{j-1,j}) + \xi_j .$$

Replacing now η_{j-1} and ξ_j by their values obtained from (3) and (4) we conclude that

$$2x_j = (1/\alpha_{j-1,j}) y_{j-1} - (\alpha_{j-1,j+1} / \alpha_{j-1,j} \alpha_{j,j+1}) y_j$$
$$+ (1/\alpha_{j,j+1}) y_{j-1}$$

(where we define $y_0 = y_n$).

Thus the system (1) can be solved by this expression, which in fact also gives the coefficients β_{ij} of the inverse $A^{-1} = (\beta_{ij})$:

$$\beta_{j,j+1} = \beta_{j+1,j} = 1/2\alpha_{j,j+1}$$

$$\beta_{j,j} = -(\alpha_{j-1,j+1}/2\alpha_{j-1,j}\alpha_{j,j+1})$$

$$\beta_{i,j} = 0 \quad \text{for all other pairs} \quad (i,j).$$

A second method for finding A^{-1} involves matrices with coefficients in R^2. We sketch the main steps.

Suppose that $\Lambda = (V_{ij})$, $\Sigma = (W_{j,k})$, etc., where U_{ij}, $W_{j,k}$, etc., belong to R^2. We can multiply Λ and Σ by the usual rule using the dot product $U_{ij} \cdot W_{jk}$ instead of number multiplication:

$$\Lambda\Sigma = (t_{ik})$$

where $t_{ik} = \Sigma_j V_{ij} \cdot W_{jk}$. The result is an ordinary real matrix. We can also multiply ΛR or $R\Lambda$ where R is a real matrix. But: this product is not associative when three vector matrices are involved: $(\Lambda\Sigma)\Gamma$ and $\Lambda(\Sigma\Gamma)$ need not coincide. However, associativity holds for all combination involving not more than two vector matrices, as in $(\Lambda\Sigma)R = \Lambda(\Sigma R)$, $(\Lambda R)\Sigma = \Lambda(R\Sigma)$, $(\Lambda R)S = \Lambda(RS)$, etc.

Using the notation introduced above, define $\Delta = (P_i \times P_j)$ and the matrix $\Theta = (U_{ij})$ with coefficients

$$U_{ii} = P_{i-1} \times P_{i+1}, \qquad 1 \le i \le n,$$

$$U_{i,i-1} = P_{i+1} \times P_i, \qquad 2 \le i \le n,$$

$$U_{i-1,i} = P_{i-1} \times P_{i-2}, \qquad 2 \le i \le n,$$

$$U_{1n} = P_2 \times P_1,$$

$$U_{n1} = P_n \times P_{n-1},$$

$$U_{ij} = 0 \quad \text{for all other pairs} \quad (i,j).$$

Denote also by $\Xi = \underset{\sim}{k}I_n$ and by E the diagonal matrix with coefficients $2\alpha_{i-1,i}\alpha_{i,i+1}$ down the diagonal.

It is easy to see that $\Xi A = A\Xi = \Delta$, and that $\Theta\Delta = E$, and therefore,

$$E = \Theta\Delta = \Theta(\Xi A) = (\Theta\Xi)A.$$

So if we denote $F = \Theta\Xi$ (a real matrix), then $I = E^{-1}FA$, whence $E^{-1}F = A^{-1}$. A direct calculation gives the values obtained above for the coefficients of A^{-1}.

We close this appendix with the remark that a direct calculation gives

$$\det A = (-1)^n 2^{n-2} \prod_{j=1}^{n} \alpha_{j,j+1} ,$$

and in particular the fact that A is invertible follows easily.

REFERENCES

1. DOR, L., Potentials and isometric embeddings in L_1, Israel J. Math., 24 (1976), 260-268.

2. FERGUSON, T., A representation of the symmetric bivariate Cauchy distribution, Ann. Math. Statist., 33 (1962), 1256-1266.

3. LINDENSTRAUSS, J., On the extension of operators with a finite-dimensional range, Ill. J. Math., 8 (1964), 488-499.

4. LINDENSTRASS, J. and TZAFRIRI, L., "Classical Banach Spaces", Lecture Notes #331, Springer, 1973.

5. PÓLYA, G. and SZEGÖ, G., "Aufgabe und Lehrsätze aus der Analysis," Berlin, 1925.

6. SCHÄFFER, J.J., "Geometry of norms in normed spaces," Lecture Notes in Pure Appl. Math., Vol. 20, Marcel Dekker, 1976.

7. SCHWARTZ, J.T., "Non-linear Functional Analysis," Gordon and Breach, 1969.

8. TORANZOS, R., Radial functions of convex and star-shaped bodies, Monthly 74 (1967), 278-280.

University of Illinois, Urbana, IL, USA.

Instituto Argentino de Matematica, CONICET, Argentina.

Functional Analysis, Holomorphy and Approximation Theory II, G.I. Zapata (ed.)

SPECTRAL THEORY FOR CERTAIN OPERATOR POLYNOMIALS

Reinhard Mennicken

Holomorphic operator functions $T \in H(\mathbb{C}, L(E,F))$ and especially operator polynomials have been studied intensively by many authors. Keldyš [29], [30] introduced the concept of eigenvalues, eigenvectors and associated vectors for holomorphic operator functions $T \in H(\mathbb{C}, L(H))$, H being a Hilbert space. For holomorphic operator bundles $T(\lambda) = I + K(\lambda)$ $(\lambda \in \mathbb{C})$ where $K(\lambda)$ is a compact operator for each $\lambda \in \mathbb{C}$ Keldyš stated [29] and proved [30] the existence of a biorthogonal system of eigenvectors and associated vectors belonging to the eigenvalues of T and the adjoint bundle T^*. He arranged this denummerable system $\{y_k^{(0)} : k \in \mathbb{N}\}$ of eigenvectors and associated vectors of T in a canonical way and defined recursively

$$y_k^{(\nu)} = \lambda_0 \, y_k^{(\nu-1)} \qquad (\nu=1,2,\ldots,n-1),$$

if $y_k^{(0)}$ is an eigenvector belonging to the eigenvalue λ_0, and

$$y_k^{(\nu)} = \lambda_0 \, y_k^{(\nu-1)} + y_{k-1}^{(\nu-1)} \qquad (\nu=1,2,\ldots,n-1),$$

if $y_k^{(0)}$ is an associated vector belonging to λ_0. Keldyš called a canonical system of eigenvectors and associated vectors n-fold complete in H if the subspace

$$\operatorname{span}\{ (y_k^{(\nu)})_{\nu=0}^{n-1} : k \in \mathbb{N}\}$$

is dense in the n-fold product H^n. For $n = 1$ this concept reduces to the concept of completeness in the normal sense. In [30] Keldyš proved a completeness theorem for holomorphic operator bundles assuming a certain smallness condition for the resolvent of T (cf.[30],p.27-31). This assumption is, as far as we know, never satisfied when we wish to apply this theorem to boundary value problems for differential equations. Other results of Keldyš in [29], [30] concern the completeness of a canonical system of eigenvectors and associated vectors of polynomial operator bundles of order n. The essential assumptions which are imposed on the coefficient operators are certain conditions of compactness, selfadjointness, normality and completeness (cf. [30], p.32). The proof is based on a linearization method, which transforms the original equation to an equivalent problem in the product space H^n, and on a Phragmen-Lindelöf argument. Applications of these results to boundary value problems for linear differential operators are stated in [29].

Allahverdiev [1], [2], [3], [4], Vizitei & Markus [47] and Markus [37], [38] strengthened and generalized the results of Keldyš concerning operator polynomials. An elaborated proof of Allahverdiev's completeness theorem in [1] can be found in the book of Gohberg & Krein [19], cf. Chap. V, sec. 9. Further completeness statements, some of them of more special character, e.g. concerning quadratic operator bundles only, are contained in the papers of Pallant [43], Gorjuk [21], Isaev [26], [27], Yakubov & Mamedov [50], Orazov [42], Kostjučenko & Orazov [32] and König [31].

In [47] Vizitei & Markus stated various conditions under which the n-fold expansions (with parantheses) in eigenvectors and associated vectors of n-th order operator polynomials were proved to be conditionally or unconditionally convergent. Their proofs are based on the above mentioned linearization method and on perturbations statements for selfadjoint or normal operators with discrete spectrum consisting of eigenvalues only and with no finite limit point in $\mathbb{C}$. The results are applied to ordinary differential operators with selfadjoint boundary conditions, or with boundary conditions which are regular in the sense of Birkhoff, and to elliptic partial differential operators which are regular in the sense of Agmon.

Monien [40] stated remarkable improvements of Markus' results but unfortunately almost all of his proofs are incorrect (cf. Bauer [5] sec. 6.2). Generalizations of Markus' expansion theorems, different from those stated by Monien, were announced by Yakubov & Mamedov in [50].

Apart from the foregoing cited results in the Russian literature (often without any proofs), completeness statements and expansion theorems similar to those of Keldyš, Allahverdiev and Markus were proved independently by Friedman and Shinbrot in [18]. This article also contains applications to linear partial differential operators.

Further results on "nonlinear" eigenvalue problems are due to Müller & Kummer [41], Kummer [33], Mittenthal [39], Turner [46], H. Langer [34] and Roach & Sleeman [44], [45]. For the present paper the cited articles of Mittenthal and Turner are of special interest: projections $P(\mu)$ onto the principal space $\xi(\mu)$, spanned by the eigenvectors and the associated vectors belonging to the eigenvalue μ, are defined in terms of the resolvent of the operator polynomial under consideration (cf. [39], p.122 and [46], p.304). It is worth mentioning that these operators in general are not pairwise biorthogonal, i.e. the equation $P(\mu_1)P(\mu_2) = 0$ for different eigenvalues μ_1, μ_2 may not be satisfied.

The assumptions of the cited completeness statements and expansion theorems are rather restrictive with respect to applications to linear differential operators. Let us consider the following simple boundary value problem:

$$(*)\quad \begin{cases} \eta''(x) + (\lambda^2+2\lambda\alpha(x)+\beta(x))\eta(x) = 0 & (x \in [0,2\pi]) \\ \eta(2\pi) = e^{2\pi i\nu}\,\eta(0), \quad \eta'(2\pi) = e^{2\pi i\nu}\,\eta'(0) \end{cases}$$

where (for simplicity) $\alpha,\beta \in \mathbb{C}^\infty([0,2\pi])$ and $\nu \in \mathbb{C}\setminus\mathbb{Z}$. According to Keldyš [30], cf. the first theorem in chap. II, or Vizitei & Markus [47], theorem 3.3 almost all eigenvalues are simple, i.e. only a finite number of associated vectors exist, and the system of eigenvectors and associated vectors is 2-fold complete in $L_2[0,2\pi]$. The theorem 4.6 of Vizitei & Markus [47], according to which the 2-fold expansion in eigenvectors and associated vectors of all functions $f_1, f_2 \in H_2[0,2\pi]$ satisfying the boundary conditions in (*) is unconditionally convergent with respect to the L_2-norm, requires by its assumptions that $\alpha(x) \equiv 0$ in $[0,2\pi]$.

The assumptions in the expansion theorems of Vizitei & Markus (and others) can be weakened to some extent by refining the perturbation conclusions in the proofs, cf. Bauer [6]. Then, in the foregoing example it is sufficient to require that the function α is small instead of zero. The obstacle to improving the results more essentially is to be found in the more or less rough linearization transforms. With respect to boundary value problems more concrete transformations, cf. Wilder [49] and Wagenführer [48], seem to be better adapted. By the transformation

$$y(x) := \begin{pmatrix} \eta_1(x) \\ \eta_2(x) \end{pmatrix} := \begin{pmatrix} 1 & 0 \\ -i(\alpha(x)+\lambda) & 1 \end{pmatrix} \begin{pmatrix} \eta(x) \\ \eta'(x) \end{pmatrix}$$

the boundary value problem (*) becomes equivalent to the linearized problem

$$(**)\quad \begin{cases} y'(x) - (A_0(x)+\lambda A_1(x))y(x) = 0 & (x \in [0,2\pi]) \\ W_1 y(0) + W_2 y(2\pi) = 0 \end{cases}$$

where

$$A_0(x) = \begin{pmatrix} i\alpha(x) & 1 \\ \alpha^2(x)-i\alpha'(x)-\beta(x) & -i\alpha(x) \end{pmatrix}, \quad A_1(x) = \begin{pmatrix} i & 0 \\ 0 & -i \end{pmatrix}$$

and

$$W_1 = -e^{2\pi i\nu} \begin{pmatrix} 1 & 0 \\ 0 & 1 \end{pmatrix}, \quad W_2 = \begin{pmatrix} 1 & 0 \\ 0 & 1 \end{pmatrix}.$$

Under appropriate assumptions n-th order differential equations with coefficients and boundary value conditions depending polynomially on the eigenvalue parameter λ can be transformed to boundary value problems for first order differential systems where the differential system is linear in λ and the boundary conditions depend polynomially on λ. Eigenvalue problems of this type have been studied by R.E. Langer [35] and Cole [8], [9] in a more classical way without using functional analytic tools.

In the present paper we are concerned with operator polynomials $T \in H(\mathbb{C},\Phi(E,F))$ where $F = F_1 \times F_2$ with Banach spaces F_1, F_2. Accordingly the operator bundle $T(\lambda)$ splits into $(T^D(\lambda),T^R(\lambda))$ and we assume that $T^D(\lambda)$ is linear in λ. Under appropriate additional assumptions we are able to define projections $P(\mu)$ onto the principal spaces $\xi(\mu)$ in terms of the resolvent of T which turn out to be pairwise biorthogonal.

The definition of these projections and the proof of their projection property and their biorthogonality are the contents of section 3. In order to prove these properties we make use of a slight generalization of Keldyš' theorem assuring the existence of a biorthogonal system of eigenvectors and associated vectors. This theorem is stated in section 2. Section 1 contains some notations and preliminary remarks.

In Section 4 we are concerned with biorthogonal expansions

$$f = \sum_j P(\mu_j)f$$

with respect to the projections $P(\mu)$ defined in section 3. To prove an expansion theorem we assume that the operator bundle T is regular which means that on certain curves in $\mathbb{C}$, tending to infinity, the resolvent $R(\lambda)$ of $T(\lambda)$ behaves like λ^p where $p \in \mathbb{N}$. The "function" f has to satisfy some "smoothness properties" and certain "boundary conditions".

In section 5 we give the applications to boundary value problems of Langer's and Cole's type. We state the existence of biorthogonal systems of eigenvectors and associated vectors and prove a general biorthogonal expansion theorem for sufficiently smooth functions f satisfying certain boundary conditions which are determined by the λ-dependent operator $T^R(\lambda)$.

Cole's sufficient conditions for the regularity of boundary value problems can be weakened to some extent. In the light of these weaker assumptions, the results of Eberhard [12], [13], Eberhard & Freiling [14], [15], [16], Freiling [17] and Heisecke [25] on boundary value problems for n-th order differential equations turn out to be very close to Langer's and Cole's theory. For details we refer to Wagenführer [48] and subsequent papers of the author.

I wish to express my thanks to G. Fiedler who worked out a major part of the proofs of Keldyš [30].

1. DEFINITIONS, NOTATIONS

Let G be a Banach space and U be a region in $\mathbb{C}$. $H(U,G)$ denotes the set of all holomorphic mappings on U to G, i.e. $a \in H(U,G)$ if and only if $a: U \to G$ is differentiable on U.

The function a is said to be meromorphic on U, $a \in M(U,G)$, if there is a region $V \subset U$ such that $a \in H(V,G)$ and if for all $\lambda \in U \setminus V$ there exists a neighbourhood W of λ and functions $b \in H(W,G)$, $g \in H(W,\mathbb{C})$ with the properties $W\setminus\{\lambda\} \subset V$, $g \neq 0$ and $ga = b$ on $W\setminus\{\lambda\}$.

If E and F are Banach spaces, $L(E,F)$ denotes the vector space of all continuous (bounded) linear operators on E to F. If $S \in L(E,F)$, $N(S)$ is its null-space or kernel, $\mathrm{Im}(S)$ its image or range. If $y \in E$, $v \in F'$, we set

$$(y \otimes v)(w) := \langle w,v\rangle y \qquad (w \in F)$$

and state that the "tensor product" $y \otimes v$ belongs to $L(F,E)$.

$T \in H(\mathbb{C},L(E,F))$ is called a holomorphic operator bundle (or operator pencil) from E to F. The "adjoint" operator bundle $T^* \in H(\mathbb{C},L(F',E'))$ is defined by $T^*(\lambda) := T(\lambda)^*$ for $\lambda \in \mathbb{C}$.

$$\rho(T) := \{\lambda \in \mathbb{C} : T(\lambda) \text{ bijective}\}$$

is called the resolvent set of T. It is well-known that $\rho(T)$ is an open subset of $\mathbb{C}$. We set $R(\lambda) := T(\lambda)^{-1}$ for $\lambda \in \rho(T)$. $R: \rho(T) \to L(F,E)$ is called the resolvent (operator bundle) of T. We know that $R \in H(\rho(T),L(F,E))$. $\sigma(T) := \mathbb{C}\setminus\rho(T)$ is called the spectrum of T, $\sigma_p(T) := \{\lambda \in \mathbb{C} : N(T(\lambda)) \neq \{0\}\}$ the point spectrum of T, each $\mu \in \sigma_p(T)$ is an eigenvalue of T, and any $y \in N(T(\mu))\setminus\{0\}$ is an eigenvector of T belonging to the eigenvalue μ.

A sequence $Y = (y_0,y_1,\ldots,y_h)$ in E is called a chain of associated vectors of the bundle T belonging to the eigenvalue μ if $y_0 \neq 0$ and

$$(1.1) \qquad \sum_{\ell=0}^{j} \frac{1}{\ell!}\, T^{(\ell)}(\mu)\, y_{j-\ell} = 0 \qquad (j=0,1,2,\ldots,h),$$

where $T^{(\ell)}(\mu)$ denotes the ℓ-th derivative of T at the point μ. Obviously y_0 is an eigenvector of T belonging to μ. We define $\#Y$ to be $h+1$ and call it the length of the chain Y. Let $K(T,\mu)$ denote the set of all chains of associated vectors belonging to the eigenvalue μ.

(1.2) $\quad \nu(y) := \sup\{\#Y : Y \in K(T,\mu),\ y_0 = y\} \qquad (y \in N(T(\mu))\setminus\{0\})$

is called the multiplicity of the eigenvector y.

$\mu \in \mathbb{C}$ is said to be an eigenvalue of finite algebraic multiplicity if

$$\mathrm{nul}(T(\mu)) := \dim N(T(\mu)) < \infty$$

and $\nu(y) < \infty$ for all $y \in N(T(\mu))\setminus\{0\}$. Since

$$L_n := \{y \in N(T(\mu))\setminus\{0\} : \nu(y) \geqq n\} \cup \{0\} \qquad (n \in \mathbb{N})$$

obviously is a subspace of $N(T(\mu))$, we conclude from

$$\dim N(T(\mu)) < \infty, \quad L_n \supset L_{n+1}, \quad \bigcap_n L_n = \{0\}$$

that $L_n = \{0\}$ for some $n \in \mathbb{N}$, i.e.

(1.3) $$\sup\{\nu(y) : y \in N(T(\mu))\setminus\{0\}\} < \infty.$$

Assume that μ is an eigenvalue of T of finite algebraic multiplicity. A system

$$\{y^k_{\nu,\mu} : \nu=0,1,2,\ldots,m_k(\mu), \quad k=1,2,\ldots,\mathrm{nul}(T(\mu))\}$$

is called a canonical system of eigenvectors and associated vectors of T belonging to the eigenvalue μ if the following relationships hold:

(1.4a) $\quad (y^{(k)}_{0,\mu}, y^{(k)}_{1,\mu}, \ldots, y^{(k)}_{m_k(\mu),\mu}) \in K(T,\mu) \qquad (k=1,2,\ldots,\mathrm{nul}\ T(\mu)),$

(1.4b) $\quad \{y^{(k)}_{0,\mu} : k=1,2,\ldots,\mathrm{nul}\ T(\mu)\}$ is a basis of $N(T(\mu))$,

(1.4c) $m_k(\mu)+1 = \max\{\nu(y) : y \in N(T(\mu)) \setminus \operatorname{span}\{y_{0,\mu}^{(i)} : i < k\}\}$

$(k = 1,2,\ldots,\operatorname{nul} T(\mu))$.

From the foregoing remarks concerning the spaces L_n we immediately conclude that each eigenvalue of finite algebraic multiplicity has a canonical system of eigenvectors and associated vectors. The numbers $m_k(\mu)$ $(k = 1,2,\ldots,\operatorname{nul} T(\mu))$ are independent of the choice of the canonical system since

$$m_k(\mu) + 1 = \max\{n \in \mathbb{N} : \dim L_n \geqq k\}.$$

For the following sections we state a reformulation of the relationships (1.1): Assume $\mu \in \sigma(T)$ and let $Y = (y_0, y_1, \ldots, y_h)$ be a sequence in E with $y_0 \neq 0$. We set

$$\xi_h(Y,\lambda) := \sum_{r=0}^{h} \frac{y_r}{(\lambda-\mu)^{h+1-r}} \tag{1.5}$$

and, for later convenience, $T^{(\ell)}(\mu) = 0$ if $\ell = -1,-2,\ldots$. Obviously $T\xi_h(Y,\cdot)$ is holomorphic in $\mathbb{C}\setminus\{\mu\}$.

(1.6) PROPOSITION. $Y \in K(T,\mu)$, i.e. Y fulfills the relationships (1.1) if and only if $T\xi_h(Y,\cdot)$ is holomorphic in μ.

PROOF. A simple calculation leads to

$$\begin{aligned} T(\lambda)\xi_h(Y,\lambda) &= \sum_{\ell=0}^{\infty} \sum_{r=0}^{h} \frac{1}{\ell!} (\lambda-\mu)^{(\ell+r)-(h+1)} T^{(\ell)}(\mu) y_r \\ &= \sum_{j=0}^{\infty} (\lambda-\mu)^{j-(h+1)} \{\sum_{\ell=j-h}^{j} \frac{1}{\ell!} T^{(\ell)}(\mu) y_{j-\ell}\}, \end{aligned}$$

from which the statement in (1.6) is evident.

2. THE EXISTENCE OF BIORTHOGONAL SYSTEMS, THE STRUCTURE OF THE RESOLVENT

$S \in L(E,F)$ is called a Fredholm operator if

$$\text{nul}(S) := \dim N(S) < \infty, \qquad \text{def}(S) := \text{codim}\, \text{Im}(S) < \infty.$$

$\Phi(E,F)$ denotes the set of all Fredholm operators on E to F. If $S \in \Phi(E,F)$, $\text{Im}(S)$ is closed (cf. e.g. Goldberg [20], Cor.IV.1.13).

In the following we always assume:

(2.1) $T \in H(\mathbb{C},\Phi(E,F))$ and $\rho(T) \neq \phi$, i.e. T is a holomorphic operator bundle with values in the set of Fredholm operators and has a non-empty resolvent set.

We state without proof:

(2.2) THEOREM. (cf. Gramsch [22], Th. 11, p.102)

(i) $\sigma(T) = \sigma_p(T)$,

(ii) $\sigma(T)$ has no finite limit point,

(iii) $R \in M(\mathbb{C},L(F,E))$.

(2.3) THEOREM. (cf. Keldyš [30], chap. I, p.19-26)

Assume $\mu \in \sigma(T)$.

Then μ is an eigenvalue of T of finite algebraic multiplicity.

Let

(2.4) $\{y_{\nu,\mu}^{(k)} : \nu = 0,1,\ldots,m_k(\mu), \quad k = 1,2,\ldots,\text{nul}\ T(\mu)\}$

be a canonical system of eigenvectors and associated vectors of T belonging to μ.

μ also is an eigenvalue of finite algebraic multiplicity of T^* and there exists a canonical system

$$(2.5)\qquad \{v_{\nu,\mu}^{(k)} : \nu = 0,1,\ldots,m_k(\mu),\ k = 1,2,\ldots,\text{nul } T(\mu)\}$$

of eigenvectors and associated vectors of T^* belonging to μ such that

$$(2.6)\qquad HT(R,\mu) = \sum_{k=1}^{\text{nul } T(\mu)} \sum_{j=0}^{m_k(\mu)} \frac{1}{(\lambda-\mu)^{m_k(\mu)+1-j}} \sum_{h=0}^{j} y_{h,\mu}^{(k)} \otimes v_{j-h,\mu}^{(k)},$$

where $HT(R,\mu)$ denotes the singular part of R at the pole μ.

Let us assume in addition that $\mu_2 \in \sigma(T)$. Then the canonical systems (2.4) and (2.5), belonging to T, μ and T^*, μ_2 respectively, fulfill the biorthogonal relationships

$$(2.7)\qquad \begin{cases} \displaystyle\sum_{n=0}^{j} \frac{1}{n!} \langle \eta^{(n)}(\mu, Y_h^{(k)})(\mu_2), v_{j-n,\mu_2}^{(i)} \rangle = \delta_{\mu,\mu_2}\, \delta_{ki}\, \delta_{m_i(\mu_2)-j,h} \\ 0 \leqq h \leqq m_k(\mu),\quad 0 \leqq j \leqq m_i(\mu_2),\quad 1 \leqq k \leqq \text{nul } T(\mu),\quad 1 \leqq i \leqq \text{nul } T(\mu_2) \end{cases}$$

where $\eta^{(n)}(\mu, Y_h^{(k)})(\mu_2)$ denotes the n-th derivative of the function

$$(2.8)\qquad \eta(\mu, Y_h^{(k)})(\lambda) := T(\lambda) \sum_{j=0}^{h} \frac{y_{j,\mu}^{(k)}}{(\lambda-\mu)^{h+1-j}}$$

at μ_2. (According to Prop. 1.6 $\eta(\mu, Y_h^{(k)})$ is analytic in $\mathbb{C}$.)

An eigenvalue μ of T is called normal if the canonical system of eigenvectors and associated vectors belonging to μ only consists of eigenvectors. The operator bundle T is said to be normal if each eigenvalue of T is normal.

(2.9) PROPOSITION. Let μ be an eigenvalue of T and $r = \text{nul } T(\mu)$. The following statements are equivalent:

(i) μ is a normal eigenvalue of T.

(ii) The resolvent R of T has a pole of order 1 in μ.

(iii) For every $y \in N(T(\mu))\backslash\{0\}$ there exists a $v \in N(T^*(\mu))$ such that $\langle T^{(1)}(\mu)y, v\rangle \neq 0$.

(iv) There exists a basis $\{y_1, y_2, \dots, y_r\}$ of $N(T(\mu))$ and a basis $\{v_1, v_2, \dots, v_r\}$ of $N(T^*(\mu))$ such that

$$\langle T^{(1)}(\mu)y_i, v_j\rangle = \delta_{ij} \qquad (i,j = 1,2,\dots,r).$$

PROOF. The implication (i) ⇒ (ii) is obvious from (2.6). The inverse implication (ii) ⇒ (i) also follows from (2.6): since the eigenvectors $y_{0,\mu}^{(k)}$ are linear independent the sum

$$\Sigma\ y_{0,\mu}^{(k)} \otimes v_{0,\mu}^{(k)},$$

where k varies over the set $\{k \in \mathbb{N} : m_k(\mu) = m_1(\mu)\}$, does not vanish which implies $m_1(\mu) = 0$. (iv) ⇒ (iii) is immediate.
We prove (iii) ⇒ (i) by contradiction: we assume that μ is not normal, i.e. $m_1(\mu) \geqq 1$ and there is a chain $y_0^{(1)}$, $y_1^{(1)}$. According to (1.1) we have

$$T(\mu)y_1^{(1)} + T^{(1)}(\mu)y_0^{(1)} = 0.$$

This equation leads to

$$\langle T^{(1)}(\mu)y_0^{(1)}, v\rangle = -\langle T(\mu)y_1^{(1)}, v\rangle = -\langle y_1^{(1)}, T^*(\mu)v\rangle = 0$$

for all $v \in N(T^*(\mu))$ which contradicts (iii). The implication (i) ⇒ (iv) remains to be proved: from (2.8) we conclude

$$\eta(\mu, Y_0^{(k)})(\lambda) = T^{(1)}(\mu)y_{0,\mu}^{(k)} + (\lambda-\mu)\sum_{j=0}^{\infty} \frac{1}{(j+2)!}(\lambda-\mu)^j T^{(j+2)}(\mu)y_{0,\mu}^{(k)}$$

and thus

$$\eta(\mu, Y_0^{(k)})(\mu) = T^{(1)}(\mu)y_{0,\mu}^{(k)} .$$

By assumption $m_k(\mu) = 0$ for $k = 1,2,\dots,r$ and therefore

$$\langle T^{(1)}(\mu)y_{0,\mu}^{(k)}, v_{0,\mu}^{(k)}\rangle = \delta_{ik} \qquad (i,k = 1,2,\dots,r)$$

according to (2.7).

3. BIORTHOGONAL PROJECTIONS

In this section we make additional assumptions with respect to the operator bundle T. We assume that $T \in H(\mathbb{C},\Phi(E,F))$ is a polynomial of degree q, i.e.

$$T(\lambda) = \sum_{i=0}^{q} \lambda^i T_i \tag{3.1}$$

with $T_q \neq 0$. Furthermore, let F_1 and F_2 be Banach spaces, $F = F_1 \times F_2$ and

$$T(\lambda) = (T^D(\lambda), T^R(\lambda))$$

where $T^D(\lambda)$ is a polynomial of order 1, i.e.

$$T^D(\lambda) = T_0^D + \lambda T_1^D J$$

and

$$T^R(\lambda) = \sum_{i=0}^{q} \lambda^i T_i^R .$$

We assume that the coefficients of these polynomials fulfill the following properties:

$T_0^D, J \in L(E,F_1)$, J is injective,

$T_1^D \in L(F_1,F_1)$ and is bijective,

$T_i^R \in L(E,F_2)$ for $i = 0,1,2,\dots,q$.

We define the linear operator $M\colon D(M) \subset F_1 \to F_1$ by

$$\left\{\begin{array}{l} D(M) := J(E) \\ My := -(T_1^D)^{-1}\, T_0^D J^{-1} y \quad (y \in D(M)) \end{array}\right. \tag{3.2}$$

and the operators M^j $(j \in \mathbb{N})$ iteratively in the usual way.

If $y \in J^{-1}(D(M^{p+1}))$ then $Jy \in D(M^{p+1})$ and thus

$$M^j Jy \in D(M) = J(E) \qquad (j = 0,1,2,\dots,p);$$

therefore

$$y^{[j]} := J^{-1}M^jJy \quad (\in E) \tag{3.3}$$

is well-defined for all $j = 0,1,2,\ldots,p$.

We set

$$T^{[j]}(\lambda) := \sum_{k=j+1}^{q} \lambda^{k-(j+1)} T_k \, , \tag{3.4}$$

$E_0 := J^{-1}(D(M^q))$ (which is a subspace of E) and, if $y \in E_0$,

$$V(\lambda)y := \sum_{j=0}^{q-1} T^{[j]}(\lambda)y^{[j]} . \tag{3.5}$$

For fixed $\lambda \in \mathbb{C}$ $V(\lambda)$ obviously is a linear operator from E_0 to F. Finally, if $\mu \in \sigma(T)$, we define

$$\begin{cases} D(P(\mu)) := E_0 \, , \\ P(\mu) := \operatorname{Res}_\mu(RV), \end{cases} \tag{3.6}$$

where $\operatorname{Res}_\mu(RV)$ denotes the residuum of the operator function $RV(\lambda) = R(\lambda)V(\lambda)$ at μ. The relationship

$$P(\mu) = \sum_{\ell=0}^{q-1} \operatorname{Res}_\mu(RT^{[\ell]})J^{-\ell}M^1J \tag{3.7}$$

is an immediate consequence from (3.3) and (3.5).

If T is an operator polynomial of the first order, i.e. $q = 1$ and $T(\lambda) = T_0 + \lambda T_1$ $(\lambda \in \mathbb{C})$, then $E_0 = E$, $V(\lambda) = T_1$ and

$$P(\mu) = \operatorname{Res}_\mu(RT_1).$$

In this "linear" case it is well-known (cf. e.g. Grigorieff [23], p. 51 or [24] p. 357-358) that the operators $P(\mu)$ fulfill the "biorthogonal" relationships

$$P(\mu_1)P(\mu_2) = \delta_{\mu_1,\mu_2} P(\mu_2)$$

for $\mu_1,\mu_2 \in \sigma(T)$. The operators $P(\mu)$ are projections onto the principal space $\xi(\mu)$ belonging to the eigenvalue μ.

In this section we are going to prove similar biorthogonal relationships for the operators $P(\mu)$ which were defined in (3.6) for operator polynomials. The introduction of these biorthogonal projections enables us to define and investigate the problem of biorthogonal expansion with respect to a system of eigenvectors and associated vectors.

(3.8) PROPOSITION. Let $y_0, y_1, \ldots, y_h$ be a chain of associated vectors of T belonging to μ. For convenience we set $y_j := 0$ if $j < 0$. We assert:

$$(3.9) \qquad Jy_k \in D(M^p),$$

$$(3.10) \qquad y_k^{[p]} = \sum_{\ell=0}^{p} \binom{p}{\ell} \mu^{p-\ell} y_{k-\ell}$$

$$(3.11) \qquad \sum_{\ell=0}^{q} T_\ell \, y_k^{[p+\ell]} = 0$$

for all $p \in \mathbb{N}$ and $k \in \{1,2,\ldots,h\}$.

We prove (3.9) and (3.10) by induction with respect to p. Both statements are true for $p = 0$. For $k \geqq 1$ we have

$$(3.12) \qquad (T_0^D + \mu T_1^D J)y_k + T_1^D Jy_{k-1} = 0$$

because of (1.1) and the linear structure of $T^D(\lambda)$. This equality also holds for $k \leqq 0$ since $y_j = 0$ for $j < 0$. Obviously $Jy_k \in D(M)$.

From (3.2) and (3.12) we conclude

$$(3.13) \qquad MJy_k = -(T_1^D)^{-1} T_0^D J^{-1} Jy_k = \mu Jy_k + Jy_{k-1}$$

so that $Jy_k \in D(M^{p+1})$ if Jy_k and Jy_{k-1} belong to $D(M^p)$. Thus (3.9) is clear.

According to (3.9) $y_k^{[p]}$ is well-defined for all $p \in \mathbb{N}$. Let us assume that (3.10) has already been proved for p. Then, using (3.13), we obtain

$$\begin{aligned} y_k^{[p+1]} &= J^{-1}M^pJ(\mu y_k + y_{k-1}) \\ &= \mu \sum_{\ell=0}^{p} \binom{p}{\ell} \mu^{p-\ell}\, y_{k-\ell} + \sum_{\ell=0}^{p} \binom{p}{\ell} \mu^{p-\ell}\, y_{k-1-\ell} \\ &= \sum_{\ell=0}^{p+1} \binom{p+1}{\ell} \mu^{p+1-\ell}\, y_{k-\ell} \end{aligned}$$

which proves (3.10).

The proof of (3.11) is a bit more complicated. We insert the following

(3.14) LEMMA. Let $j,p \in \mathbb{N}$, $\ell \in \mathbb{Z}$. Then

$$\binom{j+\ell}{p} = \sum_{m=0}^{p} \binom{j}{m} \binom{\ell}{p-m}, \quad \text{if } \ell \geqq 0, \tag{3.15}$$

$$\binom{j+\ell}{p} = \sum_{m=0}^{p} (-1)^{p-m} \binom{j}{m}\binom{p-m-\ell-1}{p-m}, \quad \text{if } \ell < 0 \text{ and } j+\ell \geqq 0, \tag{3.16}$$

$$\binom{p-j-\ell-1}{p} = \sum_{m=0}^{p} (-1)^{m} \binom{j}{m}\binom{p-m-\ell-1}{p-m}, \quad \text{if } j+\ell < 0. \tag{3.17}$$

The relationships (3.15), (3.16), (3.17) follow from the identity

$$\frac{1}{p!}\left(\frac{d}{dz}\right)^p z^{j+\ell} = \frac{1}{p!} \sum_{m=0}^{p} \binom{p}{m} \left[\left(\frac{d}{dz}\right)^m z^j\right]\left[\left(\frac{d}{dz}\right)^{p-m} z^{\ell}\right]$$

and the equality

$$\left(\frac{d}{dz}\right)^r z^s \Big|_{z=1} = \begin{cases} r!\,\binom{s}{r}, & \text{if } s \geqq 0 \\ (-1)^r\, r!\,\binom{-s+r-1}{r}, & \text{if } s < 0, \end{cases}$$

which obviously is true for arbitrary $r \in \mathbb{N}$ and $s \in \mathbb{Z}$.

Now we are ready for the proof of (3.11): If $j \in \mathbb{N}$

$$\frac{1}{j!}\, T^{(j)}(\mu) = \sum_{\ell=j}^{q} \binom{\ell}{j} \mu^{\ell-j} T_\ell\,. \tag{3.18}$$

Using (3.10) we obtain

$$\sum_{\ell=0}^{q} T_\ell y_k^{[p+\ell]} = \sum_{\ell=0}^{q} \sum_{r=0}^{p+\ell} \binom{p+\ell}{r} \mu^{p+\ell-r} T_\ell y_{k-r} .$$

A simple rearrangement of the sums on the right side and the application of the formula (3.15) leads to the relationship

$$\sum_{\ell=0}^{q} T_\ell y_k^{[p+\ell]} = \sum_{r=0}^{p+q} \sum_{\ell=r-p}^{q} \sum_{m=0}^{r} \binom{p}{m} \binom{\ell}{r-m} \mu^{p+\ell-r} T_\ell y_{k-r}$$

where, if $\ell < 0$, $T_\ell = 0$. If $m > p$ then $\binom{p}{m} = 0$ and if $m < p$ then $\binom{\ell}{r-m} = 0$ for $\ell \in \{r-p,\ldots,r-m-1\}$ whence

$$\sum_{\ell=0}^{q} T_\ell y_k^{[p+\ell]} = \sum_{r=0}^{p+q} \sum_{m=0}^{r} \binom{p}{m} \mu^{p-m} \{ \sum_{\ell=r-m}^{q} \binom{\ell}{r-m} \mu^{\ell-(r-m)} T_\ell \} y_{k-r} =$$

$$= \sum_{r=0}^{p+q} \sum_{m=0}^{r} \binom{p}{m} \mu^{p-m} \frac{1}{(r-m)!} T^{(r-m)}(\mu) \, y_{k-r}$$

where we applied (3.18) for the proof of the last equality. Another rearrangement of the sums gives

$$\sum_{\ell=0}^{q} T_\ell y_k^{[p+\ell]} = \sum_{m=0}^{p+q} \sum_{r=m}^{p+q} \binom{p}{m} \mu^{p-m} \frac{1}{(r-m)!} T^{(r-m)}(\mu) \, y_{k-r}$$

$$= \sum_{m=0}^{p+q} \binom{p}{m} \mu^{p-m} \sum_{r=0}^{p+q-m} \frac{1}{r!} T^{(r)}(\mu) \, y_{k-m-r} .$$

If $m > p$ then $\binom{p}{m} = 0$. Furthermore, if $p+q > k$ then $y_{k-m-r} = 0$ for all $k-m+1 \leqq r \leqq p+q-m$ and, if $p+q < k$ then $T^{(r)}(\mu) = 0$ for all $p+q < r \leqq k$. Thus, from the foregoing equation, we obtain

$$\sum_{\ell=0}^{q} T_\ell y_k^{[p+\ell]} = \sum_{m=0}^{p} \binom{p}{m} \mu^{p-m} \sum_{r=0}^{k-m} \frac{1}{r!} T^{(r)}(\mu) \, y_{k-m-r} ,$$

the right side of which vanishes because of the Definition (1.1) so that the proof of (3.11) is complete.

(3.19) PROPOSITION. Let $Y = (y_o, y_1, \ldots, y_h)$ be a chain of associated vectors of T belonging to μ. We assert

(3.20) $$V(\lambda) y_h = T(\lambda) \xi_h(Y,\lambda) \qquad (\lambda \in \mathbb{C})$$

where $\xi_h(Y,\lambda)$ denotes the vector function which has been defined in (1.5).

PROOF. The Definitions (3.3),(3.4),(3.5) and the relationship (3.10) imply

$$V(\lambda)y_h = \sum_{j=0}^{q-1} \Big(\sum_{r=j+1}^{q} \lambda^{r-(j+1)} T_r \Big) \Big(\sum_{\ell=0}^{j} \binom{j}{\ell} \mu^{j-\ell} y_{h-\ell} \Big)$$

$$= \sum_{j=0}^{q-1} \sum_{r=0}^{q-1-j} \sum_{\ell=0}^{j} \binom{j}{\ell} \lambda^r \mu^{j-\ell} T_{r+j+1} y_{h-\ell}$$

whence

$$V(\lambda)y_h = \sum_{j=0}^{q-1} \sum_{r=0}^{q-1-j} \sum_{\ell=0}^{j} \binom{j}{\ell} \lambda^r \mu^\ell T_{r+j+1}\, y_{h-j+\ell}$$

because $\binom{j}{\ell} = \binom{j}{j-\ell}$. A rearrangement of the sums leads to

$$(3.21)\qquad V(\lambda)y_h = \sum_{r=0}^{q-1} \sum_{\ell=0}^{q-1-r} \sum_{j=\ell}^{q-1-r} \binom{j}{\ell} \lambda^r \mu^\ell T_{r+j+1}\, y_{h-j+\ell}$$

$$= \sum_{r=0}^{q-1} \sum_{\ell=0}^{q-1-r} \sum_{j=0}^{q-1-r-\ell} \binom{j+\ell}{\ell} \lambda^r \mu^\ell T_{r+j+\ell+1}\, y_{h-j} .$$

On the other hand the Taylor series of $T(\lambda)$ at μ and the Definition (1.5) of $\xi_h(Y,\lambda)$ yield the expansion

$$T(\lambda)\xi_h(Y,\lambda) - \sum_{m=0}^{q} \sum_{j=0}^{h} (\lambda-\mu)^{m-j-1} \frac{1}{m!} T^{(m)}(\mu) y_{h-j}$$

$$= \sum_{j=0}^{h} \sum_{m=-j-1}^{q-j-1} (\lambda-\mu)^m \frac{1}{(m+j+1)!} T^{(m+j+1)}(\mu)\, y_{h-j} .$$

Since $T\xi_h(Y,\cdot)$ is holomorphic (in μ) by Proposition (1.5), we obtain

$$T(\lambda)\xi_h(Y,\lambda) = \sum_{j=0}^{h} \sum_{m=0}^{q-1-j} (\lambda-\mu)^m \frac{1}{(m+j+1)!} T^{(m+j+1)}(\mu)\, y_{h-j} .$$

If $h < q-1$ then $y_{h-j} = 0$ for $j \in \{h+1,\dots,q-1\}$ and, if $h > q-1$ then $T^{(m+j+1)}(\mu) = 0$ for $j \in \{q,\dots,h\}$ whence

$$T(\lambda)\xi_h(Y,\lambda) = \sum_{j=0}^{q-1} \sum_{m=0}^{q-1-j} (\lambda-\mu)^m \frac{1}{(m+j+1)!} T^{(m+j+1)}(\mu)\, y_{h-j} .$$

The expansion

$$(\lambda-\mu)^m = \sum_{r=0}^{m} (-1)^{m-r} \binom{m}{r} \lambda^r \mu^{m-r}$$

and the relationship (3.18) lead to

$$T(\lambda)\xi_h(Y,\lambda) = \sum_{j=0}^{q-1} \sum_{m=0}^{q-1-j} \sum_{r=0}^{m} \sum_{\ell=m+j+1}^{q} (-1)^{m-r} \binom{m}{r}\binom{\ell}{m+j+1} *$$

$$* \lambda^r \mu^{\ell-r-j-1} T_\ell y_{h-j}.$$

Because

$$\sum_{m=0}^{q-1-j} \sum_{r=0}^{m} = \sum_{r=0}^{q-1-j} \sum_{m=r}^{q-1-j}$$

we obtain

(3.22) $\quad T(\lambda)\xi_h(Y,\lambda) =$

$$= \sum_{j=0}^{q-1} \sum_{r=0}^{q-1-j} \sum_{m=0}^{q-1-j-r} \sum_{\ell=m+r+j+1}^{q} (-1)^m \binom{m+r}{r}\binom{\ell}{m+r+j+1} \lambda^r \mu^{\ell-r-j-1} T_\ell y_{h-j}$$

$$= \sum_{j=0}^{q-1} \sum_{r=0}^{q-1-j} \sum_{m=0}^{q-1-j-r} \sum_{\ell=m}^{q-1-j-r} (-1)^m \binom{m+r}{r}\binom{\ell+r+j+1}{\ell-m} \lambda^r \mu^{\ell} T_{\ell+r+j+1} y_{h-j}$$

where we used the identity

$$\binom{\ell}{m+j+1} = \binom{\ell}{\ell-m-j-1}.$$

Since, according to (3.16),

$$\sum_{m=0}^{\ell} (-1)^m \binom{\ell+r+j+1}{\ell-m}\binom{m+r}{m} = \sum_{m=0}^{\ell} (-1)^{\ell-m} \binom{\ell+r+j+1}{m}\binom{\ell-m+r}{\ell-m}$$

$$= \binom{(\ell+r+j+1)+(-r-1)}{\ell} = \binom{j+\ell}{\ell}$$

and because

$$\sum_{m=0}^{q-1-j-r} \sum_{\ell=m}^{q-1-j-r} = \sum_{\ell=0}^{q-1-j-r} \sum_{m=0}^{\ell}$$

the foregoing equation (3.22) implies that

$$T(\lambda)\xi_h(Y,\lambda) = \sum_{j=0}^{q-1} \sum_{r=0}^{q-1-j} \sum_{\ell=0}^{q-1-j-r} \binom{j+\ell}{\ell} \lambda^r \mu^\ell T_{r+j+\ell+1} y_{h-j}$$

so that, in consideration of (3.21), the proof of the relationship (3.20) finally is complete.

(3.23) THEOREM. i) If $\mu \in \sigma(T)$ then $\operatorname{Im}(P(\mu)) \subset D(P(\mu))$ which is equal to E_0 and thus independent of μ.

ii) The biorthogonal relationship

$$P(\mu_1)P(\mu_2) = P(\mu_2)P(\mu_1) = \delta_{\mu_1,\mu_2} P(\mu_1) \tag{3.24}$$

holds for all $\mu_1,\mu_2 \in \sigma(T)$.

PROOF. Let $\mu \in \sigma(T)$. Obviously

$$\operatorname{Res}_\mu (RV) = \operatorname{Res}_\mu (HT(R,\mu)V)$$

so that, by (2.6),

$$P(\mu) = \sum_{k=1}^{\operatorname{nul} T(\mu)} \sum_{j=0}^{m_k(\mu)} \sum_{h=0}^{j} \operatorname{Res}_\mu \{\frac{1}{(\lambda-\mu)^{m_k(\mu)+1-j}}(y_{h,\mu}^{(k)} \otimes v_{j-h,\mu}^{(k)})V(\lambda)\}.$$

Hence, if $y \in D(P(\mu))$ then

$$P(\mu)y = \sum_{k=1}^{\operatorname{nul} T(\mu)} \sum_{j=0}^{m_k(\mu)} \sum_{h=0}^{m_k(\mu)-j} \langle \frac{1}{j!}V^{(j)}(\mu)y, v_{m_k(\mu)-j-h,\mu}^{(k)}\rangle \, y_{h,\mu}^{(k)}$$

where $V^{(j)}(\mu)$ denotes the j-th derivative of V at the point μ. A change of the order of summation leads to

$$P(\mu)y = \sum_{k=1}^{\operatorname{nul} T(\mu)} \sum_{h=0}^{m_k(\mu)} \sum_{j=0}^{h} \langle \frac{1}{j!}V^{(j)}(\mu)y, v_{h-j}^{(k)}\rangle \, y_{m_k(\mu)-h,\mu}^{(k)} \tag{3.25}$$

whence, taking into consideration the relationship (3.9), the assertion i) is already proved.

Now we are going to prove the statement ii): According to (1.5), (2.8) and (3.20) we have

$$V(\lambda)y_{h,\mu}^{(k)} = \eta(\mu, Y_h^{(k)})(\lambda) \tag{3.26}$$

for all $h \in \{0,1,2,\ldots,m_k(\mu)\}$ and arbitrary $\lambda \in \mathbb{C}$. Therefore and because of (2.7), (3.25) we obtain

$$P(\mu_2)y^{(k)}_{m_k(\mu_1)-h,\mu_1} = \sum_{i=1}^{\text{nul}\,T(\mu_2)} \sum_{j=0}^{m_k(\mu_2)} \sum_{n=0}^{j} \frac{1}{n!} *$$

$$* \langle \eta^{(n)}(\mu_1, Y^{(k)}_{m_k(\mu_1)-h})(\mu_2), v^{(i)}_{j-n}\rangle\, y^{(i)}_{m_i(\mu_2)-j,\mu_2}$$

$$= \delta_{\mu_1,\mu_2}\, y^{(k)}_{m_k(\mu_1)-h,\mu_1}$$

for $0 \leqq h \leqq m_k(\mu_1)$, $0 \leqq k \leqq \text{nul}\,T(\mu_1)$. Thus, again in view of (3.25), the proof of ii) is complete.

4. BIORTHOGONAL EXPANSIONS

In the sequel we adopt the assumptions made in the preceding sections. We denumerate the elements $\mu_1, \mu_2, \ldots$ of the (point) spectrum $\sigma_p(T)$ in such a way that

$$|\mu_1| \leqq |\mu_2| \leqq \ldots .$$

Let $p \in \mathbb{N}$. We call the operator bundle T regular of order p if and only if the following properties are fulfilled:

i) there exist a sequence $(n_j)_{j\in\mathbb{N}}$ of natural numbers and a sequence $(\Gamma_j)_{j\in\mathbb{N}}$ of simply closed Jordan curves in $\mathbb{C}$ around 0 so that exactly the eigenvalues $\mu_1, \mu_2, \ldots, \mu_{n_j}$ are inside Γ_j for all $j \in \mathbb{N}$;

ii) $d_j := \max\{|\lambda| : \lambda \in \Gamma_j\}$ tends to infinity and there is a $\vartheta > 0$ so that

$$\vartheta d_j \leqq \text{dist}(0, \Gamma_j) \qquad (j \in \mathbb{N});$$

iii) there exists a real number $c > 0$ such that

$$\int_{\Gamma_j} |R(\lambda)|\;|d\lambda| \leqq c\, d_j^p \qquad (j \in \mathbb{N}).$$

Obviously the property iii) is fulfilled if there exist positive constants c_1, c_2 so that

$$\text{length}(\Gamma_j) \leq c_1 d_j$$

and

$$\max\{|R(\lambda)| : \lambda \in \Gamma_j\} \leq c_2 d_j^{p-1}$$

for all $j \in \mathbb{N}$.

(4.1) THEOREM. Let the operator bundle T be regular of order p. Let $f \in J^{-1}(D(M^{p+q}))$ and assume that f fulfills the "boundary conditions"

$$\sum_{j=0}^{q} T_j^R f^{[j+\ell-1]} = 0 \qquad (\ell = 1,2,\ldots,p). \tag{4.2}$$

Then we assert that

$$f = \lim_{j\to\infty} \sum_{i=1}^{n_j} P(\mu_i)f. \tag{4.3}$$

PROOF. According to (3.2) and (3.3), the definition of the $f^{[j]}$, we have

$$T_0^D f^{[\ell-1]} + T_1^D J f^{[\ell]} = 0 \qquad (\ell = 1,2,\ldots,p)$$

whence the boundary conditions (4.2) are equivalent to the conditions

$$\sum_{j=0}^{q} T_j f^{[j+\ell-1]} = 0 \qquad (\ell = 1,2,\ldots,p). \tag{4.4}$$

Let $\lambda \in \rho(T)$ and unequal to 0. We set

$$S_\ell(\lambda)f := \sum_{m=0}^{q+\ell-1} \frac{1}{\lambda^{m+1}} f^{[m]} - \sum_{m=\ell}^{q+\ell-1} \sum_{k=0}^{m-\ell} \frac{1}{\lambda^{m+1-k}} R(\lambda) T_k f^{[m]} \tag{4.5}$$

for $\ell \in \{0,1,\ldots,p\}$ and show that $S_{\ell+1}(\lambda) = S_\ell(\lambda)$: Obviously

$$S_{\ell+1}(\lambda)f = \sum_{m=0}^{q+\ell-1} \frac{1}{\lambda^{m+1}} f^{[m]} - \sum_{m=\ell}^{q+\ell-1} \sum_{k=0}^{m-\ell-1} \frac{1}{\lambda^{m+1-k}} R(\lambda) T_k f^{[m]} +$$

$$+ \frac{1}{\lambda^{q+\ell-1}} \{f^{[q+\ell]} - R(\lambda) \sum_{k=0}^{q-1} \lambda^k T_k f^{[q+\ell]}\}.$$

Using (3.1) and (4.4) we infer

$$\frac{1}{\lambda^{q+\ell+1}} \{ f^{[q+\ell]} - R(\lambda) \sum_{k=0}^{q-1} \lambda^k T_k f^{[q+\ell]} \} = \frac{1}{\lambda^{\ell+1}} R(\lambda) T_q f^{[q+\ell]}$$

$$= - \frac{1}{\lambda^{\ell+1}} R(\lambda) \sum_{m=0}^{q-1} T_m f^{[m+\ell]}$$

$$= - \sum_{m=\ell}^{q+\ell-1} \frac{1}{\lambda^{\ell+1}} R(\lambda) T_{m-\ell} f^{[m]}$$

which completes the proof.

According to (3.4), (3.5) and (4.5) we obtain

$$R(\lambda)V(\lambda)f = R(\lambda) \sum_{m=0}^{q-1} \frac{1}{\lambda^{m+1}} (T(\lambda) - \sum_{k=0}^{m} \lambda^k T_k) f^{[m]}$$

$$= \sum_{m=0}^{q-1} \frac{1}{\lambda^{m+1}} f^{[m]} - \sum_{m=0}^{q-1} \sum_{k=0}^{m} \frac{1}{\lambda^{m+1-k}} R(\lambda) T_k f^{[m]}$$

$$= S_0(\lambda) f$$

whence

$$R(\lambda)V(\lambda)f = S_p(\lambda)f \tag{4.6}$$

which, by (4.5), immediately leads to

$$\sum_{m=0}^{q+p-1} \frac{1}{\lambda^{m+1}} f^{[m]} = R(\lambda)V(\lambda)f + \sum_{m=p}^{q+p-1} \sum_{k=0}^{m-p} \frac{1}{\lambda^{m+1-k}} R(\lambda) T_k f^{[m]} .$$

If we integrate both sides of the preceding equation along Γ_j we obtain

$$f = \sum_{i=1}^{n_j} P(\mu_i) f + \sum_{m=p}^{q+p-1} \sum_{k=0}^{m-p} \frac{1}{2\pi i} \int_{\Gamma_j} \frac{1}{\lambda^{m+1-k}} R(\lambda) T_k f^{[m]} \, d\lambda$$

from the Definition (3.6). We assume without loss of generality that all $d_j \geq 1$. We estimate

$$\left| \int_{\Gamma_j} \frac{1}{\lambda^{m+1-k}} R(\lambda) T_k f^{[m]} \, d\lambda \right| \leqq \frac{c \, d_j^p}{(\vartheta d_j)^{m+1-k}} |T_k f^{[m]}|$$

$$\leqq \frac{c}{\vartheta^{m+1-k}} \frac{1}{d_j} |T_k f^{[m]}|$$

for all $0 \leqq k \leqq m-p$, $p \leqq m \leqq q+p-1$ whence the assertion of the expansion theorem is proved.

We would like to add the remark that according to (3.11) the eigenvectors and the associated vectors fulfill the conditions (4.4) and thus the boundary conditions (4.2). Therefore, in a sense, the conditions (4.2) are also necessary: if we know that all functions $f^{[j]}$ $(j = 0,1,2,...,q)$ are expandable into a series of type (4.3) then f has to fulfill the conditions (4.2). This fact is well-known for functions with (pointwise convergent) Fourier expansions: they have to satisfy the same periodicity properties as the sine and cosine functions.

It is not difficult and therefore left to the reader to state conditions under which the series (4.3) is absolutely convergent (in parantheses) which means that the sums

$$\sum_{i=1}^{n_j} |P(\mu_i)f|$$

are bounded if j tends to infinity.

We would like to point out that, on the contrary to almost all other authors, we do not require that all eigenvalues with the exception of only finitely many are normal.

5. APPLICATIONS TO DIFFERENTIAL EQUATIONS

If $s \in \mathbb{Z}$ and ω is an open subset of $\mathbb{R}$ then $H_s(\omega)$ denotes the Sobolev space of order s over ω; $H_s(\mathbb{R})$ is abbreviated by H_s. If $s \in \mathbb{N}$ we set

$$N_s := \{u \in H_s : \text{supp } u \subset \mathbb{R}\backslash(a,b)\}$$

and define

$$H_s[a,b] := H_s/N_s \; ; \qquad (5.1)$$

the corresponding quotient mapping is denoted by φ_s. It is easy to show that $H_s[a,b]$ can be identified with the vector space

$$\{u \in L_2[a,b] \; : \; \exists \; \omega \in H_s \;\; u=\omega|_{[a,b]}\}.$$

Again for $s \in \mathbb{N}$ we define

$$\text{(5.2)} \qquad H^c_{-s}[a,b] := \{v \in H_{-s} \; : \; \text{supp } v \subset [a,b]\}.$$

Obviously

$$\text{(5.3)} \qquad L_\infty[a,b] \subset L_2[a,b] = H^c_0[a,b] \subset H^c_{-1}[a,b].$$

For later use we state

$$\text{(5.4)} \qquad (H_s/N_s)' = H^c_{-s}[a,b].$$

It is well-known that the dual space of the quotient space H_s/N_s is isomorphic to $N_s^{\perp(H_s,H_{-s})}$, i.e. the orthogonal complement of N_s with respect to the dual pair (H_s,H_{-s}). Therefore we only have to prove that

$$N^{\perp(H_s,H_{-s})} = H^c_{-s}[a,b].$$

The inclusion "$\subset$" immediately follows from the definition of the support of a distribution because $C_0^\infty(\mathbb{R}\backslash[a,b]) \subset N_s$. We sketch the proof of the inverse inclusion "$\supset$":

Let $v \in H^c_{-s}[a,b]$. Since $v \in H_{-s}$ we can choose $v_i \in L_2$ for $i \in \{1,2,\dots,s\}$ such that

$$v = \sum_{i=0}^{s} v_i^{(i)}.$$

It is not difficult to show that the functions v_i can be chosen even from $L_2[a,b]$. Thus, if $u \in N_s$,

$$\langle u,v\rangle = \sum_{i=0}^{s} \langle u,v_i^{(i)}\rangle = \sum_{i=0}^{s} (-1)^i \langle u^{(i)},v_i\rangle = 0$$

where the derivatives are taken in the sense of distribution theory.

We would like to point out that in this paper $H^c_{-s}[a,b]$ is understood to be the Banach space dual, i.e. the space of continuous linear functionals, and not to be the Hilbert space dual, i.e. the space of continuous conjugate linear functionals. Consequently we have e.g.

$$\langle u,v\rangle_{(L_2,L_2)} = \int_{\mathbb{R}} u(x)v(x)dx$$

for $u,v \in L_2$.

If G is a vector space, $M_n(G)$ denotes the space of $n\times n$-matrices with elements in G.

In the following $n \in \mathbb{N}\setminus\{0\}$, $m \in \mathbb{N}$, $m \geqq 2$ and

$$a = a_1 < a_2 < \ldots < a_m = b$$

are real numbers. Furthermore for the present we only assume:

$$(5.5)\qquad \begin{cases} A,W \in H(\mathbb{C},M_n(L_\infty[a,b])), \\ W^{(j)} \in H(\mathbb{C},M_n(\mathbb{C})) \quad \text{for} \quad j \in \{1,2,\ldots,m\}. \end{cases}$$

We consider the differential operator

$$(5.6)\qquad T^D(\lambda)y := y' - A(\cdot,\lambda)y$$

and the boundary operator

$$(5.7)\qquad T^R(\lambda)y := \sum_{j=1}^{m} W^{(j)}(\lambda)y(a_j) + \int_a^b W(t,\lambda)y(t)dt,$$

both defined for $y \in H_1^n[a,b]$ where $H_1^n[a,b]$ denotes the n-fold product of $H_1[a,b]$.

For fixed $\lambda \in \mathbb{C}$, $T^D(\lambda)$ and $T^R(\lambda)$ are continuous linear operators on $H_1^n[a,b]$ to $L_2^n[a,b]$ or to $\mathbb{C}^n$ respectively. We combine both operators obtaining the "boundary value operator"

$$(5.8)\qquad T(\lambda)y := (T^D(\lambda)y, T^R(\lambda)y) \qquad (y \in H_1^n[a,b])$$

which is a continuous linear operator on $H_1^n[a,b]$ to $L_2^n[a,b] \times \mathbb{C}^n$.

(5.9) PROPOSITION. We assert:

i) $T^D \in H(\mathbb{C}, L(H_1^n[a,b], L_2^n[a,b]))$,

ii) $T^R \in H(\mathbb{C}, L(H_1^n[a,b], \mathbb{C}^n))$,

iii) $T \in H(\mathbb{C}, \Phi(H_1^n[a,b], L_2^n[a,b] \times \mathbb{C}^n))$,

iv) $\operatorname{ind} T(\lambda) := \dim N(T(\lambda)) - \operatorname{codim} \operatorname{Im}(T(\lambda)) = 0 \quad (\lambda \in \mathbb{C})$.

PROOF. i) We have

$$A(\cdot,\lambda) = \sum_{j=0}^{\infty} \lambda^j A_j$$

with $A_j \in M_n(L_\infty[a,b])$ and

$$\sum_{j=0}^{\infty} |\lambda|^j |A_j| < \infty$$

where

$$|A_j| = \max_{k,\ell=1}^{n} |(A_j)_{k,\ell}|_{L_\infty^n} .$$

If $y \in H_1^n[a,b]$ then

$$\sum_{j=0}^{\infty} |\lambda|^j |A_j y|_{L_2^n[a,b]} \leq \left(\sum_{j=0}^{\infty} |\lambda|^j |A_j| \right) |y|_{L_2^n[a,b]}$$

which already proves the first assertion.

ii) We set

$$V(\lambda) := \sum_{j=1}^{m} W^{(j)}(\lambda) \delta_{a_j} + W(\cdot,\lambda)$$

where δ_a denotes the delta distribution with support $\{a\}$ and $W(\cdot,\lambda)$ is defined to be zero outside the interval $[a,b]$. By (5.3)

$$V(\lambda) \in M_n(H_{-1}^c[a,b])$$

and thus $V \in H(\mathbb{C}, M_n(H_{-1}^c[a,b]))$ because of the smoothness assumptions with respect to W and the $W^{(j)}$'s.

If $B \in M_n(H_{-1}^c[a,b])$ and $b \in H_1^n[a,b]$ we define $\langle B,b \rangle \in \mathbb{C}^n$ to have the components

$$\langle B,b \rangle_i := \sum_{j=1}^{n} \langle B_{ij}, b_j \rangle$$

where $\langle\ ,\ \rangle$ is the canonical bilinear functional belonging to the dual pair $(H^c_{-1}[a,b], H_1[a,b])$. We infer that

(5.10) $$T^R(\lambda)y = \langle V(\lambda), y\rangle \qquad (y \in H^n_1[a,b])$$

from which the assertion ii) is immediate.

iii) Let $\lambda \in \mathbb{C}$ be fixed. The relationship

(5.11) $$T(\lambda) \in \Phi(H^n_1[a,b], L^n_2[a,b] \times \mathbb{C}^n)$$

remains to be proved. We have

$$\operatorname{codim} \operatorname{Im}(T^R(\lambda)) < \infty.$$

We will show that

$$T^D(\lambda) \in \Phi(H^n_1[a,b], L^n_2[a,b]).$$

We set

$$\tilde{T}^D(\lambda)y := y', \quad \tilde{\tilde{T}}^D(\lambda)y := A(\cdot,\lambda)y.$$

If $u \in L^n_2[a,b]$ then the function

$$y(x) := \int_a^x u(t)dt \qquad (x \in [a,b])$$

belongs to $H^n_1[a,b]$ and satisfies the equation $y' = u$ which means that $\tilde{T}^D(\lambda)$ is surjective. Moreover $N(\tilde{T}^D(\lambda)) = \mathbb{C}^n$ and thus finite-dimensional so that $\tilde{T}^D(\lambda)$ is a Fredholm operator. The embedding $H^n_1[a,b] \hookrightarrow L^n_2[a,b]$ is compact and the mapping $A(\cdot,\lambda)\colon L^n_2[a,b] \to L^n_2[a,b]$ is continuous which shows that $\tilde{\tilde{T}}^D(\lambda)$ is a compact operator. Therefore, cf. e.g. Kato [28], p.238, $T^D(\lambda) = \tilde{T}^D(\lambda) + \tilde{\tilde{T}}^D(\lambda)$ is a Fredholm operator.

iv) The operator

$$\tilde{T}(\lambda) := (\tilde{T}^D(\lambda), 0)$$

is a Fredholm operator with index 0 because

$$\dim N(\tilde{T}(\lambda)) = \operatorname{codim} \operatorname{Im}(T(\lambda)) = n.$$

Since

$$\tilde{\tilde{T}}(\lambda) := (\tilde{\tilde{T}}^D(\lambda), T^R(\lambda))$$

is a compact operator we infer, again by Kato [28], p.238, that $T(\lambda) = \tilde{T}(\lambda) + \tilde{\tilde{T}}(\lambda)$ has index 0 which completes the proof of Proposition (5.9).

The explicit form of the adjoint operator bundle T^* is stated in

(5.12) PROPOSITION. For $(v,c) \in L_2^n[a,b] \times \mathbb{C}^n$, $\lambda \in \mathbb{C}$ we have

$$T^*(\lambda)(v,c) = -v' - A(\cdot,\lambda)^t v + V(\lambda)^t c$$

where $A(\cdot,\lambda)^t$ denotes the transposed matrix of $A(\cdot,\lambda)$ and

$$V(\lambda)^t = \sum_{j=1}^{m} W^{(j)}(\lambda)^t \delta_{a_j} + W(\cdot,\lambda)^t.$$

PROOF. We define $A(x,\lambda)$ to be zero outside the interval $[a,b]$ and then

$$T_0(\lambda)y := (y' - A(\cdot,\lambda)y, \langle V(\lambda), y\rangle) \qquad (y \in H_1^n).$$

$T_0(\lambda)$ is a continuous linear mapping on H_1^n to $L_2^n \times \mathbb{C}^n$ and it is easy to show that the diagram

$$\begin{array}{ccc} H_1^n & \xrightarrow{T_0(\lambda)} & L_2^n \times \mathbb{C}^n \\ \downarrow \varphi_1^n & & \downarrow (\varphi_0^n, \mathrm{id}_{\mathbb{C}^n}) \\ H_1^n[a,b] & \xrightarrow{T(\lambda)} & L_2^n[a,b] \times \mathbb{C}^n \end{array}$$

is commutative. It is well-known that this implies that the "adjoint" diagram

$$\begin{array}{ccc} H^n_{-1} & \xleftarrow{\;T_0^*(\lambda)\;} & L_2^n \times \mathbb{C}^n \\ \Big\uparrow \varphi_1^{*n} & & \Big\uparrow (\varphi_0^{*n}, id_{\mathbb{C}^n}) \\ H^n_{-1}[a,b] & \xleftarrow{\;T^*(\lambda)\;} & L_2^n[a,b] \times \mathbb{C}^n \end{array}$$

also is commutative. Since the adjoint mappings φ_s^* are embeddings we infer

(5.13) $$T^*(\lambda) = T_0^*(\lambda)\Big|_{L_2^n[a,b]\times\mathbb{C}^n}$$

so that it will be sufficient to derive the explicit form of $T_0^*(\lambda)$:

Let $y \in H_1^n$, $(v,c) \in L_2^n \times \mathbb{C}^n$. We have

$$\langle y, T_0^*(\lambda)(v,c)\rangle = \langle T_0(\lambda)y,(v,c)\rangle$$
$$= \langle y' - A(\cdot,\lambda)y, v\rangle + \langle\langle V(\lambda),y\rangle, c\rangle$$
$$= \langle y, -v' - A(\cdot,\lambda)^t v\rangle + \langle c^t V(\lambda), y\rangle$$
$$= \langle y, -v' - A(\cdot,\lambda)^t v + V(\lambda)^t c\rangle$$

and thus

$$T_0^*(\lambda)(v,c) = -v' - A(\cdot,\lambda)^t v + V(\lambda)^t c$$

which is the desired explicit form.

(5.14) REMARK. The astriction $T_a^*(\lambda)$ of $T^*(\lambda)$ to the space $L_2^n[a,b]$ is the restriction of $T^*(\lambda)$ to the space

$$D(T_a^*(\lambda)) := \{(v,c) \in L_2^n[a,b]\times\mathbb{C}^n : v\big|_{(a_j,a_{j+1})} \in H_1^n(a_j,a_{j+1})$$
$$(j=1,\dots,m-2) \wedge v(a_j+0) - v(a_j-0) = W^{(j)}(\lambda)^t c \quad (j=1,\dots,m)\}.$$

In the foregoing defined set

$$v(a-0) = v(a_1-0) = v(a_m+0) = v(b+0) = 0$$

because $L_2[a,b] = H_0^c[a,b]$.

For the proof of Remark (5.14) let $(v,c) \in L_2^n[a,b] \times \mathbb{C}^n$. Then

$$T^*(\lambda)(v,c) \in L_2^n[a,b]$$

if and only if

$$(5.15) \qquad -v' + \sum_{j=1}^{m} W^{(j)}(\lambda)^t c\, \delta_{a_j} \in L_2^n[a,b]$$

because $A(\cdot,\lambda)^t v$ and $W(\cdot,\lambda)^t c$ belong to $L_2^n[a,b]$ anyway. Let H_a denote the Heaviside function

$$H_a(x) = \begin{cases} 0 & x < a, \\ 1 & x \geqq a. \end{cases}$$

We conclude that the relationship (5.14) holds if and only if

$$-v + \sum_{j=1}^{m} H_{a_j} W^{(j)}(\lambda)^t c \in (H_1^{loc})^n$$

which obviously is true if and only if $(v,c) \in D(T_a^*(\lambda))$.

The equation

$$(5.16) \qquad T(\lambda)y = f \qquad (y \in H_1^n[a,b],\ f \in L_2^n[a,b] \times \mathbb{C}^n)$$

defines the boundary eigenvalue problem [BEVP] which has been studied by Cole [8], [9] for $y \in C_1^n[a,b]$ and $f = (f_1,0) \in C_0^n[a,b] \times \mathbb{C}^n$. Cole referred to a famous paper of Langer [35] which considered BEVP's in the complex domain, i.e. for x varying in subsets of $\mathbb{C}$.

We call

$$(5.17) \qquad T^*(\lambda)v = g \qquad (v \in L_2^n[a,b] \times \mathbb{C}^n,\ g \in H_{-1}^c[a,b])$$

the "adjoint" BEVP of (5.16). Since, for fixed $\lambda \in \mathbb{C}$, $T^*(\lambda)$ is a continuous linear operator on the whole space $L_2^n[a,b] \times \mathbb{C}^n$ the equation (5.17) is more convenient than the adjoint BEVP

$$(5.17') \qquad T_a^*(\lambda)v = g \qquad (v \in D(T_a^*(\lambda)),\ g \in L_2^n[a,b])$$

which has been considered by Cole (and Langer) under more restrictive smoothness assumptions on v and g.

If $\rho(T) \neq \emptyset$ then, by Proposition (5.9), the boundary value operator T fulfills the assumption (2.1) so that all results stated in section 2 are applicable to the BVEP's (5.16) and (5.17). By Keldyš's theorem (2.3) the existence of biorthogonal systems of eigenvectors and associated vectors is guaranteed. For "normal" boundary value operator bundles T this statement has been proved by Langer, cf. also Cole, without using any functional analytic argument. Indeed, for Langer and Cole, already the definition of associated vectors would have been impossible as they considered the astriction operator bundle $T_a^*(\lambda)$ the domain of which is dependent on λ.

In the sequel we need additional assumptions in order that the boundary value operator bundle T will satisfy the assumptions of the sections 2 and 3 and the results stated there becomes applicable.

We assume that

$$(5.18)\begin{cases} V(\lambda) = \sum_{j=0}^{q} \lambda^j V_j & \text{where without loss of generality } q \geqq 1, \\ A(\cdot,\lambda) = A_0 + \lambda A_1 & \text{with } A_0, A_1 \in M_n(L_\infty[a,b]). \end{cases}$$

We set $E = H_1^n[a,b]$, $F_1 = L_2^n[a,b]$, $F_2 = \mathbb{C}^n$ and

$$(5.19)\qquad \begin{cases} T_0^D y := y' + A_0 y & (y \in E) \\ T_1^D := A_1 \text{ on } F_1 . \end{cases}$$

J denotes the canonical embedding of $H_1^n[a,b]$ into $L_2^n[a,b]$. We define

$$H_{j,\infty}[a,b] := \{f \in H_j[a,b] : f^{(j)} \in L_\infty[a,b]\}$$

for $j \in \mathbb{N}$ and infer

$$M_n(H_{j,\infty}[a,b])\cdot H_j^n[a,b] \subset H_j^n[a,b]$$

from the Leibniz-formula since $f^{(i)} \in C_0[a,b]$ for $i < j$ if $f \in H_j[a,b]$. Let $r \in \mathbb{N}$. In addition to (5.18) we assume that

$$(5.20) \quad \begin{cases} A_0, A_1 \in M_n(H_{r,\infty}[a,b]), \\ \det A_1(x) \neq 0 \text{ almost everywhere in } [a,b], \\ A_1^{-1} \in M_n(L_\infty[a,b]). \end{cases}$$

If $r > 0$ then $A_1 \in M_n(C_0[a,b])$. Therefore, in this case, the second and the third condition are fulfilled if and only if $\det A_1(x) \neq 0$ everywhere in $[a,b]$.

We infer that the boundary value operator bundle T fulfills all the assumptions made in section 3 so that the results stated there and in the subsequent section 4 become available for this operator bundle. Here the operator M, defined in (3.2), has the following concrete form:

$$(5.21) \quad \begin{cases} D(M) = H_1^n[a,b], \\ My = -A_1^{-1}(y' + A_0 y) \end{cases}$$

where we could omit J^{-1} because we understand $H_1^n[a,b]$ to be a subspace of $L_2^n[a,b]$.

(5.22) PROPOSITION. Under the assumption (5.20)

$$D(M^\ell) = H_\ell^n[a,b]$$

for all $\ell \leq r+1$.

PROOF. The assertion is true for $\ell = 0$ because $D(M^0) = L_2^n[a,b]$. Let $y \in H_{\ell+1}^n[a,b]$. According to (5.20), (5.21) $My \in H_\ell^n[a,b]$ and thus, by the induction hypothesis, $My \in D(M^\ell)$, i.e. $y \in D(M^{\ell+1})$. Conversely, let $y \in D(M^{\ell+1})$. From

$$y' = -A_1 My - A_0 y$$

and $My \in D(M^\ell) = H^n_\ell[a,b]$ it follows that $y' \in H^n_\ell[a,b]$ and thus $y \in H^n_{\ell+1}[a,b]$ which already completes the proof.

According to (3.3)

$$\begin{cases} y^{[0]} := y \\ y^{[j]} := -A_1^{-1}(y^{[j-1]'} + A_0 y^{[j-1]}) \qquad (j=1,2,\dots,\ell) \end{cases}$$

is recursively defined if $y \in H^n_\ell[a,b]$.

An explicit form of the resolvent R of the boundary value operator bundle T can be derived in the usual way using Green's matrix. For this purpose it is useful to know that, for fixed $\lambda \in \mathbb{C}$, $V(\lambda)$ does not only belong to $M_n(H^c_{-1}[a,b])$ but that it is also given by a Lebesgue-Stieltjes-measure: if we define

$$F(x,\lambda) := \sum_{a_j<x} W^{(j)}(\lambda) + \int_a^x W(t,\lambda)dt \tag{5.23}$$

then $V(\lambda) = dF(\cdot,\lambda)$. Let $Y(\cdot,\lambda) \in M_n(H_1[a,b])$ be a fundamental matrix of the equation

$$y' + A(\cdot,\lambda)y = 0$$

for each $\lambda \in \mathbb{C}$. We may suppose that Y is continuous in $(x,\lambda) \in [a,b] \times \mathbb{C}$ and is holomorphic with respect to λ for each $x \in [a,b]$. Let $\lambda \in \rho(T)$. Then

$$(T^R Y)(\lambda) := T^R(\lambda)Y(\cdot,\lambda) \qquad (\lambda \in \mathbb{C})$$

is invertible and the Green's matrix of T, i.e.

$$G(x,s,\lambda) := \begin{cases} Y(x,\lambda)((T^R Y)(\lambda))^{-1} \int_a^s dF(t,\lambda)Y(t,\lambda)Y^{-1}(s,\lambda) & \text{if } s < x \\ -Y(x,\lambda)((T^R Y)(\lambda))^{-1} \int_s^b dF(t,\lambda)Y(t,\lambda)Y^{-1}(s,\lambda) & \text{if } s > x \end{cases}$$

is well-defined. We conclude that

$$G(\cdot,\cdot,\lambda) \in M_n(L_2([a,b]\times[a,b])) \qquad (\lambda \in \mathbb{C})$$

and that

$$(5.24) \quad R(\lambda)(g,c)(x) = \int_a^b G(x,s,\lambda)g(s)ds + Y(x,\lambda)((T^R Y)(\lambda))^{-1}c$$

for $(g,c) \in L_2^n[a,b] \times \mathbb{C}^n$ and $\lambda \in \mathbb{C}$.

The boundary value problem (5.16) is called regular of order p if and only if the corresponding boundary value operator T has this property which has been defined in section 4. Langer [35] and Cole [8], [9] stated sufficient criteria for the regularity of (5.16), cf. especially Cole [9], p. 541, Theorem 8 which he proved by a profound analysis of the corresponding Green's matrix. Langer and Cole made rather incisive assumptions with respect to the coefficients A_0 and A_1. These assumptions can be weakened to some extent as we are going to show in a subsequent paper. We would like to point out that Langer's and Cole's regularity criteria only concern the resolvent $R(\lambda)$ restricted to $L_2^n[a,b] \times \{0\}$ so that they did not have to estimate the boundary part

$$Y(x,\lambda)((T^R Y)(\lambda))^{-1}c.$$

However, their criteria remain true for $R(\lambda)$ itself since the additional boundary term can be estimated in a similar manner.

We summarise our result in

(5.25) THEOREM. Assume that the coefficients of the boundary value operator T, defined in (5.8), satisfy the conditions (5.5), (5.18) and (5.20) (with a fixed $r \in \mathbb{N}$). Let $\rho(T) \neq \emptyset$ and assume that T is regular of order p. Suppose that $p+q \leq r$, $f \in H_{p+q}^n[a,b]$ and that it fulfills the boundary conditions

$$(5.26) \qquad \sum_{j=0}^{q} T_j^R f^{[j+\ell-1]} = 0 \qquad (\ell=1,2,\ldots,p).$$

Under these assumptions we have the following biorthogonal

expansion in parentheses:

$$\left| f - \sum_{i=1}^{n_j} P(\mu_i)f \right|_{H_1^n[a,b]} \to 0 \qquad (j \to \infty)$$

where the natural numbers n_j are given by the regularity of T and the projectors $P(\mu)$ have been defined in (3.6), see also (3.7) or (3.25).

The proof is immediate from the Theorem (4.1).

The expansion theorems proved by Langer and Cole (cf. e.g. [9], p.547, Theorem 9) only apply to normal boundary value operator bundles T. Furtheremore, Cole's expansion of the function f in general is not biorthogonal in the sense defined in the present paper. And Cole did not require that f satisfies boundary conditions. Indeed he seemed to be a bit astonished that, according to his convergence theorem, in most cases not f but a "modified" function $f - a_0$ is expandable with respect to the eigenvectors of T, cf. [9], p.549, section 14.

REFERENCES

1. ALLAHVERDIEV, Dž. E., On the completeness of the systems of eigenvectors and associated vectors of nonselfadjoint operators close to normal ones, Dokl. Akad. Nauk SSSR 115 (1957), 207-210 (Russian).

2. ALLAHVERDIEV, Dž. E., On the completeness of the system of eigenelements and adjoint elements of nonselfadjoint operators, Dokl. Akad. Nauk SSSR 160 (1965), 503-506 (Russian), Engl. transl.: Soviet Math. Dokl. 6 (1965), 102-105.

3. ALLAHVERDIEV, Dž. E., On the completeness of the system of eigenelements and adjoint elements of a class of nonselfadjoint operators depending on a parameter λ, Dokl. Akad. Nauk SSSR 160 (1965), 1231-1234 (Russian), Engl. transl.: Soviet Math. Dokl. 6 (1965), 271-275.

4. ALLAHVERDIEV, Dž. E., Multiply complete systems and nonselfadjoint operators depending on a parameter λ, Dokl. Akad. Nauk SSSR 166 (1966), 11-14 (Russian), Engl. transl.: Soviet Math. Dokl. 7 (1966), 4-8.

5. BAUER, G., Über Entwicklungsfragen bei Operatorenbüscheln, Dipolomarbeit Regensburg 1979.

6. BAUER, G., Störungstheorie für diskrete Spektraloperatoren und Anwendungen in der nichtlinearen Spektraltheorie, to be published.

7. COLE, R.H., Reduction of an n-th order linear differential equation and m-point boundary conditions to an equivalent matrix system, Amer. J. Math. 68 (1946), 179-183.

8. COLE, R.H., The expansion problem with boundary conditions at a finite set of points, Can. J. Math. 13 (1961), 462-479.

9. COLE, R.H., General boundary condtions for an ordinary linear differential system, Trans. Amer. Math. Soc. 111 (1964), 521-550.

10. COLE, R.H., The two-point boundary problem, Amer. Math. Monthly 72 (1965), 701-711.

11. Di PRIMA, R.C. and HABETLER, G.J., A completeness theorem for nonselfadjoint eigenvalue problems in hydrodynamic stability, Arch. Rat. Mech. Anal. 34 (1969), 218-227.

12. EBERHARD, W., Über das asymptotische Verhalten von Lösungen der linearen Differentialgleichung $M[y] = \lambda N[y]$ für grosse Werte von λ, Math. Z. 119 (1971), 160-170.

13. EBERHARD, W., Über Eigenwerte und Eigenfunktionen einer Klasse von nichtselbstadjungierten Randwertproblemen, Math. Z. 119 (1971), 171-178.

14. EBERHARD, W. und FREILING, G., Nicht-S-hermitesche Rand- und Eigenwertprobleme, Math. Z. 133 (1973), 187-202.

15. EBERHARD, W. und FREILING, G., Das Verhalten der Greenschen Matrix und der Entwicklungen nach Eigenfunktionen N-regulärer Eigenwertprobleme, Math. Z. 136 (1974), 13-30.

16. EBERHARD, W. und FREILING, G., Stone-reguläre Eigenwertprobleme, Math. Z. 160 (1978), 139-161.

17. FREILING, G., Reguläre Eigenwertprobleme mit Mehrpunkt-Integral-Randbedingungen, Math. Z. 171 (1980), 113-131.

18. FRIEDMAN, A. and SHINBROT, M., Nonlinear eigenvalue problems, Acta Math. 121 (1968), 77-125.

19. GOHBERG, I.C. and KREIN, M.G., Introduction to the theory of linear nonselfadjoint operators, Providence, Amer. Math. Soc. 1969.

20. GOLDBERG, S., Unbounded linear operators: theory and applications, New York-Toronto-London-Sydney, McGraw-Hill 1966.

21. GORJUK, I.V., A certain theorem on the completeness of the system of eigen- and associated vectors of the operator bundle $L(\lambda) = \lambda^2 C+\lambda B+E$, Vestnik Moscov Univ. Ser. I Mat. Meh. 25 (1970), 55-60 (Russian, Engl. summary).

22. GRAMSCH, B., Meromorphie in der Theorie der Fredholmoperatoren mit Anwendungen auf elliptische Differentialoperatoren, Math. Ann. 188 (1970), 97-112.

23. GRIGORIEFF, R.D., Approximation von Eigenwertproblemen und Gleichungen zweiter Art im Hilbertschen Raume, Math. Ann. 183 (1969), 45-77.

24. GRIGORIEFF, R.D., Diskrete Approximation von Eigenwertproblemen, I. Qualitative Konvergenz, Num. Math. 24 (1975), 355-374.

25. HEISECKE, G., Rand-Eigenwertprobleme N(y) = λM(y) bei λ-abhängigen Randbedingungen, ZAMM 61 (1981), 242-244.

26. ISAEV, G.A., The completeness of a certain part of the eigen- and associated vectors of polynomial operator pencils, Uspehi Mat. Nauk 28 (1973), 241-242 (Russian).

27. ISAEV, G.A., The numerical range of operator pencils and multiple completeness in the sense of M.V. Keldyš, Funkcional Anal. i Priložen 9 (1975), 31-34 (Russian), Engl. transl.: Functional Anal. Appl. 9 (1975), 27-30.

28. KATO, T., Perturbation theory for linear operators, Berlin-Heidelberg-New York, Springer Verlag 1976.

29. KELDYŠ, M.V., On the eigenvalues and eigenfunctions of certain classes of nonselfadjoint equations, Dokl. Akad. Nauk SSSR 77 (1951), 11-14 (Russian).

30. KELDYŠ, M.V., On the completeness of eigenfunctions of some classes of nonselfadjoint linear operators, Uspehi Mat. Nauk 26 (1971), 15-41 (Russian), Engl. transl.: Russian Math. surveys 26 (1971), 15-44.

31. KÖNIG, H., A trace theorem and a linearization method for operator polynomials, preprint 1981.

32. KOSTJUČENKO, A.G. and ORAZOV, M.B., The completeness of the root vectors of certain selfadjoint quadratic pencils, Funkcional Anal. i Priložen 11 (1977), 85-87 (Russian), Engl. transl.: Functional Anal. Appl. 11 (1978), 317-319.

33. KUMMER, H., Zur praktischen Behandlung nichtlinearer Eigenwertaufgaben abgeschlossener linerer Operatoren, Mitteilung Math. Sem. Giessen 62 (1964), 1-56.

34. LANGER, H., Zur Spektraltheorie polynomialer Scharen selbstadjungierter Operatoren, Math. Nachr. 65 (1975), 301-319.

35. LANGER, R.E., The boundary problem of an ordinary linear differential system in the complex domain, Trans. Amer. Math. Soc. 46 (1939), 151-190.

36. MAMEDOV, K.Š., The convergence of multiple expansions in the system of eigen- and associated vector-valued functions of a polynomial differential-operator pencil, Izv. Akad. Nauk Azerbaĭdžan. SSR Ser. Fiz.-Tehn. Mat. Nauk 1 (1977), 36-40 (Russian, Engl. summary).

37. MARKUS, A.S., Some criteria for completeness of the system of root vectors of a linear operator in a Banach space, Mat. Sb. 70 (1966), 526-561 (Russian), Engl. transl.: Amer. Math. Soc. Transl. 85 (1970), 51-91.

38. MARKUS, A.S., The completeness of a part of the eigen- and associated vectors for certain nonlinear spectral problems, Funkcional Anal. i Priložen 5 (1971), 78-79 (Russian), Engl. transl.: Functional Anal. Appl. 5 (1971), 334-335.

39. MITTENTHAL, L., Operator valued analytic functions and generalizations of spectral theory, Pacific J. Math. 24 (1968), 119-132.

40. MONIEN, B., Entwicklungssätze bei Operatorbüschel, Dissertation Hamburg 1968.

41. MÜLLER, P.H. und KUMMER, H., Zur praktischen Bestimmung nichtlinear auftretender Eigenwerte, Anwendung des Verfahrens auf eine Stabilitätsuntersuchung (Kipperscheinung), ZAMM 40 (1960), 136-143.

42. ORAZOV, M.B., The completeness of the eigen- and associated vectors of a selfadjoint quadratic pencil, Funkcional Anal. y Priložen 10 (1976), no.2, 82-83 (Russian), Engl. transl.: Functional Anal. Appl. 10 (1976), 153-155.

43. PALLANT, Ju. A., A test for the completeness of a system of eigenvectors and associated vectors of a polynomial bundle of operators, Dokl. Akad. Nauk SSSR 141 (1961), 558-560 (Russian), Engl. transl.: Soviet Math. Dokl. 2 (1961), 1507-1509.

44. ROACH, G.F. and SLEEMAN, B.D., On the spectral theory of operator bundles, Applicable Anal. 7 (1977), 1-14.

45. ROACH, G.F. and SLEEMAN, B.D., On the spectral theory of operator bundles II, Applicable Anal. 9 (1979), 29-36.

46. TURNER, R.E.L., A class of nonlinear eigenvalue problems, J. Functional Anal. 2 (1968), 297-322.

47. VIZITEI, V.N. and MARKUS, A.S., On convergence of multiple expansions in eigenvectors and associated vectors of an operator bundle, Mat. Sb. 66 (1965), 287-320 (Russian), Engl. transl.: Amer. Math. Soc. Transl. 87 (1970), 187-227.

48. WAGENFÜHRER, E., Transformations of n-th order linear differential equations whose coefficients are nonlinear in ρ to systems of n first order equations which are linear in the parameter, to be published.

49. WILDER, C.E., Reduction of the ordinary linear differential equation of the n-th order whose coefficients are certain polynomials in a parameter to a system of n first order equations which are linear in the parameter, Trans. Amer. Math. Soc. 29 (1927), 497-506.

50. YAKUBOV, S. Ya. and Mamedov, K.Š., Multiple completeness of the system of a eigen- and associated elements of a polynomial operator bundle and multiple expansions with respect to this system, Funkcional Anal. i Priložen 9 (1975), 91-93 (Russian), Engl. transl.: Functional Anal. Appl. 9 (1975), 90-92.

Functional Analysis, Holomorphy and
Approximation Theory II, G.I. Zapata (ed.)

INTEGRO-DIFFERENTIAL OPERATORS AND THEORY OF SUMMATION

M. Mikolás

I. HISTORICAL REMARKS

The theory of differential and integral operators of arbitrary complex order which originates in the 19th century with some results of Abel, Liouville, Riemann and Heaviside, has been property established since the turn of the 20th century by certain fundamental works of Hadamard, Hardy and Littlewood, M. Riesz, H. Weyl. At the end of the fourties, Marcel Riesz summarized the results of the field due to him and his school, discussing in the work [13] of more than 200 pages also some applications of the theory in modern mathematical physics, connected with the so-called Riesz potentials.

The kernel of the Riesz method is the analytic continuation of the Riemann-Liouville "fractional" integral:

(1) $$_{x_o}I_x^s f = \frac{1}{\Gamma(s)} \int_{x_o}^{x} f(t)(x-t)^{s-1}dt \quad \begin{pmatrix} f \text{ bounded, integrable} \\ \mathrm{Re}\ s > 0 \end{pmatrix}$$

which is nothing else but a natural extension of Cauchy's solution of the initial value problem

(2) $$y^{(m)}(x) = f(x); \quad y(x_o) = y'(x_o) = \dots = y^{(m-1)}(x_o) = 0,$$

m corresponding to the complex parameter s in (1). Note that in case of a Lebesgue integrable function f and any fixed s with $\mathrm{Re}\ s > 0$, the existence of the integral (1) is assured for almost all x and the operator ${}_{x_0}I_x^s$ satisfies the so-called index law (or semigroup property):

$$(3) \qquad {}_{x_0}I_x^{s_1}({}_{x_0}I_t^{s_2}) = {}_{x_0}I_x^{s_2}({}_{x_0}I_t^{s_1}) = {}_{x_0}I_t^{s_1+s_2} \qquad \left(\begin{array}{c} \mathrm{Re}\ s_1 > 0, \quad \mathrm{Re}\ s_2 > 0 \\ x_0 < t \leq x \end{array}\right)$$

The analytic continuation of ${}_{x_0}I_x^s$ is meant of course with respect to the order of integration s and depends on the fact that the integral (1) is a holomorphic function of s in its domain of existence. On the same basis, supplemented by the observation that the widening of the semigroup of fractional integral operators of the type (1) to an Abelian group is equivalent to the introduction of fractional differential operators in the Hardy-Littlewood sense [2] - this has been recently the starting-point of a newly developed branch of semigroup theory, namely the "operational calculus of fractional powers". The results are also interlinked with differential equations in Banach spaces and a few modern topics in functional analysis, e.g. about certain operators in abstract Hilbert spaces. (Cf. e.g. [1] and [15].)

In the pertinent works [3]-[10] of the author which have been published since 1958 in several periodicals in English, French or German and partly also in the Hungarian-written book [11], some complex analytical procedures, the idea of Riesz about analytic continuation, furthermore certain deeper tools from summation theory and operator theory are linked, in order to get the most general, unified treatment of integro-differential operators for arbitrary Lebesgue integrable functions, based mainly on Weyl's concept [16] of fractional integration. Let us mention that the words "integro-differential operators" are utilized in what follows for the sake of brevity in a much wider sense than it's usual. By this term, integral or differential operators of arbitrary complex order will be meant, referring so to the fact that it is about a common generalization of integration and differentiation of any positive integral order. (A similar meaning has the word "diffintegration" in a recent book of Oldham and Spanier [12].) There exists also a longer work in English which yields a survey on all main topics on integro-differential operators published during the last decades. This is author's summarizing report [10], which was given at the first international conference on the field held in New Haven (Conn., USA) in 1974, and was published in the Proceedings of that congress edited by Springer-Verlag as Vol. 457 of the series "Lecture Notes in Mathematics". (Cf. [14].)

It may be pointed out that the theory of integro-differential operators is becoming now a new branch of analysis, between the classical and functional one, whose applications reach from the theory of functions, integral transformations, theory of approximation and from a large scale of differential and integral equations to the modern operator theory and the theory of generalized functions.

II. EXPOSITION OF THE METHODS

In the sequel, according to the character of this seminar, we will elucidate the close connection between integro-differential operators of arbitrary complex order and some strong summation methods.

The one side of the inherence in question is that the analytic continuation of the Weyl fractional integral and herewith the introduction of integro-differential operators can be realized by means of powerful processes of summation, due to Abel-Poisson, Borel, Le Roy and Lindelöf etc. As an illustration, let us formulate only the central results of this theory, given in [3] and [10].

Consider Weyl's fractional integral of order s of a function $f \in L(0,1)$ with the period 1 in the form

$$(4)\qquad f_{[s]}(x) = \cos\frac{\pi s}{2}\sum_{n=1}^{\infty}\frac{2a_n(x)}{(2n\pi)^s} + \sin\frac{\pi s}{2}\sum_{n=1}^{\infty}\frac{2b_n(x)}{(2n\pi)^s},$$

where $\mathrm{Re}\ s > 1$ and

$$(5)\qquad \begin{cases} a_n(x) = (\alpha_n-\alpha_o)\cos 2n\pi x + \beta_n \sin 2n\pi x, \\ b_n(x) = (\alpha_n-\alpha_o)\sin 2n\pi x - \beta_n \cos 2n\pi x, \end{cases}$$

α_n, β_n denoting the corresponding Fourier coefficients of f. If we apply the most effective of the above-mentioned summation methods, namely a suitable extension of the so-called Mittag-Leffler summation* (due to M. Riesz) to the series in (4), then it can be determined completely the characterizing Mittag-Leffler star for both of these series, so that we get an explicit expression of $f_{[s]}(x)$, holding everywhere in the common part of the mentioned domains.

This means that the method yields about the maximal information on the holomorphy domain and the singularities of $f_{[s]}(x)$ as a function of s which can be hoped in full generality. By the way, we have the integral representation:

$$(6)\qquad f_{[s]}(x) = \int_0^1 f(x-t)\,[\mathfrak{z}_s(t) - \mathfrak{z}_s(x)]\,dt \qquad (\mathrm{Re}\ s > 1)$$

where

$$(7)\qquad \mathfrak{z}_s(u) = \sum_{n=1}^{\infty}\frac{2}{(2n\pi)^s}\cos\left(2n\pi u - \frac{\pi s}{2}\right) \qquad (u \neq 0, \pm 1, \pm 2, \dots)\ .$$

---

*This process is defined for an arbitrary series Σu_m by

$$(\mathrm{ML})\qquad \sum_{m=0}^{\infty} u_m = \lim_{\delta\to+0}\sum_{m=0}^{\infty}\Gamma(1+\delta m)^{-1}\,u_m\ .$$

Thus the result depends also upon the properties of the generalized (Hurwitz) zeta-function $\zeta(s,u)$, defined by analytic continuation of the series $\sum_{m=0}^{\infty} (u+m)^{-s}$. It holds namely for the kernel function (7) the formula:

$$\beta_s(u) = \Gamma(s)^{-1} \zeta(1-s,u) \qquad (0 < u \leq 1). \tag{8}$$

The other side of the exposed connexion will be discussed more in detail, i.e. that certain function series can be handled very effectively by means of integro-differential operators. It is about a <u>new summation method</u> which was introduced in its simplest form in the works [4]-[6] of the author and has been developed further since the sixties in several directions, e.g. in [9]-[11].

Let $\varphi_n(x)$ $(n=1,2,\ldots)$ a sequence of functions bounded and Lebesgue integrable in an interval (x_0,x_1) and suppose that the series $\sum {}_{x_0}I_x^{\nu} \varphi_n$ converges at a point $x \in (x_0,x_1)$ for any $\nu > 0$. Then the limit

$$^{(W)}\sum \varphi_n(x) = \lim_{\nu \to +0} \sum {}_{x_0}I_o^{\nu} \varphi_n \tag{9}$$

is called the (W)-<u>sum</u> of the series $\sum \varphi_n$; and in case of the existence of (9), $\sum \varphi_n$ is said to be (W)-<u>summable</u> at x.

This (W)-method leads to sharp results e.g. for trigonometric series and ordinary Dirichlet series. In particular, putting $x_o = -\infty$ and using certain properties of the Hurwitz zeta-function, a simple - necessary and sufficient - summability condition for trigonometric Fourier series can be deduced. Moreover we find that the local "strength" of the (W)-method is beyond that of any classical summation process.

By formal grounds, it is reasonable for these applications to define two variants of the method:

We say that the series

$$(10)\qquad a_0 + \sum_{n=1}^{\infty} (a_n \cos nx + b_n \sin nx)$$

is (W_+)-or (W_-)-summable at a point x, if there exists, $\vartheta > 0$ being chosen sufficiently small, the limit for $\vartheta \to +0$ of the sum of the series

$$a_0 + \sum_{n=1}^{\infty} n^{-\vartheta}[a_n \cos(nx + \frac{\pi\vartheta}{2}) + b_n \sin(nx + \frac{\pi\vartheta}{2})]$$

or

$$a_0 + \sum_{n=1}^{\infty} n^{-\vartheta}[a_n \cos(nx - \frac{\pi\vartheta}{2}) + b_n \sin(nx - \frac{\pi\vartheta}{2})],$$

respectively. These limits are called the (W_+)- resp. (W_-)-sum of (10).

III. MAIN RESULTS

On the above mentioned lines, the following theorems can be obtained.

1. The trigonometric Fourier series of a bounded function f at a point x is (W_+)-summable if and only if the limit

$$(11)\qquad f\langle x+0\rangle = \lim_{\vartheta\to+0} [\vartheta \int_0^{\delta} f(x+t)t^{\vartheta-1}dt]$$

exists, where δ is an arbitrary positive number (but a fixed one). $f\langle x+0\rangle$ does not depend on δ and in case of its existence, it yields also the (W_+)-sum of the Fourier series.

2. We have especially $f\langle x+0\rangle = f(x+0)$ at every point where the function has a limit from the right. Furthermore, the (W_+)-summability holds uniformly in each closed interval where f is continuous. (Two-side continuity at the end-points being assumed.)

3. Analogous statements hold also for the (W_-)-method. We have only to put $x-0$ instead of $x+0$.

4. There exist such trigonometrical series which are not summable neither by any (C,r)- nor by the (A)-method, yet are $(W_\pm)$-summable.

5. In case of a related summation method, defined for (10) by

(12) $$a_0 + \lim_{\vartheta\to+0} \sum_{n=1}^{\infty} n^{-\vartheta}(a_n \cos nx + b_n \sin nx),$$

all our theorems are valid with $\varphi_x(t) = \frac{1}{2}[f(x+t)+f(x-t)]$ instead of $f(x\pm t)$ and with

(13) $$f\langle\langle x\rangle\rangle = \lim_{\vartheta\to+0} [\vartheta \int_0^\delta \varphi_x(t)t^{\vartheta-1}dt]$$

instead of $f\langle x\pm 0\rangle$. In addition, we obtain for any bounded function f the result, that the set of points at which $f\langle\langle x\rangle\rangle$ exists is wider than that of the points where the so-called Lebesgue condition holds i.e. the limit

(14) $$\lim_{\ell\to+0} \frac{1}{2\ell} \int_{x-\ell}^{x+\ell} f(t)dt$$

exists. As it is well-known, this last one is the most general summability condition of practical use.

We remark that these results are based essentially on the connection between the closed form of the series occuring in the definition of the $(W_\pm)$-method and the Hurwitz zeta-function. By certain propositions of Tauberian type it can be shown still: the effectiveness of the $(W_\pm)$-methods in case of bounded functions exceeds not only the effectiveness of the Abel-Poisson method but also that of a more general class of processes, namely the so-called Abel-Cartwright methods.

The latters are defined for an arbitrary series $\Sigma\, u_n$ by

(15) $$\overset{(A_q)}{\Sigma}\, u_n = \lim_{\vartheta\to+0} \Sigma\, u_n\, e^{-n^q\vartheta}$$

where q is a fixed positive number.

Recent investigations indicate wide application possibilities to boundary asymptotics of power series in Hadamard's sense, namely concerning improvement and localization of the results in question.

IV. PROOF OF A TYPICAL SUMMATION THEOREM

For illustration, we will consider the above-mentioned theorem 5. which can be formulated in detail as follows:

The trigonometric Fourier series of a bounded function f is summable in the sense (12) at a point x if and only if the limit (13) (with an arbitrarily small but fixed $\delta > 0$) exists. In particular we have $f\langle\langle x\rangle\rangle = [f(x+0)+f(x-0)]/2$ whenever $f(x\pm 0)$ exist, and the summability is uniform in each closed continuity interval of the function. The domain of effectiveness of our summation process is greater than that of any Cesàro method or of the Abel-Poisson summation.

PROOF. We start with some elementary lemmas which can be deduced easily from the classical theory of the Hurewitz zeta-function [cf. (8)].

LEMMA 1. $\zeta(s,u)$ satisfies the formula

$$(16)\qquad \zeta(1-\vartheta,x) = 2(2\pi)^{-\vartheta}\,\Gamma(\vartheta)\sum_{n=1}^{\infty} n^{-\vartheta}\cos(2n\pi x - \frac{\pi\vartheta}{2})$$

$$(0 < \vartheta < 1,\quad 0 < x < 1)$$

and the inequality

$$(17)\qquad |\zeta(\theta,x)| \leq x^{-\theta} + (1-\theta)^{-1} + 1 \quad (0 < \theta < 1,\quad 0 < x \leq 1).$$

LEMMA 2. We have the representation

(18)
$$\sum_{n=1}^{\infty} n^{-\vartheta} \cos n\tau =$$
$$= \frac{1}{4} (2\pi)^{\vartheta} \left(\cos \frac{\pi\vartheta}{2}\right)^{-1} \Gamma(\vartheta)^{-1} [\zeta(1-\vartheta, \frac{\tau}{2\pi}) + \zeta(1-\vartheta, 1-\frac{\tau}{2\pi})],$$

and for the partial sums of the left-hand series the estimation

(19)
$$|\sum_{n=1}^{N} n^{-\vartheta} \cos n\tau| <$$
$$< (1 + \frac{1}{1-\vartheta})\pi^{1-\vartheta}[\tau^{\vartheta-1} + (2\pi-\tau)^{\vartheta-1}]$$
$$(N = 2,3,\ldots; \quad 0 < \tau < 2\pi).$$

These propositions imply that the sum of the series

(20)
$$\alpha_0 + \sum_{n=1}^{\infty} n^{-\vartheta} (\alpha_n \cos nx + \beta_n \sin nx)$$
$$(0 < \vartheta < 1, \quad 0 \leq x < 2\pi),$$

with $\alpha_0 = \frac{1}{2\pi} \int_0^{2\pi} f(t)dt$, $\alpha_n = \frac{1}{\pi} \int_0^{2\pi} f(t)\cos nt\, dt$, $\beta_n = \frac{1}{\pi} \int_0^{2\pi} f(t)\sin nt\, dt$ $(n=1,2,\ldots)$ and a bounded function f, can be written in the form:

(21)
$$\frac{1}{2\pi} \int_0^{2\pi} \varphi_x(v) Z_{\vartheta}(v) dv$$

where

(22)
$$Z_{\vartheta}(v) = 1 + (2\pi)^{\vartheta} \left(\cos \frac{\pi\vartheta}{2}\right)^{-1} \Gamma(\vartheta)^{-1} \zeta(1-\vartheta, \frac{v}{2\pi}).$$

Furthermore the kernel function $Z_{\vartheta}(v)$ has the only singular point $v = 0$ and obeys the inequality

(23)
$$|Z_{\vartheta}(v) - 2\pi\left(\cos \frac{\pi\vartheta}{2}\right)^{-1} \Gamma(\vartheta)^{-1} v^{\vartheta-1}| \leq$$
$$\leq |1 - (2\pi)^{\vartheta} \left(\cos \frac{\pi\vartheta}{2}\right)^{-1} \Gamma(\vartheta+1)^{-1}| +$$
$$+ (2\pi)^{\vartheta} \left(\cos \frac{\pi\vartheta}{2}\right)^{-1} \frac{\vartheta}{\Gamma(\vartheta+1)},$$

so that the difference on the left-hand side of (23) tends in $0 < v \leq 2\pi$ uniformly to 0 as $\vartheta \to +0$.

Let now split (21) into three parts:

$$(24)\qquad \frac{1}{2\pi}\int_0^{2\pi} \varphi_x(v)Z_\vartheta(v)dv =$$

$$= \frac{1}{2\pi}\int_0^{2\pi} \varphi_x(v)[Z_\vartheta(v) - 2\pi(\cos\frac{\pi\vartheta}{2})^{-1}\Gamma(\vartheta)^{-1} v^{\vartheta-1}]dv +$$

$$+ (\cos\frac{\pi\vartheta}{2})^{-1}\Gamma(\vartheta)^{-1}\int_0^{\delta} \varphi_x(v)v^{\vartheta-1}dv +$$

$$+ (\cos\frac{\pi\vartheta}{2})^{-1}\Gamma(\vartheta)^{-1}\int_\delta^{2\pi} \varphi_x(v)v^{\vartheta-1}dv =$$

$$= J_1 + J_2 + J_3$$

with a fixed $\delta \in (0,1)$.

As regards the first term, with any given $\varepsilon > 0$ a number $\vartheta'_\varepsilon < 1$ can be associated such that

$$(25)\qquad |J_1| \leq \frac{1}{2\pi}\int_0^{2\pi} |\varphi_x(v)|\cdot\varepsilon\, dv = \varepsilon\cdot k \qquad (\vartheta < \vartheta'_\varepsilon)$$

where $K = \sup_{t\in[0,2\pi]} |f(t)|$.

On the other hand, using simple properties of the gamma-function we get

$$(26)\qquad |J_2 - \vartheta\int_0^{2\pi} \varphi_x(v)v^{\vartheta-1}dv| =$$

$$= \vartheta|(\cos\frac{\pi\vartheta}{2})^{-1}\Gamma(\vartheta+1)^{-1} - 1||\int_0^{\delta}\varphi_x(v)v^{\vartheta-1}dv| <$$

$$< |(\cos\frac{\pi\vartheta}{2})^{-1}\Gamma(\vartheta+1)^{-1} - 1|K < K\varepsilon\ ,$$

provided that $\vartheta < \vartheta''_\varepsilon$.

Finally, there exists a number $\vartheta'''_\varepsilon > 0$ such that for $\vartheta > \vartheta'''_\varepsilon$ the inequality

(27) $$|J_3| < (\cos \tfrac{\pi\vartheta}{2})^{-1}\,\Gamma(\vartheta)^{-1}\cdot K(2\pi)^{\vartheta-1}\int_{\delta}^{2\pi}(\tfrac{v}{2\pi})^{-1}\,dv$$

$$< \vartheta(\cos \tfrac{\pi\vartheta}{2})^{-1}\,\Gamma(\vartheta+1)^{-1}\,K\cdot 2\pi \log \tfrac{2\pi}{\delta} < K\varepsilon$$

holds.

By (24)-(27), we see that

$$\left|\frac{1}{2\pi}\int_0^{2\pi}\varphi_x(v)Z_\vartheta(v)\,dv - \vartheta\int_0^{\delta}\varphi_x(v)v^{\vartheta-1}dv\right| < 3K\varepsilon,$$

whenever ϑ is sufficiently small. This is equivalent to the statement that the limits

$$\lim_{\vartheta\to+0}\left[\frac{1}{2\pi}\int_0^{2\pi}\varphi_x(v)Z_\vartheta(v)\,dv\right],$$

$$\lim_{\vartheta\to+0}\left[\vartheta\int_0^{\delta}\varphi_x(v)v^{\vartheta-1}dv\right]$$

can exist only simultaneously and in case of existence, they are equal.

If both of the limits $f(x\pm0)$ exist, then we have for $0 < \eta < \delta < 1$:

$$\left|\vartheta\int_0^{\delta}\varphi_x(v)v^{\vartheta-1}dv - \tfrac{1}{2}[f(x+0)+f(x-0)]\right| \le$$

$$\le \frac{\vartheta}{2}\int_0^{\delta}[|f(x+v)-f(x+0)| + |f(x-v)-f(x-0)|]v^{\vartheta-1}\,dv +$$

$$+ \frac{1}{2}|f(x+0)+f(x-0)|\,\left|\vartheta\int_0^{\delta}v^{\vartheta-1}dv - 1\right| \le$$

$$\le \frac{1}{2}\Big[\sup_{v\in[0,\eta]}\{|f(x+v)-f(x+0)|\} +$$

$$+ \sup_{v\in[0,\eta]}\{|f(x-v)-f(x-0)|\}\Big] +$$

$$+ \frac{\vartheta}{2}\int_{\eta}^{\delta}v^{-1}[|f(x+v)-f(x+0)| + |f(x-v)-f(x-0)|]\,dv +$$

$$+ \frac{1}{2}|f(x+0)+f(x-0)|(1-\delta^{\vartheta});$$

and the last bound becomes as small as we please, if first η, thereafter (η being fixed) ϑ is chosen in a suitable manner. Since the bounds figuring in (25)-(27) are independent of x, it follows also the assertion on uniform summability.

In order to show that the method (12) is more effective than any Cesàro or the Abel-Poisson summation process, we refer to the well-known fact that the series $\sum_{n=1}^{\infty} n^{-(1+i\tau)}$ $(\tau \neq 0)$, by a Tauberian theorem of Hardy and Littlewood, is not summable by any of these methods. Nevertheless it is plainly summable in (12) sense, because the continuity of $\zeta(\theta,u)$ as a function of θ implies for $\vartheta \to +0$:

$$\sum_{n=1}^{\infty} n^{-1-i\tau}\, n^{-\vartheta} = \lim_{M\to\infty} \left| \sum_{m=0}^{M} (1+m)^{-(1+\vartheta+i\tau)} + \frac{M^{-(\vartheta+i\tau)}}{\vartheta+i\tau} \right|$$

$$\to \lim_{M\to\infty} \left| \sum_{m=0}^{M} (1+m)^{-1-i\tau} + \frac{M^{-i\tau}}{i\tau} \right| =$$

$$= \zeta(1+i\tau,\ 1) = \zeta(1+i\tau).$$

Thus the verification of the theorem is completed.

REFERENCES

1. BUTZER, P.L., TREBELS, W., Hilberttransformation, gebrochene Integration und Differentiation. Köln-Opladen 1968, 81 pp.

2. HARDY, G.H., LITTLEWOOD, J.E., Some properties of fractional integrals I.-II. Math. Zeitschrift 27 (1928), 565-606; 34 (1932), 403-439.

3. MIKOLÁS, M., Differentiation and integration of complex order of functions represented by trigonometric series and generalized zeta-functions. Acta Math. Acad. Sci. Hung. 10 (1959), 77-124.

4. MIKOLÁS, M., Sur la sommation des séries de Fourier au moyen de l'intégration d'ordre fractionnaire. Comptes Rendus Acad. Sci. Paris 251 (1960), 837-839.

5. MIKOLÁS, M., Application d'une nouvelle méthode de sommation aux séries trigonométriques et de Dirichlet. Acta Math. Acad. Sci. Hung. 11 (1960), 317-334.

6. MIKOLÁS, M., Über die Dirichlet-Summation Fourierscher Reihen. Annales Univ. Sci. Budapest, Sectio Math. 3-4 (1960-61), 189-195.

7. MIKOLÁS, M. Generalized Euler sums and the semigroup property of integro-differential operators. Annales Univ. Sci. Budapest, Section Math. 6 (1963), 89-101.

8. MIKOLÁS, M., Sur la propriété principale des opérateurs différentiels généralisés. Comptes Rendus Acad. Sci. Paris 258 (1964), 5315-5317.

9. MIKOLÁS, M., Procédés de sommation (A,λ_n) dans l'analyse de Fourier. Communications CIM Nice, 1970, 132 p.

10. MIKOLÁS, M., On the recent trends in the development, theory and applications of fractional calculus. Lecture Notes in Math., vol. 457 (1975), Springer-Verlag.

11. MIKOLÁS, M., Real Function Theory and Orthogonal Series. (In Hungarian.) Budapest 1978, 494 pp.

12. OLDHAM, K.B., SPANIER, J., The Fractional Calculus. Academic Press, New York-London 1974, 234 pp.

13. RIESZ, M., L'intégrale de Riemann-Liouville et le problème de Cauchy. Acta Mathematica, 81 (1949), 1-223.

14. ROSS, B. (ed.), Fractional Calculus and Its Applications. Berlin-Heidelberg-New York 1975, Springer-Verlag, 381 pp.

15. WESTPHAL, U., Ein Kalkül für gebrochene Potenzen infinitesimaler Erzeuger von Halbgruppen und Gruppen von Operatoren I.-II. Compositio Math. 22 (1970), 67-103; 104-136.

16. WEYL, H., Bemerkungen zum Begriff des Differentialkoeffizienten gebrochener Ordnung. Vierteljahrschr. Naturforsch. Ges. Zürich, 62 (1917), 296-302.

Technical University of Budapest, I.
Dept. of Math. and Mathematical Research Institute of the Hungarian Academy of Sciences

Budapest, Hungary

Functional Analysis, Holomorphy and
Approximation Theory II, G.I. Zapata (ed.)

APPROXIMATION-SOLVABILITY OF SOME NONCOERCIVE NONLINEAR EQUATIONS AND SEMILINEAR PROBLEMS AT RESONANCE WITH APPLICATIONS(*)

P.S. MILOJEVIĆ

1. INTRODUCTION

In the first section of the paper, we study the (approximation) solvability of operator equations of the form $f \in Tx$ with T A-proper or strongly A-closed and such that $\|u\| + (u,v)/\|v\| \to \infty$ as $\|x\| \to \infty$ for $u \in Tx$, $v \in Kx$, where $K: X \to 2^{Y^*}$. Applications to operator equations involving monotone like and other classes of mappings and to elliptic BV propblems are also given. The results of this section complement the earlier ones of the author [20-22].

In Section 2, we first prove several new abstract results on the (approximation) solvability of semilinear equations of the form $Ax + Nx = f$, where A is a Fredholm mapping of index zero and N a given quasibounded nonlinear mapping such that $A + N$ is A-proper or strongly A-closed. The results are then applied to studying semilinear elliptic equations of the form $Au + F(x,u,Du,\dots,D^{2m}u) = f$ exhibiting (double) resonance with F having a linear growth. The obtained results extend earlier ones of Dancer [11] and Berestycki and Figueiredo [1] involving nonlinearities that depend on x and u.

(*) The work was partially supported by a CNPq Grant, 1980/81.

2. SOLVABILITY OF CERTAIN NONLINEAR NONCOERCIVE OPERATOR EQUATIONS

Let X and Y be normed linear spaces, $\{E_n\}$ and $\{F_n\}$ two sequences of finite dimensional spaces with $\dim E_n = \dim F_n$, V_n and W_n continuous linear mappings of E_n into X and of Y onto F_n respectively, V_n injective, $\delta = \sup \|V_n\| < \infty$ and $\mathrm{dist}(x, V_nE_n) \to 0$ for each $x \in X$. Then $\Gamma = \{E_n, V_n; F_n, W_n\}$ is an admissible scheme for (X,Y). If $P_n: X \to X_n \subset X$ and $Q_n: Y \to Y_n \subset \subset Y$ are linear projections, $P_nx \to x$, $Q_ny \to y$ for each x,y, $\Gamma_o = \{X_n, P_n; V_n, Q_n\}$ is a projectionally complete scheme for (X,Y). Let $K(Y)$ and $BK(Y)$ be the families of all nonempty closed convex and bounded closed and convex subsets of Y respectively. Let $V \subset X$ be a subspace, $D \subset X$, Γ a scheme for (V,Y), $D_n = V_n^{-1}(D)$, $T: D \cap V \to 2^Y$ and $T_n \equiv W_nTV_n: D_n \to K(F_n)$. We shall always assume that T_n is upper demicontinuous. The classes of mappings to be studied are introduced next.

DEFINITION 2.1. (a) We say that $T: D \cap V \to 2^Y$ is <u>approximation-closed</u> (<u>A-closed</u>) with respect to a scheme Γ for (V,Y) if, whenever $D \cap V \ni V_{n_k}u_{n_k} \to x$ and $\|y_{n_k} - W_{n_k}f\| \to 0$ for some $y_{n_k} \in T_{n_k}u_{n_k}$ and $f \in Y$, then $x \in D \cap V$ and $f \in Tx$.

(b) $T: D \cap V \to 2^Y$ is <u>strongly</u> A-<u>closed</u> w.r.t. Γ for (V,Y) if, whenever $V_{n_k}u_{n_k} \in D \cap V$ is bounded and $\|y_{n_k} - W_{n_k}f\| \to 0$ for some $y_{n_k} \in T_{n_k}u_{n_k}$ and $f \in Y$, then $f \in Tx$ for some $x \in D \cap V$.

(c) $T: D \cap V \to 2^Y$ <u>is</u> A-<u>proper</u> <u>w.r.t.</u> Γ_o for (V,Y) if, whenever $\{V_{n_k}u_{n_k} \in D \cap V\}$ is bounded and $\|y_{n_k} - W_{n_k}f\| \to 0$ for some $y_{n_k} \in T_{n_k}u_{n_k}$ and $f \in Y$, then some subsequence $V_{n_{k(i)}}u_{n_{k(i)}} \to x$. The upper semi-(demi-) continuity of T together with some restrictions on Γ and/or (X,Y) imply its A-closedness (cf. [28]).

Monotone like mappings are examples of strongly A-closed ones and, as will be seen later on, many classes of nonlinear mappings are of the A-proper type. Theory of A-proper mappings, their uniform limits and of strongly A-closed (i.e., pseudo A-proper) mappings unifies and extends theories of compact and ball-condensing vector fields, of monotone and accretive like and other classes of mappings an is, more importantly, useful in studying various classes of mappings to wich the other existing theories are not applicable. The theory is constructive in case of A-proper mappings and was initiated by Pertyshyn and we refer to [30] for a survey of the theory until 1975. Our terminology is somewhat different from the usual one.

In this section we shall study the (approximation) solvability of operator equations

$$(2.1) \qquad f \in Tx, \qquad (x \in X, \quad f \in Y)$$

where T is either A-proper or strongly A-closed w.r.t. Γ for (X,Y) and satisfies a rather weaker than coercivity condition (see (2.5)-(2.6), or (2.8) below). In most of the cases we shall assume that T is K-quasi-bounded for some $K: X \to 2^{Y^*}$, i.e., whenever $\{x_n\} \subset X$ is bounded and $y_n \in Tx_n$, $f_n \in Kx_n$ are such that $(y_n, f_n) \leq c\|x_n\|$ for each n and some $c > 0$ then $\{y_n\}$ is bounded. For our study of Eq. (2.1) we need introduce a multivalued bounded mapping $G: X \to K(Y)$ such that $0 \notin Gx$ for $\|x\|$ large and which satisfies the following conditions:

(2.2) For each large $r > 0$, $\deg(W_n G V_n, B_n(0,r), 0) \neq 0$ for each large n, where $B_n = V_n^{-1}(B(0,r))$ and the degree is that defined in [17,18].

(2.3) $(u,v) = \|u\|\,\|v\|$ for each $u \in Gx$, $v \in Kx$, $x \in X$.

(2.4) There exists $K_n: V_n(E_n) \to 2^{K_n^*}$ for each n such that $(W_n y, u) = (y,v)$ for each $u \in K_n x$, $v \in Kx$, $x \in V_n E_n$, $y \in Y$.

When $Y = X$ or $Y = Y^*$ there are natural choices for G, K, K_n that satisfy (2.2)-(2.4). If $Y = X$ is a Π_1-Banach space and $\Gamma_o = \{X_n, P_n\}$ is a projectionally complete scheme for X, $\|P_n\| = 1$, then choosing $G = I$, $K = J: X \to 2^{X^*}$, the normalized duality mapping, and $K_n = K|_{X_n}$ we have (2.2)-(2.4) since $P_n^* Jx \subset Jx$ for each $x \in X_n$, $n \geq 1$. If X is reflexive, $Y = X^*$ and $\Gamma_o = \{X_n, P_n; R(P_n^*), P_n^*\}$ is a projectionally complete scheme for (X, X^*), then taking $G = J: X \to 2^{X^*}$, $K = I$ and $K_n = I|_{X_n}$ we see that they satisfy (2.2)-(2.4). For a given $G: X \to Y$ it is always possible to find a $K: X \to 2^{Y^*}$ such that (2.3) holds; for example, we can take $K = JG$, where $J: Y \to 2^{Y^*}$ is the normalized duality mapping. Then (2.4) holds for $\Gamma = \{X_n, V_n, Y_n, Q_n\}$ with $K_n = K|_{X_n}$ if $Q_n^* Kx = Kx$ for $x \in X_n$. In applications it is often possible to construct other types of mappings K, K_n, G and a scheme Γ which satisfy (2.2)-(2.4).

Let $J: X \to 2^{X^*}$ be the normalized duality mapping and define a semi-inner product $(\cdot,\cdot)_-: X \times X \to R$ by $(x,y)_- = \inf\{\phi(x) \mid \phi \in Jy\}$ $(= \min\{\phi(x) \mid \phi \in Jy\})$. We are in a proposition to give our basic approximation-solvability result for Eq. (2.1) with T being A-proper which was announced first in [28].

THEOREM 2.1. Let $T: X \to 2^Y$ be K-quasibounded, $K = JG$ and G bounded and $G, T + \mu G$ A-proper and A-closed w.r.t. Γ for $\mu \geq 0$. Suppose that (2.2)-(2.4) hold and that for each f in Y there exists an $r_f > 0$ such that

(2.5) $$\|u - tf\| \geq \gamma > 0 \quad \text{for } u \in Tx, \ \|x\| = r_f, \ t \in [0,1];$$

(2.6) $$(u,v)_- \geq -\|u\|\|v\| \quad \text{for } u \in Tx, \ v \in Gx, \ \|x\| = r_f.$$

Then Eq. (2.1) is feebly approximation-solvable for each f in Y (i.e., there exists a solution $u_n \in E_n$ of $W_n f \in W_n T V_n u$ for each large n and a subsequence $V_{n_k} u_{n_k} \to x$ with $f \in Tx$).

PROOF. Let f in Y be fixed. It is easy to see that the A-properness of T and (2.5) imply that there exists an $n_o \geq 1$ such that

$$tW_n f \notin W_n TV_n u \quad \text{for} \quad u \in \partial B_n(0,r_f), \quad t \in [0,1], \quad n \geq n_o .$$

Therefore, $\deg(W_nTV_n - W_nf, B_n, 0) = \deg(W_nTV_n, B_n, 0)$ for $n \geq n_o$. To show that this degree is nonzero, we define on $[0,1] \times \bar{B}_n$ a homotopy $H_n(t,u) = tW_nTV_nu + (1-t)W_nGV_nu$. We need show that for each $n \geq n_1$ with some n_1 fixed

$$(2.7) \qquad 0 \notin H_n(t,u) \quad \text{for} \quad u \in \partial B_n , \quad t \in [0,1].$$

If (2.7) did not hold, then there would exist $t_k \in [0,1]$, $t_k \to t_o$, and $u_{n_k} \in \partial B_n$ such that $0 \in H_{n_k}(t_k,u_{n_k})$ for each k. Suppose first that $t_o \in (0,1)$ and let $v_k \in TV_{n_k}u_{n_k}$ and $w_k \in GV_{n_k}u_{n_k}$ be such that $t_kW_{n_k}v_k + (1-t_k)W_{n_k}w_k = 0$, $k \geq 1$. Then for each large k

$$W_{n_k}v_k + (1-t_o)t_o^{-1} W_{n_k}w_k = -(1-t_k)t_k^{-1} W_{n_k}w_k + (1-t_o)t_o^{-1} W_{n_k}w_k =$$

$$= [(1-t_o)t_o^{-1} - (1-t_k)t_k^{-1}]W_nw_k \to 0 \quad \text{as} \quad k \to \infty,$$

and, by the A-properness and A-closedness of $T + \mu G$, some subsequence $\{V_{n_k}u_{n_k}\} \supset \{V_mu_m\} \to x$ with $0 \in t_oTx + (1-t_o)Gx$ and $\|x\| = r_f$. This leads to a contradiction with (2.6) since, if $u \in Tx$ and $v \in Gx$ are such that $t_ou + (1-t_o)v = 0$ and $\lambda = (1-t_o)t_o^{-1}$, then $\|u\|\|v\| + (u,v)_- = \lambda\|v\|^2 - \lambda(v,v)_- = 0$. Therefore, it remains to consider the case $t_o = 0,1$. Let $t_o = 1$. Then $W_{n_k}v_k = -(1-t_k)t_k^{-1} W_{n_k}w_k \to 0$ as $k \to \infty$ and, by the A-properness of T, some subsequence $V_{n_{k(i)}}u_{n_{k(i)}} \to x$ with $0 \in Tx$ and $\|x\| = r_f$, in contradiction with (2.5). Finally, let $t_o = 0$. Then for $x_k \in KV_{n_k}u_{n_k}$ and $y_k \in K_{n_k}V_{n_k}u_{n_k}$ we obtain using (2.3) and (2.4) and nothing that $t_k \neq 0$ for infinitely many k

$$(v_k,x_k) = (W_{n_k}v_k,y_k) = -(1-t_k)t_k^{-1}(W_{n_k}w_k,y_k) = -(1-t_k)t_k^{-1}(w_k,x_k) =$$
$$= -(1-t_k)t_k^{-1}\,\|w_k\|\,\|x_k\| \to -\infty \quad \text{as} \quad k \to \infty,$$

which implies that $\{v_k\}$ is bounded by the K-quasiboundedness of T. Hence, $\|W_{n_k}v_k\| = (1-t_k)^{-1}\, t_k\|W_{n_k}w_k\| \to \infty$, a contradiction. This completes the proof of (2.7).

Now, it follows from (2.7) that $\deg(W_nTV_n - W_nf, B_n, 0) \neq 0$ for each $n \geq n_1$ and consequently there exists a $u_n \in B_n$ such that $W_nf \in W_nTV_nu_n$ for $n \geq n_1$. Since T is A-proper and A-closed, some subsequence $V_{k_n}u_{n_k} \to x$ with $f \in Tx$. □

REMARK 2.1. (a) Conditions (2.5) and (2.6) are implied by

(2.8) $\|u\| + \dfrac{(u,v)}{\|v\|} \to \infty$ as $\|x\| \to \infty$ for $u \in Tx$, $v \in Gx$.

(b) Analysing the proof of Theorem 2.1, we see that condition (2.6) can be replaced by

(2.9) For each $f \in Y$ there exists an $r_f > 0$ such that

$$T(x) \cap \lambda G(x) = \emptyset \quad \text{for} \quad x \in \partial B(0,r_f) \quad \text{and} \quad \lambda < 0.$$

Let us now extend Theorem 2.1 to the case when T is just strongly A-closed or satisfies the following condition:

(*) Whenever $\{x_n\} \subset X$ is bounded and $u_n \to f$ for some $u_n \in Tx_n$ and $f \in Y$, then there exists an $x \in X$ such that $f \in Tx$.

THEOREM 2.2. Suppose that all the hypotheses of Theorem 2.1 hold with the A-properness and A-closedness of T replaced by either condition (*) for T or the strong A-closedness of T and that for each $r > 0$ large and each $\mu > 0$ small, $\deg(\mu W_nGV_n, B_n(0,r), 0) \neq 0$ for each large n. Then $T(X) = Y$. Moreover, the conclusion remains valid if (2.6) is replaced by (2.9).

PROOF. Let f in Y be fixed. By condition (2.5) there exists a $\mu_o > 0$ such that for each $\mu \in (0,\mu_o)$

$$\|u+\mu v-tf\| \geq \gamma/2 \quad \text{for} \quad u \in Tx, \quad v \in Gx, \quad \|x\| = r_f, \quad t \in [0,1].$$

Let $\mu \in (0,\mu_o)$ be fixed. Since $T + \mu G$ is A-proper and A-closed, there exists an $n_o \geq 1$ such that $tW_nf \notin W_nTV_nu + \mu W_nGV_nu$ for $u \in \partial B_n(0,r_f)$, $t \in [0,1]$ and $n \geq n_o$. Hence, $\deg(W_nTV_n+\mu W_nGV_n - W_nf, B_n, 0) = \deg(W_nTV_n + \mu W_nGV_n, B_n, 0)$ for each $n \geq n_o$. Next, define on $[0,1] \times \bar{B}_n$ a homotopy $H_{\mu n}(t,u) = tW_nTV_nu + \mu W_nGV_nu$. We claim that there exists an $n_1 = n_1(\mu) \geq n_o$ such that for each $n \geq n_1$

$$0 \notin H_{\mu n}(t,u) \quad \text{for} \quad u \in \partial B_n, \quad t \in [0,1] \tag{2.10}$$

and $n_1(\mu_1) \leq n_2(\mu_2)$ whenever $\mu_2 < \mu_1 < \mu_o$. If not, then there exist $t_k \in [0,1]$, $t_k \to t_o$, $u_{n_k} \in \partial B_{n_k}$ such that $0 \in H_{\mu n_k}(t_k,u_{n_k})$ for each k. Let $v_k \in TV_{n_k}u_{n_k}$ and $w_k \in GV_{n_k}u_{n_k}$ be such that $t_kW_{n_k}v_k + \mu W_{n_k}w_k = 0$. Suppose first that $0 < t_o \leq 1$. Then

$$W_{n_k}v_k + \mu t_o^{-1} W_{n_k}w_k = \mu(t_o^{-1} - t_k^{-1})W_{n_k}w_k \to 0 \quad \text{as} \quad k \to \infty$$

and by the A-properness of $T + \mu G$ some subsequence $V_{n_{k(i)}}u_{n_{k(i)}} \to x$ with $0 \in t_oTx + \mu Gx$ and $\|x\| = r_f$. As in Theorem 2.1, this leads to a contradiction with (2.6).

Suppose now that $t_o = 0$. Since $t_k \neq 0$ for infinitely many k, we have that $W_{n_k}v_k = -\mu t_k^{-1} W_{n_k}w_k$ for such k, and for any $x_k \in KV_{n_k}u_{n_k}$

$$(v_k,x_k) = (W_{n_k}v_k,y_k) = -\mu t_k^{-1}(W_{n_k}w_k,y_k) = -\mu t_k^{-1}(w_k,x_k) =$$

$$= -\mu t_k^{-1} \|w_k\| \, \|x_k\| \to -\infty \quad \text{as} \quad k \to \infty.$$

Since T is K-quasibounded, we get that $\{v_k\}$ is bounded and consequently $\|W_{n_k}v_k\| = \mu t_k^{-1}\|W_{n_k}w_k\| \to \infty$ as $k \to \infty$ by the A-properness of G. This contradicts the boundedness of $\{v_k\}$ and therefore $t_o \neq 0$. Hence, (2.10) is valid.

Now, it follows from (2.10) that $\deg(W_nTV_n+\mu W_nGV_n-W_nf,B_n,0) = \deg(\mu W_nGV_n,B_n,0)$ for each $n \geq n_1$, and therefore, the equation $f \in Tx + \mu Gx$ is solvable in $\bar{B}(0,r_f)$. Taking $\mu_k \in (0,\mu_o)$ with $\mu_k \to 0$ and $x_k \in \bar{B}(0,r_f)$ with $f \in Tx_k + \mu_k Gx_k$ and using condition (*) we obtain an $x \in X$ such that $f \in Tx$.

Let us now suppose that T is strongly A-closed and let $\mu_k \in (0,\mu_o)$ be such that $\mu_k \to 0$ decreasingly. For each k fixed, condition (2.10) for $H_{\mu_k n}$ holds for each $n \geq n_k$ and therefore $\deg(W_{n_k}TV_{n_k} + \mu_k W_{n_k}GV_{n_k} - W_{n_k}f,B_{n_k},0) \neq 0$. Hence, there exists a $u_{n_k} \in B_{n_k}$ such that $W_{n_k}f \in W_{n_k}TV_{n_k}u_{n_k} + \mu_k W_{n_k}GV_{n_k}u_{n_k}$ for each k. Let $v_k \in TV_{n_k}u_{n_k}$ and $w_k \in GV_{n_k}u_{n_k}$ be such that $W_{n_k}v_k + \mu_k W_{n_k}w_k = W_{n_k}f$. Since $\|W_{n_k}v_k - W_{n_k}f\| \to 0$ as $k \to \infty$, there exists $x \in X$ such that $f \in Tx$ by the strong A-closedness of T. □

Eq. (2.1) with T A-proper or strongly A-closed and satisfying (2.5)-(2.6) or (2.8) has been earlier studied by the author under other additional conditions on T. So, in [20] have announced the following results (see also Note added in proof).

THEOREM 2.3. Let $T: X \to 2^Y$, $K = JG$ and G be bounded and (2.2)-(2.3) and (2.5)-(2.6) hold. Then

(a) Eq. (2.1) is feebly approximation-solvable for each f in Y if, in addition, $H(t,x) = tTx + (1-t)Gx$ is an A-proper and A-closed homotopy on $[0,1] \times X$.

(b) $T(X) = Y$ if, in addition, $H_\mu(t,x) = tTx + \mu Gx$ is an A-proper and A-closed homotopy at 0 on $[0,1]\times X\setminus B(0,R)$ for some large

R and all $\mu \in (0,\beta)$ for some $\beta > 0$, T either satisfies condition (*) or is strongly A-closed and for each $r \geq R$ and $\mu \in (0,\beta)$, $\deg(\mu W_n G V_n, B_n(0,r), 0) \neq 0$ for each large r.

The following special case is useful in applications ([20-22])

COROLLARY 2.1. If G and $T + \mu G$ are A-proper and A-closed for $\mu \geq 0$ ($\mu > 0$, resp.) and either T is bounded or $(u,v) \geq -c\|v\|$ for $u \in Tx$, $v \in Kx$ with $\|x\| \geq R_o$ and some $c > 0$, then $H(t,x)$ ($H_\mu(t,x)$, resp.) in an A-proper and A-closed homotopy on $[0,1] \times X$ (at 0 on $[0,1] \times X \setminus B(0,R)$, $R \geq R_o$, resp.). Therefore, the conclusions of Theorem 2.3 (a)-(b) are valid provided the other its hypotheses hold.

We note that in the above results we have not assumed (2.4). Detailed proofs of Theorem 2.3 and Corollary 2.1 and their applications to various special classes of nonlinear mappings and BVP for partial differential equations can be found in the author's papers [21,22] (cf. also [28]). Analysing the proofs of Theorem 2.3 and Corollary 2.1 we see that (2.6) can be replaced by (2.9). We have also proven there the following

PROPOSITION 2.1. Suppose that (2.3), (2.4) hold and that $(M_n u, v) > 0$ for $v \in K_n u$ and $0 \neq u \in E_n$ and some linear isomorphism $M_n : E_n \to F_n$. Then, for each r and large $\mu > 0$, $\deg(\mu W_n G V_n, B_n(0,r), 0) \neq 0$ for n large.

The above results are applicable to studying perturbed equations

$$f \in Tx + Fx \tag{2.11}$$

with F such that

(2.12) $(u,v)_- \geq -\alpha\|v\|$ for $u \in Fx$, $v \in Gx$, $\|x\| \geq R$ and some $\alpha > 0$.

If T is K-coercive, then (2.12) can be weaken as in the next result. We recall that $A: X \to 2^Y$ satisfies <u>condition</u> (+) if $\{x_n\}$ is bounded whenever $y_n \to f$ for some $y_n \in Ax_n$ and f in Y.

THEOREM 2.4 ([28]). Let $T: X \to 2^Y$ be K-coercive, $K = JG$, i.e. $(u,v)_- \geq c(\|x\|)\|v\|$ for $u \in Tx$, $v \in Gx$, $x \in X$, and some $c: R^+ \to R^+$ such that $c(r) \to \infty$ as $r \to \infty$, $F: X \to 2^Y$ be bounded and $(u,v)_- \geq$ $\geq -c(\|x\|)\|v\|$ for $u \in Fx$, $v \in Gx$, $x \in X\backslash B(0,R)$ and some $R > 0$. Suppose that G, K and K_n are as in Theorem 2.2 and $T + \beta F + \mu G$ is A-proper and A-closed for each $\mu > 0$ and $\beta \in (\beta_o, 1)$ with some $\beta_o > 0$. Then, if $T + \beta F$ is K-quasibounded and $T + F$ satisfies conditions (+) and (*), $T + F$ is surjective, i.e. $(T+F)(X) = Y$.

PROOF. Let $f \in Y$ be given. By condition (+), there exists an $r_f \geq R$ and $\gamma > 0$ such that $\|u+v-tf\| \geq \gamma$ for $u \in Tx$, $v \in Fx$, $\|x\| = r_f$ and $t \in [0,1]$. Since F is bounded, there exists a $\beta_1 \in (\beta_o, 1)$ such that $(1-\beta_1)\|v\| \leq \gamma/2$ for $v \in Fx$, $\|x\| = r_f$. Therefore, for each $\beta \in (\beta_1, 1)$, $u \in Tx$, $v \in Fx$, $\|x\| = r_f$ and $t \in [0,1]$ we have that $\|u+\beta v-tf\| = \|u+v-tf-(1-\beta)v\| \geq \gamma/2$, i.e. $T + \beta F$ satisfies (2.5) on $\partial B(0,r_f)$ for each $\beta \in (\beta_1, 1)$. Moreover, for each $u \in Tx$, $v \in Fx$, $w \in Gx$, with $\|x\| = r_f$ and $\beta \in (\beta_1, 1)$ we obtain that $(u+\beta v, w)_- \geq c(\|x\|)\|w\| - \beta c(\|x\|)\|w\| > 0$, and therefore $T + \beta F$ satisfies (2.6) on $B(0,r_f)$ for each $\beta \in (\beta_1, 1)$. Hence, by Theorem 2.2, for a given $\beta \in (\beta_1, 1)$ there exists $x \in \bar{B}(0,r_f)$ such that $f \in Tx + \beta Fx$. Let $\beta_k \in (\beta_1, 1)$ be such that $\beta_k \to 1$ and $x_k \in \bar{B}(0,r_f)$ such that $f \in Tx_k + \beta_k Fx_k$. Since F is bounded, it follows from condition (*) that there exists a $x \in X$ such that $f \in Tx + Fx$. □

We continue our exposition by deriving solvability results for Eq. (2.1) involving various special classes of nonlinear mappings. Recall that $A: X \to Y$ is said to be c-strongly K-monotone for some $K: X \to 2^{Y^*}$ if $(Ax-Ay,z) \geq c\|x-y\|^2$ for $x, y \in X$, $z \in K(x-y)$ and some $c > 0$. If $c = 0$, A is called K-monotone. We shall always assume that K is such that $Q_n^*Kx \subset Kx$ for $x \in X_n$ and a given scheme $\Gamma_o = \{X_n, P_n; Y_n, Q_n\}$ for (X,Y), and $\|u\| \leq \alpha\|x\|$ for $u \in Kx$ and some $\alpha > 0$. When $Y = X^*$ ($Y = X$, resp.) such mappings are called c-strongly monotone (c-strongly accretive with $K = J$, the normalized duality mapping, resp.). The ball measure of non-compactness of a bounded subset $D \subset X$ is defined by $\chi(D) = \inf\{r > 0 \mid D \subset \bigcup_{i=1}^{n} B(x_i, r),\ x_i \in X,\ n \in N\}$. A mapping $T: D \subset X \to BK(Y)$ is said to be k-ball contractive if $\chi(T(Q)) \leq k\chi(Q)$ for each $Q \subset D$; it is ball-condesing if $\chi(T(Q)) < \chi(Q)$ whenever $Q \subset D$ and $\chi(Q) \neq 0$. We have

EXAMPLE 2.1 (a) ([19,25]) Let $A: X \to Y$ be c-strongly K-monotone and $F: X \to BK(Y)$ k-ball-contractive with $k\delta < c$, or ball-condensing if $\delta = c = 1$, where $\delta = \sup\|Q_n\|$. Then $A+F: X \to BK(Y)$ is A-proper w.r.t. a projectionally complete scheme $\Gamma_o = \{X_n, P_n; Y_n, Q_n\}$.
(b) ([24,28]) $A + F$ of part (a) is A-closed w.r.t. Γ_o if F is upper semicontinuous (u.s.c.) and A is either continuous, or demicontinuous with X reflexive.

In Example 2.1 we could have assumed more generally that A is a-stable w.r.t. Γ_o. By Theorem 2.3 in [19] and the above abstract results we obtain the following extension of Theorem 2.7 in [19] and Corollary 1 in [21].

THEOREM 2.5. Let A and F be as in Example 2.1 (a)-(b).
(a) Let $T = A + F$ satisfy condition (+) and either one of the following conditions holds:

(2.13) T is odd on $X \backslash B(0,R)$ for some $R > 0$;

(2.14) A is bounded and (2.9) holds;

(2.15) $(u,z) \geq -\alpha\|x\|^2$ for $u \in Tx$, $z \in Kx$, $\|x\| \geq R$ and some $\alpha > 0$.

Then the equation $f \in Ax + Fx$ is feebly approximation-solvable for each f in Y. If $k\delta = c$ and T satisfies also condition (*), then $T(X) = Y$.

(b) Let T be K-quasibounded and satisfy (2.5) and (2.9). Then the equation $f \in Ax + Fx$ is feebly approximation-solvable for each f in Y. If $k\delta = c$ and T satisfies also condition (*), then $T(X) = Y$.

(c) Let T be K-quasibounded, satisfy conditions (+) and (*), $k\delta = c$ and

(2.16) $(u,z) \geq -\alpha\|x\|^2$ for $u \in Tx$, $z \in Kx$, $\|x\| \geq R$ and some $\alpha > 0$.

Then $T(X) = Y$.

If A is just K-monotone, then $A + cK$ is c-strongly K-monotone and, using Theorem 2.5 with F compact for each $c > 0$, we obtain the following extension of Theorem 2.7 in [19] for such mappings.

THEOREM 2.6. Let $A: X \to Y$ be K-monotone, with either $Y = X$ and $K = J$ or $Y = X^*$ and $K = I$, and $F: X \to BK(Y)$ be u.s.c. and compact. Suppose that $T = A + F$ satisfies conditions (+) and (*) and A is either continuous or demicontinuous with X reflexive. Suppose that T satisfies either one of conditions (2.13)-(2.15) or T is K-quasibounded and satisfies (2.16). Then $(T+F)(X) = Y$.

Theorem 2.5 with $F = 0$ yields the following extension of a result of F. Browder [6]. Unlike his approach based on differential equations, our proof is much simpler.

COROLLARY 2.2 Let X be a Π_1-space and $A: X \to X$ accretive and either continuous or demicontinuous with $X = X^{**}$. Then A is m-accretive, i.e. $\lambda I + A$ is surjective for each $\lambda > 0$.

The following result is useful in applications of Theorem 2.6 (cf. [9]).

LEMMA 2.1 Let X be reflexive, have normal structure, X and X^* strictly convex and $A: X \to X$ be m-accretive. Then A satisfies condition (*). If, in addition, A is strongly demiclosed, i.e. whenever $x_n \rightharpoonup x$ and $Ax_n \to f$ then $Ax = f$, and F is compact, then $A + F$ satisfies (*).

Let us now discuss more general than c-strongly K-monotone mappings. Let $C_b(D,Y)$ denote the normed linear space with the supremum norm of all continuous bounded functions from the topological space D into the normed linear space Y.

DEFINITION 2.2 ([28]). A mapping $T: X \to BK(Y)$ is said to be <u>semi</u> a-<u>stable</u> w.r.t. a projectionally complete scheme $\Gamma_o = \{X_n, P_n; Y_n, Q_n\}$ for (X,Y) if there exists a $U: X\times X \to BK(Y)$ such that $Tx = U(x,x)$, $x \in X$, and

(i) the mapping $x \to (x,\cdot)|\bar{D}$ is compact from $\bar{D}$ into $C_b(\bar{D},Y)$ for each bounded subset $D \subset X$.

(ii) For each $x \in X$, $U(x,\cdot)$ is a-stable w.r.t. Γ_o, i.e., for some $c > 0$ and each large n

$$\|Q_n u - Q_n v\| \geq c\|y-z\| \quad \text{for} \quad u \in U(x,y),\ v \in U(x,z),\ y,z \in X_n.$$

In particular, $T: X \to BK(Y)$ is <u>semi</u> c-<u>strongly</u> K-<u>monotone</u> if there exists a $U: X\times X \to BK(Y)$ and $K: X \to 2^{Y^*}$ such that $Tx = U(x,x)$, (i) of Definition 2.1 holds and for each $x \in X$, $U(x,\cdot)$ is c-strongly K-monotone. As before, we assume that $\|u\| \leq \alpha\|x\|$ for $u \in Kx$ and some $\alpha > 0$ and $Q_n^* Kx \subset Kx$, $x \in X_n$. It is clear that such mappings are semi a-stable w.r.t. Γ_o. If $c = 0$, we have the class of semi K-monotone mappings. For such mappings we have proven in [25,28] the following result whose part (a) answers positively the question raised by Browder [7] and extends some of his results (cf. [6,7]).

THEOREM 2.7 (a) If $T: X \to BK(Y)$ is semi a-stable, then T is A-proper w.r.t. Γ_o. If, also, Q_nT is injective in X_n, T satisfies (+) and is either u.s.c. or demicontinuous with $Y = Y^{**}$, then it is also A-closed and the equation $f \in Tx$ is feebly approximation-solvable for each f in Y. If T is continuous a-stable and singlevalued, it is an A-proper and A-closed homeomorphism.

(b) If T is as in (a) and $F: X \to BK(Y)$ is u.s.c. and k-ball contractive with $k\delta < c$ or ball-condensing if $\delta = c = 1$, then $T + F$ is A-proper and A-closed w.r.t. Γ_o.

Now, in view of Theorem 2.7 (b), our general results yield ([28])

THEOREM 2.8 (a) The conclusions of Theorem 2.5 remain valid if we assume in it that A is semi c-strongly K-monotone with $K = JG$.

(b) The conclusion of Theorem 2.6 remains valid if we assume in it that A is semi K-monotone.

REMARK 2.2 In Theorem 2.8 (a) one can assume that F is semi k-ball-contractive with $k\delta < c$ (i.e. $F(x) = U(x,x)$ with U satisfying (i) of Definition 2.1 and $U(x,\cdot)$ is k-ball-contractive for each x), or semi ball-condensing if $\delta = c = 1$ (cf. [28]). We note that (i) of Definition 2.1 holds if X is reflexive and $U(\cdot,x): X \to Y$ is completely continuous uniformly for x in a bounded set.

In [31], Pohožaev studied a class of A-proper mappings $T:X \to X^*$

such that $(Tx-Ty, x-y) \geq c(\|x-y\|) - \phi(x-y)$, $x,y \in X$, where $\phi: X \to R$ is weakly upper semicontinuous at 0 and $\phi(0) = 0$ and $c: R^+ \to R^+$ is continuous, $c(0) = 0$ and $r \to 0$ whenever $c(r) \to 0$. A slightly more general class is given by

PROPOSITION 2.2 Let $T: X \to X^*$ be such that for each $r > 0$, some constant $q(r) > 0$ and x,y with $\|x-y\| \geq r$,

$$(2.17) \qquad (Tx-Ty, x-y) \geq q(r)c(\|x-y\|) - \phi(x-y).$$

Then T is of type (S_+) (i.e., whenever $x_n \rightharpoonup x$ and $\limsup (Tx_n, x_n - x) \leq 0$, then $x_n \to x$), and is therefore A-proper w.r.t. $\Gamma_a = \{X_n, V_n; X_n^*, V_n^*\}$ or Γ_o.

PROOF. Let $x_n \rightharpoonup x$ and $\limsup (Tx_n, x_n - x) \leq 0$. If $x_n \not\to x$, then there exists $r > 0$ such that $\|x_n - x\| \geq r$ for infinitely many n. Hence, (2.17) holds for such n's leading to a contradiction. □

Partial differential equations studied recently by Hetzer [15], in a much more complicated way using generalized degree theory of Browder [6], satisfy (2.17) modulo a compact mapping. Let $Q \subset R^n$ be a bounded region. We are interested in a generalized solution $u \in V$ of

$$(2.18) \qquad \sum_{|\alpha|,|\beta|\leq m} (-1)^{|\alpha|} D^\alpha(a_{\alpha\beta}(x)D^\alpha u) + \sum_{|\alpha|\leq m} (-1)^{|\alpha|} D^\alpha A_\alpha(x,u,Du,\ldots,D^m u) = f(x)$$

with $f \in L_2(Q)$, where $V \subset W_2^m$ is a closed subspace with $\mathring{W}_2^m \subset V$. Let $s_m = \#\{\alpha \mid |\alpha| \leq m\}$. Suppose that $a_{\alpha\beta} \in L_\infty(Q)$ for $|\alpha|,|\beta| \leq m$, $c_o > 0$ and

$$(2.19) \qquad \sum_{|\alpha|,|\beta|=m} a_{\alpha\beta}(x) z_\alpha z_\beta \geq c_o \sum_{|\alpha|=m} z_\alpha^2 \quad \text{for } x \in Q(\text{a.e.}),$$
$$\text{all } (z_\alpha) \in R^{s_m - s_{m-1}}.$$

(2.20) For each $|\alpha| \leq m$, $A_\alpha : Q \times R^{s_m} \to R$ satisfies the Carathéodory condtions and $|A_\alpha(x,y)| \leq \mu|y| + \psi(x)$ for $x \in Q$(a.e.), and some $\mu \in R$, $\psi \in L_2(Q)$.

(2.21) There exists a function $p: R^+\setminus\{0\} \to [0,c)$ such that ([15])

$$\sum_{|\alpha|=m} [A_\alpha(x,y,z)-A_\alpha(x,y,z')](z_\alpha - z'_\alpha) \geq -p(r) \sum_{|\alpha|=m} |z_\alpha - z'_\alpha|^2$$

for $x \in Q$(a.e.), $y \in R^{s_{m-1}}$, $r > 0$, and $z,z' \in R^{s_m - s_{m-1}}$ with $|z-z'| \geq r$.

Define continuous and bounded mappings $A_1, A_2, N_1, N_2 : V \to V$ by $(A_1u,v) = \sum_{|\alpha|,|\beta|=m} (a_{\alpha\beta}D^\alpha u, D^\beta v)_{L_2}$, $(A_2u,v) = \sum_{|\alpha|,|\beta|<m} (a_{\alpha\beta}D^\alpha u, D^\beta v)_{L_2}$, $(N_1u,v) = \sum_{|\alpha|=m} (A_\alpha(x,u,Du,\ldots,D^m u), D^\alpha v)_{L_2}$ and $(N_2u,v) = \sum_{|\alpha|<m} (A_\alpha(x,u,Du,\ldots,D^m u), D^\alpha v)_{L_2}$ for $u,v \in V$. It is a well known fact that A_2 and N_2 are compact. It was shown in Hetzer [15] that, for each $r > 0$, and $u,v \in V$ with $\|u-v\| \geq r$, $(N_1u - N_1v, u-v) \geq -q(r)c(\|u-v\|) - \phi(u-v)$, where $q(r) = (c_o + 3p(\tau(r)))/4$, $\tau(r) = r/4|Q|^{1/2}$, c as in Proposition 2.1 and $\phi(u-v) = \sum_{|\alpha|=m} (A_\alpha(x,v,Dv,\ldots,D^m v) - A_\alpha(x,u,Du,\ldots,D^{m-1}u,D^m v), D^\alpha(u-v))_{L_2}$. Therefore, $A_1 + N_1$ satisfies (2.17) and so $T \equiv A_1 + A_2 + N_1 + N_2$ is A-proper and A-closed w.r.t. Γ_o. Since $A = A_1 + A_2$ is Fredholm of index zero, the theory of solvability of $Ax + Nx = f$ developed in [24,26-28] is applicable to (2.18) assuming (2.20),(2.21). In particular, the result of Hetzer [15] is deducible by this theory (see [27] for a more general case when $A = A^*$). When $N(A) = \{0\}$, we have:

THEOREM 2.9 Suppose that (2.19)-(2.21) hold.

(a) If T satisfies condition (+) and either $A_\alpha(x,-y) = -A_\alpha(x,y)$ for $x \in Q$(a.e.) and $y \in R^{s_m}$, $|\alpha| \leq m$, or T satisfies (2.9), then Eq. (2.18) is feeby approximation-solvable in the variational sense in V for each $f \in L_2$.

(b) If $N(A) = \{0\}$ and μ is sufficiently small, the conclusion of (a) holds.

(c) If $N(A) = \{0\}$, μ is sufficiently small, p takes on also the value c and T satisfies condition (*), then Eq. (2.18) has a generalized solution $u \in V$.

PROOF. Part (a) follows from a result in [30,31], Theorem 2.1 and Remark 2.1, respectively. Part (b) follows from the generalized first Fredholm theorem for A-proper mappings in [23]. Let us prove (c). Now A_1+N_1 is monotone on $V\backslash B(0,r)$ for each $r > 0$ and $A_1 + N_1 + \alpha I$ satisfies (2.17) for $\alpha > 0$. Hence, $T + \alpha I$ is A-proper and A-closed w.r.t. $\Gamma_o = \{X_n, P_n\}$ for V for each $\alpha > 0$ and the conclusion now follows from Theorem 4.5.2 in [28] (or Corollary 1 in [24]). □

Theorems 2.1-2.4 and Corollary 2.1 are applicable to many other classes of A-proper and strongly A-closed mappings discussed in detail in [27,28]. We conlcude the section with a couple of more applications proven in [28] (cf. also Corollary 3.1N in [30]).

THEOREM 2.10 Let $X = X^{**}$, $\Gamma_o = \{X_n, P_n\}$, $\|P_n\| = 1$, $S: X \to C(X)$ a generalized contraction (in the sense of Belluce and Kirk) and $C: X \to C(X)$ u.s.c., compact and $(u,v) \geq -c\|x\|^2$ for $u \in Cx$, $v \in Jx$, $\|x\| \geq R$, $c > 0$. Then the equation $f \in x - Sx - Cx$ is feebly approximation-solvable for each $f \in X$.

The proof follows from Theorem 2.4. For monotone like mappings we have ([28]):

THEOREM 2.11 Let X be reflexive, $T: X \to 2^{X^*}$ quasibounded and either pseudo monotone and demiclosed, or generalized pseudo monotone, or quasi-monotone, and $F: X \to 2^{X^*}$ such that $T + F$ is strongly A-closed w.r.t. $\Gamma_a = \{X, V_n; X_n^*, V_n^*\}$ (e.g., F could be completely continuous, or, for the first two types of T, quasi-bounded and generalized pseudo monotone with $(u,x) \geq -c(\|x\|)\|x\|$ for some continuous $c: R^+ \to R^+$, or the sum of two such mappings). Then, if $T + F$ satisfies (2.5)-(2.9), $(T+F)(X) = X^*$.

DEFINITION 2.3 $T: X \to BK(X^*)$ is of semibounded variation if for $R > 0$ and $\|x\|, \|y\| \leq R$, $(u-v, x-y) \geq -c(R, \|x-y\|')$ for $u \in Tx$, $v \in Ty$, where $\|\cdot\|'$ is a norm on X compact relative to $\|\cdot\|$ and $c(R,r) \geq 0$ is continuous in R and r and $c(R,tr)/t \to 0$ as $t \to 0^+$ for fixed R and r.

COROLLARY 2.2 Let T be hemicontinuous, of semibounded variation and satisfy (2.5)-(2.9). Then $T(X) = X^*$.

Theorem 2.11 follows from Theorem 2.2 and extends the known surjectivity results for monotone like mappings. It includes the surjectivity results of Brezis [2] for bounded coercive pseudo monotone mappings, of Browder and Hess [8] for generalized pseudo monotone mappings, of Hess [14], Calvert and Webb [10] and Fitzpatrick [13] for quasimonotone mappings, and of Wille [34] and Browder [5] for maximal monotone and bounded generalized pseudo monotone mappings that satisfy (2.8) and $(Tx,x) \geq -\|x\| \geq R$, respectively. Corollary 2.2 follows from Theorem 2.11 since T is pseudo monotone, demiclosed and quasibounded by a result in [29]. It extends the earlier results of Browder [4] and Dubinskii [12]. Both of these results are valid for mappings between X and Y under suitable restrictions on the spaces.

Finally we shall look at intertwined perturbations of mappings of semibounded variation. Let X and X_o be separable reflexive Banach spaces, X continuously and densily embedded in X and let the injection $I: X \to X_0$ be compact. Denote by $\|\cdot\|_0$ the norm of X_0.

DEFINITION 2.4 ([33]). $T: X \to X^*$ is called a <u>Gårding mapping</u> if there exists a mapping $U: X \times X \to X^*$ such that $Tx = U(x,x)$ for $x \in X$ and

(i) For each $y \in X$, $U(y,\cdot): X \to X^*$ is completely continuous;

(ii) For each $x \in X$, $U(\cdot,x): X \to X^*$ is hemicontinuous;

(iii) For each $R > 0$ and $\|x\| \le R$, $\|y\| \le R$,

$$(U(x,x)-U(y,x),x-y) \ge -c(R,\|x-y\|_0)$$

where $c: R^+ \times R^+ \to R^+$ is continuous and $\lim_{t\to 0^+} c(r,tR)/t = 0$ for each $r > 0$, $R > 0$.

Since such mappings are pseudo monotone (see [33]) and satisfy condition (*), Theorem 2.2 gives the following extension of the result Oden [33].

COROLLARY 2.3 Let T be a Gårding mappings and (2.5) hold. Then, if T is quasibounded and either one of conditions (2.6) and (2.9) holds, $T(X) = X^*$.

We shall now apply Corollary 2.3 to finding a generalized solution $u \in V$ of

$$(2.22) \qquad \sum_{|\alpha|\le m} (-1)^{|\alpha|} D^\alpha(x,u,Du,\ldots,D^m u) = f, \quad f \in L_q(Q),$$

where $Q \subset R^n$ is a bounded domain with the smooth boundary and V is closed subspace of $W_p^m(Q)$ with $\mathring{W}_p^m \subseteq V$ and $p \in (-1,\infty)$. Assume

(2.23) For each $|\alpha| \le m$, $A_\alpha: Q \times R^{s_m} \to R$ satisfies the Carathéodory conditions and there exist $K > 0$ and $k(x) \in L_q(Q)$ such that

$$|A_\alpha(x,y)| \le K(|y|^{p-1} + k(x)) \quad \text{for} \quad x \in Q \text{ a.e.}, \quad y \in R^{s_m}.$$

As before, a generalized form associated with (2.22) induces a bounded and continuous mappings $T: V \to V^*$. Let $w_f \in V^*$ be such that $(w_f, v) = \int_Q fv\,dx$ for each $v \in V$. Then finding generalized solutions of (2.22) is equivalent to solving the operator equation

$$(2.24) \qquad Tu = w_f, \quad u \in V.$$

Let $D^\gamma u = \{(D^\alpha u) \mid |\alpha| \le m-1\}$. We need require the following conditions.

(2.25) Let $c_1: R^+ \to R^+$ be continuous and $r \to 0$ if $c_1(r) \to 0$, $c: R^+ x R^+ \to R^+$ be such that $c(R,\cdot)$ is weakly upper-semicontinuous at 0 and $c(R,0) = 0$ for each R and for each $v \in V$, $R > 0$ and $\|u\|_{m,p} \le R$, $\|w\|_{m,p} \le R$ we have for some $k \le m-1$

$$\sum_{|\alpha|=m} \int_Q [A_\alpha(x,D^\gamma v,D^m u) - A_\alpha(x,D^\gamma v,D^m w)]D^\alpha(u-w) \ge$$

$$\ge c_1(\|u-w\|_{m,p}) - c(R,\|u-w\|_{k,p}).$$

(2.26) Let $c: R^+ x R^+ \to R^+$ be as in Definition 2.4 and suppose that for each $v \in V$, $R > 0$ and $\|u\|_{m,p} \le R$, $\|w\|_{m,p} \le R$ the integral inequality in (2.25) holds with $c_1 \equiv 0$.

Some algebraic conditions that imply a stronger version of (2.25) can be found in [12]. The following algebraic conditions imply (2.25) and (2.26), respectively.

(2.27) There are constant $c_1 > 0$, $c \ge 0$ such that for each $y \in R^{s_{m-1}}$, each $R > 0$ and $(y^i,z^i) \in R^{s_{m-1}} x R^{s_m - s_{m-1}}$ with $|y^i| + |z^i| \le R$, $i = 1,2$, and $x \in Q$ a.e. we have

$$\sum_{|\alpha|=m} [A_\alpha(x,y,z^1) - A_\alpha(x,y,z^2)](z_\alpha^1-z_\alpha^2) \geq$$

$$\geq c_1 \sum_{|\alpha|=m} |z_\alpha^1-z_\alpha^2|^p - c \sum_{|\alpha|\leq m-1} |y_\alpha^1-y_\alpha^2|^p .$$

(2.28) For each $y \in R^{s_{m-1}}$, $z^1, z^2 \in R^{s_m - s_{m-1}}$ and $x \in Q$ a.e. we have

$$\sum_{|\alpha|=m} [A_\alpha(x,y,z^1) - A_\alpha(x,y,z^2)](z_\alpha^1-z_\alpha^2) \geq 0.$$

THEOREM 2.11 Let (2.23) hold and T satisfy condition (2.5). Suppose that either one of the following conditions holds:

(2.29) T is odd on $V\backslash B(0,r)$ for some $r > 0$, i.e. $A_\alpha(x,-\xi) = -A_\alpha(x,\xi)$ for $x \in Q$ a.e., and all $|\xi|$ large and $|\alpha| \leq m$.

(2.30) For each $h \in V^*$ there exists an $r_h > 0$ such that

$$Tu \neq \lambda Ju \quad \text{for} \quad u \in \partial B(0,r_h), \quad \lambda < 0,$$

where $J: V \to V^*$ is the normalized duality mapping. Then

(a) If (2.25) holds, the generalized boundary value problem (2.22) is feebly approximation-solvable in V for each $f \in L_q$.

(b) If (2.26) holds, (2.22) has a generalized solution in V for each $f \in L_q$.

PROOF. Define the mappings $U: V\times V \to V$ and $C: V \to V^*$ by

$$(U(u,v),w) = \sum_{|\alpha|=m} \int_Q A_\alpha(x,D^\gamma v,D^m u)D^\alpha w\, dx, \quad \text{for } u,v,w \in V,$$

$$(Cu,w) = \sum_{|\alpha|<m} \int_Q A_\alpha(x,D^\gamma u,D^m u)D^\alpha w\, dx, \quad \text{for } u,w \in V.$$

Then C is known to be completely continuous and $Tu = U(u,u) + Cu$ is A-proper w.r.t. an injective scheme $\Gamma_I = \{X_n,V_n;X_n^*,V_n^*\}$ for (V,V^*)

if (2.25) holds by Example 1.4.5 in [28] (i.e., Example 1.4 in [27]) since T is of type (S_+). If (2.26) holds, then $U_1(u,v) = U(u,v) + Cv$ is a Gårding mapping and therefore T is strongly A-closed w.r.t. Γ_1 and satisfies condition (*). Consequently, the conclusions follow from Theorem 2.1 and Remark 2.1 and Theorem 2.2, respectively when (2.30) holds, and from the corresponding results in [31,30] when T is odd. □

Theorem 2.11 extends the earlier results of Pohozaev [31], Browder [4], Dubinskii [12], involving mappings of type (S_+) and of semi-bounded variation. Bondary value problems satisfying (2.27) with $c = 0$ or (2.28) have been also studied earlier by the author in [21,26,28] (cf. also [29]).

3. SOLVABILITY OF SEMILINEAR EQUATIONS AT RESONANCE

Throughout the section we shall always assume that H is a Hilbert space which contains the Banach space X as a vector sub-space. We shall study the (approximation-) solvability of equations of the form

(3.1) $$Ax + Nx = f, \qquad (x \in D(A), \quad f \in H)$$

where $A\colon D(A) \equiv V \subset X \to H$ is a linear densily defined closed mapping such that its null space $X_o = N(A)$ has finite dimension, the range $R(A)$ is closed with $R(A) = X_o^{\perp}$ and $N\colon V \to H$ is a nonlinear mapping of certain type. Here $X_o^{\perp}$ denotes the orthogonal complement in H of $X_o \subset H$. Let $A_1 = A$ restricted to $V \cap R(A)$ and since $A_1^{-1}\colon R(A) \to V$ is closed, it is continuous by the closed graph theorem. Let $c > 0$ be such that $\|A_1^{-1}x\| \le \frac{1}{c}\|x\|$ for $x \in R(A)$. Then $(Ax,x) \ge -\|Ax\|\,\|x\| \ge -c^{-1}\|Ax\|^2$ for $x \in V$. Let α be the supremum of all such c. Then $\alpha \in [0,\infty]$ and

$$(3.2) \qquad (Ax,x) \geq -\frac{1}{\alpha}\|Ax\|^2 \quad \text{for} \quad x \in V.$$

Equation (3.1) with $X = H$ and A as above and such that $A_1^{-1}: R(A) \to R(A)$ is compact has been studied by Brézis and Nirenberg [3], Berestycki and de Figueiredo [1] and many others under various conditions on the perturbation N (cf. [1,3] for the bibliography on these problems). Our research has feen motivated by [1] and deals with Eq. (3.1) such that $A + N$ is A-proper or strongly A-closed. Using the degree theory for multivalued mappings instead of the Brouwer's degree, we see that the results of this section are also valid for multivalued nonlinearities of the same type.

We begin with the following result proven in [1] with $X = H$ and $A_1^{-1}: R(A) \to R(A)$ compact. However, analysing its proof one sees that the compactness of A_1^{-1} is not needed and we give its proof for the sake of completeness.

LEMMA 3.1 Let $A: V \subset X \to H$ be such that $\alpha \in (0,\infty)$ and $A_1^{-1} + \alpha^{-1}I: R(A) \to R(A)$ be strongly monotone (i.e., $(A_1^{-1}x + \alpha^{-1}x,x) \geq c\|(A_1^{-1} + \alpha^{-1}I)x\|^2$ for each $x \in R(A)$ and some constant $c > 0$). If equality holds in (3.2), i.e.

$$(3.3) \qquad (Ax,x) = -\frac{1}{\alpha}\|Ax\|^2 \quad \text{for some} \quad x \in V,$$

then $x \in N(A) \oplus N(A+\alpha I)$ in H.

PROOF. Let $x \in V$ be such that (3.3) holds. Then $x = x_o + x_1$ uniquely, where $x_o \in N(A)$ and $x_1 \in R(A)$. If $x_1 = 0$, the conclusion follows. Suppose that $x_1 \neq 0$ and then (3.3) becomes $(Ax_1,x_1) = -\alpha^{-1}\|Ax_1\|^2$, or, setting $u = Ax_1$, we get $(A_1^{-1}u + \alpha^{-1}u,u) = 0$. The strong monotonicity of $A_1^{-1} + \alpha^{-1}I$ implies that $A_1^{-1}u + \alpha^{-1}u = 0$, or $Ax_1 + \alpha x_1 = 0$. □

REMARK 3.1 Lemma 3.1 includes the case when $X = H$ and A is selfadjoint. Assuming additionaly in Lemma 3.1 that $A_1^{-1}: R(A) \to R(A)$ is compact then as in [1] one obtains that $-\alpha$ is an eigenvalue of A.

Introduce in V a new norm $\|\cdot\|_o$. Then throughout the section we shall assume that the nonlinear mapping $N: V \to H$ is quasibounded, i.e.

$$(3.4) \qquad |N| = \limsup_{\|x\|_o \to \infty} \frac{\|Nx\|}{\|x\|_o} < \infty .$$

We say that A has Property I if it has the properties discussed at the begining of the section and if $A_1^{-1}: R(A) \to R(A)$ is compact. Let Q be the orthogonal projection of H onto $N(A)$ and $\Gamma_a = \{X_n, V_n; Y_n, Q_n\}$ an admissible scheme for (V,H) with V equipped with the norm $\|\cdot\|_o$. We are ready now to prove various surjectivity results for Eq. (3.1). Our first result is

THEOREM 3.1 Let $A: V \subset X \to H$ have Property I, $0 < \alpha < \infty$; $A_1^{-1} + \alpha I$ strongly monotone, and $C,N: V \to H$ quasibounded and such that

(3.5) Whenever $\|x_n\|_o \to \infty$, $u_n = \frac{x_n}{\|x_n\|_o} \to u$ in H and $\frac{Nx_n}{\|x_n\|_o} \rightharpoonup v$, $\frac{Cx_n}{\|x_n\|_o} \rightharpoonup w$, then (i) $\|v\|^2 \le \alpha(u,v)$, (ii) $\|w\|^2 \le \alpha(u,w)$, (iii) $(2w,v) \le \alpha(u,v)$ and one of the inequalities in (i) and (ii) is strict.

(3.6) Whenever $\|x_n\|_o \to \infty$, $u_n = \frac{x_n}{\|x_n\|_o} \to u_o + u_1$ in H with $u_o \in N(A)$, $u_1 \in N(A+\alpha I)$ and $Nx_n/\|x_n\|_o \rightharpoonup v$, then $v \ne \alpha u_1$; in particular, this is so when $(v,u_1) < \alpha\|u_1\|^2$ if $u_1 \ne 0$ and $(v,u_o) > 0$ if $u_1 = 0$.

Suppose that A and $H(t,x) = Ax + (1-t)Nx + tCx$ are A-proper and A-closed w.r.t. Γ_a for (V,H) and for each large R,

$\deg(Q_n H_1, B(0,R) \cap X_n, 0) \neq 0$ for each large n. Then Eq. (3.1) is feebly approximation-solvable for each $f \in H$.

PROOF. Let $f \in H$ be fixed. Then there exists an $R = R(f)$ such that $\|x\|_o < R$ whenever $H(t,x) = (1-t)f$ for some $x \in V$ and $t \in [0,1]$. If not, then there would exist $\{x_n\} \subset V$ and $t_n \in [0,1]$ such that $t_n \to t_o$, $\|x_n\|_o \to \infty$ and

$$H(t_n, x_n) = (1-t_n)f \quad \text{for each} \quad n. \tag{3.7}$$

Set $u_n = x_n/\|x_n\|_o$. Since $u_n = u_{on} + u_{1n}$ uniquely with $u_{on} \in N(A)$ and $u_{1n} \in R(A)$, dividing (3.7) by $\|x_n\|_o$ we obtain

$$Au_{1n} + (1-t_n)\frac{Nx_n}{\|x_n\|_o} + t_n\frac{Cx_n}{\|x_n\|_o} - (1-t_n)\frac{f}{\|x_n\|} = 0,$$

or

$$u_{1n} + (1-t_n)A_1^{-1}(I-Q)\frac{Nx_n}{\|x_n\|_o} + t_nA_1^{-1}(I-Q)\frac{Cx_n}{\|x_n\|_o} + \tag{3.8}$$

$$+ (1-t_n)A_1^{-1}(I-Q)\frac{f}{\|x_n\|} = 0$$

$$(1-t_n)Q\frac{Nx_n}{\|x_n\|_o} + t_nQ\frac{Cx_n}{\|x_n\|_o} \quad (1-t_n)Q\frac{f}{\|x_n\|} = 0. \tag{3.9}$$

Since $\{Nx_n/\|x_n\|_o\}$ and $\{Cx_n/\|x_n\|_o\}$ are bounded in H by the quasiboundedness of N and C, we may assume that they converge weakly to v and w, respectively, and by the complete continuity of A_1^{-1} we obtain from (3.8) passing to the limit that $u_{1n} \to u_1$ in H, $u_1 \in R(A)$, and

$$u_1 + (1-t_o)A_1^{-1}(I-Q)v + t_oA_1^{-1}(I-Q)w = 0,$$

or

$$Au_1 + (1-t_o)(I-Q)v + t_o(I-Q)w = 0. \tag{3.10}$$

Since $\dim N(A) < \infty$, we may assume that $u_{on} \to u_o$ and so $u_n \to u = u_o + u_1$ in H. Passing to the limit in (3.9) we obtain $(1-t_o)Qv + t_oQw = 0$, and by (3.10) it follows that

$$(3.11) \qquad Au + (1-t_o)v + t_o w = 0.$$

Using (3.5)-(iii), it follows from (3.2) and (3.11) that
$0 \leq (Au, Au + \alpha u) \leq (1-t_o)^2 [\|v\|^2 - \alpha(u,v)] + t_o^2\|w\|^2 - \alpha t_o(w,u)$.
If $t_o \neq 0$, this leads to a contradiction in view of (3.5)-(i) and (ii). If $t_o = 0$, we get that $(Au, Au + \alpha u) = 0$, and therefore $u \in N(A) \oplus N(A+\alpha I)$ by Lemma 3.1. Since $u = u_o + u_1$ and $N(A+\alpha I) \subset R(A)$, $u_1 \in N(A+\alpha I)$ and $Au = Au_1 = -\alpha u_1 = -v$ by (3.11). Hence, $v = \alpha u_1$ in contraction to (3.6). Therefore, our claim is valid and let $R > 0$ be a such one.

Now, since N and C are bounded on $\partial B(0,R)$ and A and $H(t,x)$ are A-proper and A-closed for each t, it is easy to see that there exist an $n_o \geq 1$ and a $\gamma > 0$ such that for each $n \geq n_o$ $\|Q_nH(t,x) - (1-t)Q_nf\| \geq \gamma$ for $x \in \partial B(0,R) \cap X_n$, $t \in [0,1]$. Consequently, by the homotopy theorem for the Brouwer's degree we obtain that for each $n \geq n_o$

$$\deg(Q_nA+Q_nN-Q_nf, B(0,R) \cap X_n, 0) = \deg(Q_nH_1, B(0,R) \cap X_n, 0) \neq 0 .$$

Hence, there exists a $x_n \in B(0,R) \cap X_n$ for each $n \geq n_o$ such that $Q_nAx_n + Q_nNx_n = Q_nf$, and therefore, some subsequence $x_{n_k} \to x$ and $Ax + Nx = f$. □

The following special case is suitable in many applications.

COROLLARY 3.1 Let $(V, \|\cdot\|)_o$ be compactly embedded in H, $0 < \alpha < \infty$, $A_1^{-1} + \alpha I$ strongly monotone and $N: V \to H$ quasibounded and satisfy (3.6) and

(3.12) Whenever $\|x_n\|_o \to \infty$, $u_n = \frac{x_n}{\|x_n\|_o} \to u$ in H and $\frac{Nx_n}{\|x_n\|_o} \rightharpoonup v$, then $\|v\|^2 \leq \alpha(u,v)$.

Suppose that $A + tN$ is A-proper and A-closed w.r.t. Γ_a for (V,H) for each $t \in [0,1]$. Then Eq. (3.1) is feebly approximation-solvable for each f in H.

PROOF. We shall first show that $A_1^{-1}: R(A) \to R(A)$ is compact. Since $A: V \subset X \to H$ is closed and $\|x\| \leq c\|x\|_o$ for some c, $A: (V,\|\cdot\|_o) \to H$ is also closed and therefore $A_1^{-1}: R(A) \cap H \to$ $\to (V \cap R(A), \|\cdot\|_o)$ is continuous. Since V is compactly embedded in H, it follows that $A_1^{-1}: R(A) \to R(A)$ is compact. Set $C = \frac{\alpha}{2} I$. Then, in view of (3.12), it is easy to see that (3.5) holds with the strict inequality in (ii). Since C is compact, we see that $A + (1-t)N + tC$ is A-proper and A-closed w.r.t. Γ_a for each $t \in [0,1]$ and the conclusion of the corollary follows from Theorem 3.1. □

When a Landesman-Lazer type condition holds instead of (3.6), we have

THEOREM 3.2 Suppose that all conditions of Theorem 3.1 hold with $w \neq 0$ and (3.6) replaced by

(3.13) Whenever $\|x_n\|_o \to \infty$, $u_n = \frac{x_n}{\|x_n\|_o} \to u_o + u_1$ in H with $u_o \in N(A)$, $u_1 \in N(A+\alpha I)$ and $Nx_n/\|x_n\|_o \rightharpoonup v$, then $(v,u_1) < \alpha\|u_1\|^2$ if $u_1 \neq 0$, and, $\liminf\,(Nx_n,u_o) > (f,u_o)$ for some f in H, if $u_1 = 0$.

Then the equation $Ax + Nx = f$ is feebly approximation-solvable.

PROOF. As in Theorem 3.1 it suffices to show that there exists an

$R > 0$ such that $\|x\|_o < R$ whenever $H(t,x) \equiv Ax + (1-t)Nx + tCx = (1-t)f$ for some $t \in [0,1]$. Again, arguing by contradiction, suppose that there exist $t_n \to t_o$ and $x_n \in V$ with $\|x_n\|_o \to \infty$ such that $H(t_n,x_n) = (1-t_n)f$ for each n. Set $u_n = x_n/\|x_n\|_o$ and write $u_n = u_{on} + u_{1n}$ uniquely with $u_{on} \in N(A)$ and $u_{1n} \in R(A)$. Then, as in Theorem 3.1 we see that $u_{on} \to u_o$, $u_{1n} \to 0$ in H and $t_o = 0$. Taking the inner product of the equation $H(t_n,x_n) = (1-t_n)f$ with u_o, we obtain for each

$$(1-t_n)(Nx_n,u_o) + t_n(Cx_n,u_o) = (1-t_n)(f,u_o),$$

since $(Ax_n,u_o) = 0$. Therefore

$$\lim[(1-t_n)(Nx_n,u_o) + t_n(Cx_n,u_o)] = (f,u_o)$$

and since by (3.5)(ii), $(Cx_n/\|x_n\|_o,u_o) \to (w,u_o) \geq \alpha^{-1}\|w\|^2 > 0$ and $t_n\|x_n\|_o > 0$, we see that

$$\lim\inf(Nx_n,u_o) \leq (f,u_o)$$

in contradiction to (3.13). Therefore, such an $R > 0$ exists. □

As in the case of Corollary 3.1, we obtain

COROLLARY 3.2 Suppose that $(V,\|\cdot\|_o)$ is compactly embedded in H, $\alpha \in (0,\infty)$, $A_1^{-1} + \alpha I$ strongly monotone and $N: V \to H$ quasibounded and satisfies (3.12)-(3.13). Suppose that $A + tN$, $t \in [0,1]$, is A-proper and A-closed w.r.t. Γ_a for (V,H). Then the equation $Ax + Nx = f$ is feebly aporximation-solvable.

In case when $H_o = A + N$ is not A-proper but just strongly A-closed instead, we have the following extensions of the above results.

THEOREM 3.3 Let A and $H(t,x) = Ax + (1-t)Nx + tCx$ be A-proper and A-closed w.r.t. Γ_a for (V,H) for each $t \in (0,1]$ and $A + N$

be strongly A-closed w.r.t. Γ_a and satisfy condition (*). Suppose also that N and C are bounded. Then

(a) If all other conditions of Theorem 3.1 hold, $R(A+N) = H$.

(b) If all other conditions of Theorem 3.2 hold, the equation $Ax + Nx = f$ is solvable.

PROOF. Let $f \in H$ be given. Then, as in the proofs of Theorems 3.1 and 3.2, there exists an $R = R(f) > 0$ such that

$$H(t,x) \neq (1-t)f \quad \text{for} \quad \|x\| = R, \qquad t \in [0,1].$$

Let $\varepsilon \in (0,1)$ be fixed. Since $H(t,\cdot)$ is A-proper and A-closed on $\partial B(0,R)$ for each $t \in (0,1]$, there exists an $n(\varepsilon) \geq 1$ such that for each $n \geq n(\varepsilon)$

$$Q_n H(t,x) \neq (1-t)Q_n f \quad \text{for} \quad x \in \partial(0,R) \cap X_n, \qquad t \in [\varepsilon,1]$$

and $n(\varepsilon_1) \geq n(\varepsilon_2)$ whenever $\varepsilon_1 < \varepsilon_2$. Therefore, there exists a $x_\varepsilon \in B(0,R)$ such that $Ax_\varepsilon + (1-\varepsilon)Nx_\varepsilon + \varepsilon Cx_\varepsilon = (1-\varepsilon)f$. Let $\varepsilon_k \in (0,1)$ be such that $\varepsilon_k \to 0$ decreasingly and $x_k \in B(0,R)$ such that $H(\varepsilon_k,x_k) = (1-\varepsilon_k)f$. Then $Ax_k + Nx_k = (1-\varepsilon_k)f + \varepsilon_k(Nx_k - Cx_k) \to f$ by the boundedness of N and C. Finally, by condition (*) there exists a $x \in \bar{B}(0,R)$ such that $Ax + Nx = f$. □

COROLLARY 3.3 Let $(V,\|\cdot\|_o)$ be compactly embedded in H, $0 < \alpha < \infty$, $A_1^{-1} + \alpha I$ strongly monotone, $N: V \to H$ bounded and quasibounded and $A + tN$ A-proper and A-closed w.r.t. Γ_a for (V,H) for each $t \in [0,1)$. Suppose that A + N is strongly A-closed w.r.t. Γ_a and satisfies condition (*). Then

(a) If (3.6) and (3.12) hold, $R(A+N) = H$.

(b) If (3.12) and (3.13) hold, the equation $Ax + Nx = f$ is solvable.

In our last result we show that the A-properness requirement of

$H(t,x)$ can be relaxed if a scheme $\Gamma = \Gamma_o = \{X_n, P_n; Y_n, Q_n\}$ is projectionally complete (i.e. $P_n x \to x$ and $Q_n y \to y$ for each $x \in V$ and $y \in H$), $Q_n Ax = Ax$ for each $x \in X_n$, $n \geq 1$, and $Q_n^* y \to y$ for $y \in H$. Such schemes have been extensively used in our works [24,26-28] and when X has a nice approximation property such a scheme can be always constructed using the properties of A. For example, suppose that X is such that

(i) There are finite dimensional subspaces X_n of X such that $X_o \equiv N(A) \subset X_1 \subset X_2 \subset \ldots$ with the inclusions being proper.

(ii) There are projections P_n of X onto X_n such that $P_n x \to x$ for each $x \in X$ and $AP_n x \to Ax$ for each $x \in D(A)$.

(iii) $P_o P_n = P_n P_o$ and $P_n(D(A)) \subset D(A)$ for $n = 0,1,2,\ldots$.

Define a linear mapping $Q_n: H \to H$, $n \geq 1$, as follows: since each $y \in H$ has the unique representation $y = y_o + y_1$ with $y_o \in N(A)$, $y_1 \in R(A)$, it suffices to define $Q_n y_o = y_o$ and $Q_n y_1 = AP_n A_1^{-1} y_1$. Set $Y_n = R(Q_n)$. Then we have that $\Gamma_o = \{X_n, P_n; Y_n, Q_n\}$ so constructed is projectionally complete and $Q_n Ax = Ax$ for $x \in X_n$. Now, without requiring the A-properness of $H(t,x)$ for each t, we have with $X = (V, \|\cdot\|_o)$,

THEOREM 3.4 (a) Let $A+N: V \to H$ be A-proper and A-closed w.r.t. Γ_o. Then Eq. (3.1) is feebly approximation-solvable for each $f \in H$ if all other conditions of Theorem 3.1 hold, and just for a given f for which (3.13) and all other conditions of Theorem 3.2 hold.

(b) If $A+N: V \to H$ is strongly A-closed w.r.t. Γ_o, then the conclusions of (a) hold existencially.

The proof of Theorem 3.4 consists in showing, for a given $f \in H$, that there exist an $R > 0$ and $n_o \geq 1$ such that $\|x_n\|_o < R$ whenever $Q_n H(t,x_n) \equiv Q_n Ax_n + (1-t)Q_n Nx_n + tQ_n Cx_n = (1-t)Q_n f$ for some $t \in [0,1]$, and $x_n \in X_n$ with $n \geq n_o$. To that end one argues by contradiction using the properties of Γ_o and arguments similar to those in the previous results. We omit the details.

Let us now look at some applications of the abstract results to the solvability of

$$(3.15) \qquad Au + F(x,u,Du,\ldots,D^{2m}u) = f, \qquad f \in L_2(Q)$$

where $Q \subset R^n$ is a bounded domain, $A: D(A) \subset L_2(Q) \to L_2(Q)$ a linear mapping and $F: Q\times R^{s_{2m}} \to R$ such that

(3.16) F satisfies the Carathéodory conditions and $|F(x,y)| \leq \leq \mu|y| + \Psi(x)$ for $x \in Q$(a.e.), $y \in R^{s_{2m}}$ and some $\mu \in R$, $\Psi \in L_2(Q)$;

(3.17) $\ell_{\pm}(x) = \liminf\limits_{y\to\pm\infty} \dfrac{F(x,y,z)}{y}$, $k_{\pm}(x) = \limsup\limits_{y\to\pm\infty} \dfrac{F(x,y,z)}{y}$ exist for $x \in Q$(a.e.), uniformly in $z \in {}^{s_{2m-1}}$ and satisfy

$$(3.18) \qquad 0 \leq \ell_+(x) \leq k_+(x) \leq \alpha, \qquad 0 \leq \ell_-(x) \leq k_-(x) \leq \alpha.$$

Let $V = D(A)$ with $\|\cdot\|_{2m}$, $V \subset W_2^{2m}(Q)$ and $u^{\pm}(x) = \max(\pm u(x),0)$. Our first result for (3.15) is a consequence of Corollaries 3.1 and 3.3.

THEOREM 3.5 Let A have Property I, $0 < \alpha < \infty$, $\dim N(A) < \infty$, $A_1^{-1} + \alpha I: R(A) \to R(A) \subset L_2$ be strongly monotone and (3.16)-(3.18) hold. Suppose that $A+tN: V \to L_2$ is A-proper w.r.t. Γ_o for (V,L_2) for each $t \in [0,1)$ and

$$(3.19) \qquad \ell_+(x)u^+(x) + \ell_-(x)u^-(x) \not\equiv 0 \quad \text{for} \quad 0 \neq u \in N(A),$$

$$(3.20) \qquad [\alpha-k_+(x)]u^+(x) + [\alpha-k_-(x)]u^-(x) \not\equiv 0 \quad \text{for} \quad 0 \neq u \in N(A+\alpha I).$$

Then Eq. (3.15) is feebly approximation-solvable for each $f \in L_2$ if also $A + N$ is A-proper; it is just solvable for each $f \in L_2$

if also $A + N$ satisfies condition (*) and is strongly A-closed.

If zero is an eigenvalue of A and if for any $0 \neq u \in N(A)$ either $u > 0$ or $u < 0$ a.e. in Q, then (3.19) holds if and only if ℓ_+ and ℓ_- are $\neq 0$ on subsets of Q with positive measure. A second order self-adjoint elliptic operator with 0 as its eigenvalue is such an example. We now consider the case when this does not happen, i.e. we assume ([1])

(3.21) $\quad \ell_+(x) \equiv 0; \quad 0 \leq \ell_-(x), \quad k_+(x), k_-(x) \leq \alpha; \quad \ell_-(x) \not\equiv 0.$

In (3.21) one could have ℓ_-, k_- or k_+ equal to α on subsets of Q with positive measure. Hence, problem (3.15) exibits double resonance in this case. An analogous problem could be studied with the roles of ℓ_+ and ℓ_- interchanged. Double resonance boundary value problems have been earlier studied by Dancer [11] and Berestycki and Figueiredo [1]. As an application of Corollary 3.2, we obtain

THEOREM 3.6 Let A have Property I, $\alpha \in (0,\infty)$, $\dim N(A) < \infty$, $A_1^{-1} + \alpha I: R(A) \to R(A) \subset L_2$ be strongly monotone and (3.16), (3.17), (3.20) and (3.21) hold. Suppose that any $0 \neq u \in N(A)$ does not change sign in Q and:

(3.22) There exists a $c(x) \in L_2(Q)$ such that $F(x,y,z) \geq c(x)$ for $x \in Q$(a.e.), $y \geq 0$ and $z \in R^{s_{2m-1}}$.

(3.23) $F_+(x) = \liminf_{y \to +\infty} F(x,y,z)$ exists for $x \in Q$(a.e.) uniformly in $z \in R^{s_{2m-1}}$ and

$$\int_Q fu < \int_Q F_+ u \quad \text{for} \quad 0 \neq u \in N(A), \quad u > 0.$$

(3.24) Whenever $\|x_n\|_0 \to \infty$, $u_n = x_n/\|x_n\|_0 \to u \in N(A)$ in H with $u > 0$, then there exists an n_0 such that $u_n(x) > 0$ a.e. in Q for each $n \geq n_0$.

Then, if $A + tN: V \to L_2$ is A-proper w.r.t. Γ_o for (V,L_2) for $t \in [0,1]$ ($t \in [0,1)$ and $A+N$ satisfies (*) and is strongly A-closed), Eq. (3.15) is feebly approximation-solvable (solvable, resp.).

Let us now discuss some examples of A and F in Eq. (3.15) to which the above results apply. Since $A: V = D(A) \to L_2$ is Fredholm of index zero, we have shown in [24] that $A+tN:(V,\|\cdot\|) \to L_2$ is A-proper and A-closed w.r.t. various schemes for $t \in [0,1]$ provided $N: V \to L_2$ is k-ball-contractive with k sufficiently small. A typical example of A is a strongly uniformly elliptic operator $\Sigma_{|\alpha|,|\beta|\le m} (-1)^{|\alpha|} D^\alpha(a_{\alpha\beta}(x)D^\beta u)$ with $D(A) = W_2^{2m} \cap \mathring{W}_2^m$. The following condition implies $N: D(A) \to L_2$ is k-ball-contractive:

(3.25) $|F(x,y,z)-F(x,y,z')| \le \beta \sum_{|\alpha|=2m} |z-z'_\alpha|$ for $x \in Q$(a.e.), $y \in R^{s_{2m-1}}$ $z,z' \in R^{s'_{2m}}$, $s'_{2m} = s_{2m}-s_{2m-1}$, and β sufficiently small.

More generally, let $A = \sum_{|\alpha|\le 2m} a_\alpha(x)D^\alpha$ be strongly uniformly elliptic with real coefficients $a_\alpha \in C^{o,\lambda}(\bar{Q})$ for $|\alpha| \le 2m$, $0 < \lambda < 1$, $\partial Q \in C^\infty$, $V = D(A) = W_2^{2m}(Q) \cap \mathring{W}_2^m(Q)$ with $\|\cdot\|_{2m}$-norm and $H = L_2(Q)$. By a result of Skrypnik [32] there exist a linear strongly uniformly elliptic operator $K = \sum_{|\alpha|\le 2m} b_\alpha(x)D^\alpha$ with C^∞-real coefficients and constants $c > 0$, $c_o > 0$ such that $(Au,Ku)_o \ge c\|u\|_{2m}^2 - c_o\|u\|_o^2$ and $((-1)^m Ku,u)_o \ge c_o\|u\|_m^2$ for each $u \in V$. Then $A: V \to L_2$ is Fredholm of index zero and, if (3.25) holds with $\beta < c$, $A+tN: V \to L_2$ is A-proper w.r.t. Γ_o for (V,L_2) for $t \in [0,1]$. Now, (3.25) can be relaxed to

(3.26) There exists a function $p: R^+\backslash\{0\} \to [0,c)$ such that

$$\sum_{|\alpha|=2m} b_\alpha(x) \big(F(x,y,z)-F(x,y,z')\big)(z_\alpha - z'_\alpha) \ge -p(r) \sum_{|\alpha|=2m} |z_\alpha - z'_\alpha|^2$$

for $x \in Q$(a.e.), $y \in R^{s_{2m-1}}$, $r > 0$, and $z,z' \in R^{s'_{2m}}$ with $|z-z'|\ge r$.

Then $A+tN: V \to L_2$ is A-proper w.r.t. Γ_o for $t \in [0,1]$ by a variant of Proposition 2.2 and, if p takes on also the value c, then $A + N$ is just strongly A-closed. Other conditions on N that imply the A-properness of $A + tN$ or the strong A-closedness of $A + N$ can be found in [24,27,28]. In view of this discussion, we see that Theorems 3.5 and 3.6 extend some results of Dancer [11] and Berestycki and Figueiredo [1] involving nonlinearities that depend only on x and u, and, in contrast to theirs, our results are also constructive when $A+N$ is A-proper.

REFERENCES

1. H. BERESTYCKI and D.G. de FIGUEIREDO, Double resonance in semilinear elliptic problems, Comm. Part. Diff. Eq. 6 (1) (1981), 91-120.

2. H. BRÉZIS, Equations et inequations nonlinear dans les espaces vectorielles en dualité, Ann. Inst. Fourier (Grenoble) 18 (1968), 115-175.

3. H. BRÉZIS and L. NIRENBERG, Characterization of the ranges of some nonlinear operators and applications to boundary value problems, Ann. Sc. Norm. Sup. Pisa, Serie IV, 5 (1978), 225-326.

4. F.E. BROWDER, Nonlinear elliptic boundary value problems, Bull. Amer. Math. Soc. 69 (1963), 862-874.

5. F.E. BROWDER, Existence theory for boundary value problems for quasilinear elliptic systems with strongly nonlinear lower order terms, Proc. Symp. Pure Math., AMS v. 23 (1973), 269-286.

6. F.E. BROWDER, Nonlinear operators and nonlinear equations of evaluation in Banach spaces, Proc. Symp. Pure Math., AMS, vol. 18, Part 2, 1976.

7. F.E. BROWDER, Problems of present day mathematics, Proc. Symp. Pure Math., v. 28, 1976.

8. F.E. BROWDER and P. HESS, Nonlinear mappings of monotone type in Banach spaces, J. Funct. Anal. 11 (1972), 251-294.

9. B.D. CALVERT and C.P. GUPTA, Nonlinear elliptic boundary value problems in L^p-spaces and sums of ranges of accretive operators, Nonlinear Analysis, Theory, Methods, Applications, 2 (1), (1978), 1-26.

10. B.D. CALVERT and J.R.L. WEBB, An existence theorem for quasimonotone operators, Rend. Acad. Naz. Lincei 8 (1971), 362-368.

11. E.N. DANCER, On the Dirichlet problem for weakly nonlinear elliptic partial differential equations, Proc. Royal Soc. Edinburg, 76 (1977), 283-300.

12. Yu.A. DUBINSKII, Nonlinear elliptic and parabolic equations, Itogi Nauki: Sovremennye Problemy Mat., vol. 9, VINITI, Moscow, 1976, pg. 5-130; English transl. in J. Soviet Math. 12 (1979), nº 5.

13. P.M. FITZPATRICK, Surjectivity results for nonlinear mappings from a Banach space to its dual, Math. Ann. 204 (1973), 177-188.

14. P. HESS, On nonlinear mappings of monotone type homotopic to odd operators, J. Funct. Anal. 11 (1972), 138-167.

15. G. HETZER, A continuation theorem and the variational solvability of quasilinear elliptic boundary value problems at resonance, Nonlinear Analysis; TMA 4 (4) (1980), 773-780.

16. S. KANIEL, Quasi-compact non-linear operators in Banach spaces and applications, Arch. Rational Meach. Anal. 20 (1965), 259-278.

17. J.M. LASRY and R. ROBERT, Degree theorems de point fixe pour les fonctions multivoques; applications. Seminaire Goulaonic-Lions-Schwartz, 1974, École Polytechnique, Paris cedex 05.

18. T.W. MA, Topological degrees for set valued compact vector fields in locally convex spaces, Dissertationes Math., 92 (1972), 1-43.

19. P.S. MILOJEVIĆ, A generalization of Leray-Schauder theorem and surjectivity results for multivalued A-proper and pseudo A-proper mappings, Nonlinear Analysis, TMA, 1 (3) (1977), 263-276.

20. P.S. MILOJEVIĆ, Surjectivity results for A-proper, their uniform limits and pseudo A-proper maps with applications, Notices Amer. Math. Soc. January 1977, 77T-B27.

21. P.S. MILOJEVIĆ, On the solvability and continuation type results for nonlinear equations with applications, I, Proc. Third Intern. Symp. on Topology and its Applic., Belgrade, 1977.

22. P.S. MILOJEVIĆ, On the solvability and continuation type results for nonlinear equations with applications II, Canadian Math. Bull. 25(1) (1982), 98-109.

23. P.S. MILOJEVIĆ, Some generalizations of the first Fredholm theorem to multivalued A-proper mappings with applications to nonlinear elliptic equations, J. Math. Anal. Applic. 65 (2) (1978), 468-502.

24. P.S. MILOJEVIĆ, Approximation-solvability results for equations involving nonlinear perturbations of Fredholm mappings with applications to differential equations, Proc. Intern. Sem. Functional Analysis, Holomorphy and Approx. Theory, Rio de Janeiro, August 1979, Lecture Notes in Pure and Applied Math., Marcel Dekker, N.Y. (Ed. G. Zapata), Vol. 83.

25. P.S. MILOJEVIĆ, Fredholm alternatives and surjectivity results for multivalued A-proper and condensing mappings with applications to nonlinear integral and differential equations, Czechoslovak Math. J. 30 (105), (1980), 387-417.

26. P.S. MILOJEVIĆ, Continuation theorems and solvability of equations with nonlinear noncompact perturbations of Fredholm mappings, Atas 12º Seminário Brasileiro de Análise, ITA, São José dos Campos, São Paulo, October, 1980, 163-189.

27. P.S. MILOJEVIĆ, Continaution theory for A-proper and strongly A-closed mappings and their uniform limits and nonlinear perturbations of Fredholm mappings, Proc. Intern. Sem. Functional Analysis, Holomorphy and Approx. Theory, Rio de Janeiro, August 1980, North Holland Publ. Comp. (Ed. J.A. Barroso), Mathematics Studies, No 71, (1982), 299-372.

28. P.S. MILOJEVIĆ, Theory of A-proper and pseudo A-closed mappings, Habitation Memoir, Universidade Federal de Minas Gerais, Belo Horizonte, December 1980, 1-208.

29. P.S. MILOJEVIĆ and W.V. PETRYSHYN, Continuation and surjectivity theorems for uniforms limits of A-proper mappings with applications, J. Math. Anal. Appl. 62 (2) (1978), 368-400.

30. W.V. PETRYSHYN, On the approximation-solvability of equations involving A-proper and pseudo A-proper mappings, Bull. AMS, 81 (1975), 223-312.

31. S.I. POHOŽAEV, The solvability of nonlinear equations with odd operators, Funct. Anal. i Prilozenia 1 (1967), 66-73.

32. I.V. SKRYPNIK, On the coercivity inequalitites for pairs of linear elliptic operators, Soviet Math. Doklad, 19 (2) (1978), 324-327.

33. J.T. ODEN, Existence theorems for a class of problems in nonlinear elasticity, J. Math. Anal. Appl. 69 (1979), 51-83.

34. F. WILLE, On monotone operators with perturbations, Arch. Rational Mech. Anal. 46 (1972), 269-288.

Departamento de Matemática
Universidade Federal de Minas Gerais
Belo Horizonte, MG, Brasil

Functional Analysis, Holomorphy and Approximation Theory II, G.I. Zapata (ed.)

HOLOMORPHIC FUNCTIONS ON HOLOMORPHIC INDUCTIVE LIMITS AND ON THE STRONG DUALS OF STRICT INDUCTIVE LIMITS

Luiza Amália Moraes

INTRODUCTION

The concepts of holomorphically bornological (hbo), holomorphically barreled (hba), holomorphically infrabarreled (hib) and holomorphically Mackey (hM) spaces have been introduced by Barroso, Matos and Nachbin in [1]. In this note, we will survey results concerning this holomorphic classification of the spaces. More explicitly, we will be concerned with the following situations:

Let E be a inductive limit of locally convex spaces E_i $i \in I$.

1. Find sufficient conditions on E_i $i \in I$ and on E so that

   (a) E_i hbo for every $i \in I \Rightarrow E$ hbo

   (b) E_i hba for every $i \in I \Rightarrow E$ hba

   (c) E_i hib for every $i \in I \Rightarrow E$ hib

   (d) E_i hM for every $i \in I \Rightarrow E$ hM

2. Find sufficient conditions on E_i $i \in I$ and E so that E' is a hbo space.

PRELIMINARIES

Let us make a brief review of what will be needed here. Unless stated otherwise, E and F denote complex locally convex spaces, U is a non void open subset of E and F^U is the set of all mappings of U into F. If I is a set and F is a seminormed space, we denote by $l^\infty(I;F)$ the seminormed space of all bounded mappings of I into F. $\mathcal{H}(U;F)$ is the vector space of all holomorphic mappings of U into F; and $H(U;F)$ is the vector space of all mappings of U into F which are holomorphic when considered as mappings of U into a fixed completion $\hat{F}$ of F. We will say that a mapping $f: U \to F$ is holomorphic iff f belongs to $\mathcal{H}(U;F)$. A mapping $f: U \to F$ is algebraically holomorphic (equivalently G-holomorphic) if the restriction $f|U \cap S$ is holomorphic, for every finite dimensional vector subspace S of E meeting U, where S carries its natural topology. We denote by $P(^kE;F)$, $P_M(^kE;F)$ and $P_{HY}(^kE;F)$, respectively, the vector space of all k-homogeneous continuous polynomials $P: E \to F$, the vector space of all k-homogeneous polynomials $P: E \to F$ that are bounded on bounded subsets of E and the vector space of all k-homogeneous polynomials $P: E \to F$ that are continuous on the compact subsets of E; $\mathcal{H}_M(U;F)$ (respectively: $\mathcal{H}_{HY}(U;F)$) will denote the vector space of all G-holomorphic mappings of U into F that are bounded on the compact subsets of U (respectively: that are continuous on the compact subsets of U). When $F = \mathbb{C}$, it is not included in the notation for function spaces; so, we will write $P(^kE)$ for $P(^kE;\mathbb{C})$, $\mathcal{H}(U)$ for $\mathcal{H}(U;\mathbb{C})$, $l^\infty(I)$ for $l^\infty(I;\mathbb{C})$, and so on. The set of all continuous seminorms on E is denoted by $CS(E)$. A mapping $f: U \to F$ is amply bounded if $\beta \circ f$ is locally bounded for every $\beta \in SC(F)$; more generally, a collection $\mathfrak{X}$ of mappings of U into F is amply bounded if the collection $\beta \circ \mathfrak{X} = \{\beta \circ f \mid f \in \mathfrak{X}\}$ is locally bounded for every $\beta \in CS(F)$. A given E is a holomorphically bornological space if, for every U and F, we have $\mathcal{H}(U;F) = \mathcal{H}_M(U;F)$. A given E is a holomorphically barreled space (respectively: a holomorphically infrabarreled space) if for every U and every F, we have that each collection $\mathfrak{X} \subset \mathcal{H}(U;F)$ is amply bounded if, and always only if, $\mathfrak{X}$ is bounded on every finite dimensional compact subset of U (respectively: on every compact subset of U). A given E is a holomorphically Mackey space if for every U and every F, we have that each mapping $f: U \to F$ belongs to $H(U;F)$ iff $\psi \circ f \in \mathcal{H}(U)$ for every $\psi \in F'$. We represent by τ_o the topology of uniform convergence on the compact subsets and by τ_{of} the topology of uniform convergence on the finite dimensional compact subsets. A net $(x_\alpha)_{\alpha \in A}$ of elements in E is very strongly convergent if $(\lambda_\alpha x_\alpha)_{\alpha \in A}$ converges to zero for any net $(\lambda_\alpha)_{\alpha \in A}$ of scalars. An inductive limit of $(E_i)_{i \in I}$ by $(\rho_i)_{i \in I}$ is said to be a holomorphic inductive limit if for every U, F, $f: U \to F$ we have $f \in \mathcal{H}(U;F)$ iff $f \circ \rho_i \in \mathcal{H}(U_i;F)$ for every $i \in I$.

SECTION 1: STRONG DUALS OF STRICT INDUCTIVE LIMITS OF FRÉCHET-MONTEL SPACES

Barroso, Matos and Nachbin prove in [1] that Silva spaces are holomorphically bornological spaces. We infer from the results of Dineen [4] that the strong duals of Fréchet-Montel spaces (DFM-spaces) are also holomorphically bornological spaces and this improves the result of Barroso, Matos and Nachbin as every Silva space is a DFM-space and there are DFM-spaces that are not Silva spaces. Boland and Dineen give in [2] an example of a strict inductive limit of Fréchet-Montel spaces (FM-spaces) that is not holomorphically bornological (see Proposition 14(a) of [2]). We are going to prove in this section that if E is a strict inductive limit of FM-spaces and there exists a continuous norm on E, then E' is a holomorphically bornological space. This result improves Proposition 3 of [8]: we have now $\mathcal{H}_M(U;F) = \mathcal{H}(U;F)$ (and consequently $\mathcal{H}_{HY}(U;F) = \mathcal{H}(U;F)$) for every U and for every F. As $\mathcal{D}'$ satisfies our conditions, we have in particular $\mathcal{H}_M(U;F) = \mathcal{H}(U;F)$

for every $U \subset \mathcal{D}'$ and for every F. This improves Proposition 14 of [2].

The results contained in Lemmas 1 & 3 are well-known and we have stated then here only for the sake of completeness.

LEMMA 1. If E is the strict inductive limit of FM-spaces E_n, E' is a bornological space.

PROOF. Let $T: E' \to F$ be a linear mapping that is bounded on every compact subset of E'. We know from Proposition 2.8 of [5] that $E' = \varprojlim_{\pi_n} E_n$, where $\pi_n: E' \to E'_n$ is the canonical surjection. So, for every $n \in \mathbb{N}$, there exists a linear mapping $T_n: E'_n \to F$ such that $T = T_n \circ \pi_n$. Let K_n be a compact subset of E'_n. By Example 2.10 of [5], there exists a compact subset K of E' such that $K = \pi_n(K)$. By hypothesis, T is bounded on K and so, T_n is bounded on K_n. So, T_n is bounded on every compact subset of E'_n; as E'_n is a DFM-space, we infer from Corollary 11 of [4] that T_n is continuous. So, $T = T_n \circ \pi_n$ is continuous and this proves that E' is a bornological space.

DEFINITION 2. A set X is determining for $f: U \to \mathbb{C}$ if $X \cap U \neq \emptyset$ and $f|X \cap U = 0$ imply $F \equiv 0$.

LEMMA 3. Let E be a Montel space such that E' is bornological. The following are equivalent:

(a) There exists a continuous norm p on E.

(b) If $V_p = \{\varphi \in E : p(\varphi) \leq 1\}$ for each $p \in SC(E)$, then there exists at least a $p \in SC(E)$ such that V_p^o is determining for every $f \in \mathcal{H}_M(W)$, where W is any convex balanced open subset of E'; this is true for every continuous norm p on E.

(c) There exists a subset K of E' such that K is determining for every $\varphi \in E = (E')'$.

PROOF. (a) $\Rightarrow$ (b): Let p be a continuous norm on E. Then p is the Minkowski functional of $V_p = \{\varphi \in E : p(\varphi) \leq 1\}$. Since E' is Montel, $K = V_p^o$ is a compact subset of E' and we claim it satisfies (b).

We show that if $f \in \mathcal{H}_M(W)$ where W is convex balanced open in E' and $f|K \cap W = 0$, then $f \equiv 0$. Since K is convex and balanced, it is clear that $K \cap W \neq \emptyset$ and if $f|K \cap W = 0$ then $\frac{d^n f(0)}{n!}|K \cap W = 0$ for each $n \in N$. Hence, to show $f \equiv 0$ it suffices to show that K is determining for the elements of $P_M({}^nE')$ for every $n \in N$. We prove this by induction. It is clear for $n = 1$ as E' is a bornological space (Lemma 1) and p is a continuous norm on E: $T \in (E')' = E \ \forall\, T \in P_M({}^1E')$; so $T(x) = 0 \ \forall\, x \in K \Rightarrow \lambda T \in V^{oo} = V \ \forall\, \lambda \in \mathbb{C} \Rightarrow p(\lambda T) \leq 1 \ \forall\, \lambda \in \mathbb{C} \Rightarrow p(T) \leq \frac{1}{|\lambda|} \forall\, \lambda \in \mathbb{C} \Rightarrow p(T) = 0 \Rightarrow T \equiv 0$.
Now let $n > 1$ and suppose we have shown that K is determining for all elements of $P_M({}^kE')$ for every $k < n$. Let $P \in P_M({}^nE')$ be such that $P/K = 0$, and let L denote the symmetric n-linear form corresponding to P. From the polarization formula we see that L is bounded on the bounded subsets of $(E')^n = E' \times E' \times \ldots \times E'$ (n factors) and vanishes on $K \times \ldots \times K$. Now fix $x \in K$. Then $L_x : E' \to \mathbb{C}$ defined by $L_x(z) = L(x, z, \ldots, z)$ is an element of $P_M({}^{n-1}E')$ which vanishes on K, and hence $L_x \equiv 0$. Next let $y \in E'$ be arbitrary. Then $L^y : E' \to \mathbb{C}$ defined by $L^y(z) = L(z, y, \ldots, y)$ is a linear form that is continuous on the bounded subsets of E' and $L^y/K = 0$. By the induction hypothesis $L^y \equiv 0$ on E'. But then in particular $L^y(y) = L(y, \ldots, y) = P(y) = 0$ for every $y \in E'$. This shows that $P \equiv 0$ on E' and therefore by induction K is determining for the elements of $P_M({}^kE')$ for every $k \in \mathbb{N}$. This completes the prove of (a) $\Rightarrow$ (b).

(b) $\Rightarrow$ (c): obvious.

(c) $\Rightarrow$ (a): Let K be a compact subset of E' such that if $\varphi \in E =$

$= (E')'$ and $\varphi/K = 0$, then $\varphi \equiv 0$. We claim that K^o is the unitary ball of a continuous norm on E. If this is not true, there exists $\varphi \in (E')' = E$, $\varphi \neq 0$, such that $\alpha\varphi \in K^o$ for every $\alpha \in \mathbb{C}$. So $\|\alpha\varphi\|_K \leq 1$ for every $\alpha \in \mathbb{C}$ and consequently $\varphi/K = 0$. By hypothesis $\varphi \neq 0$. Contradiction. This completes the prove of (c) $\Rightarrow$ (a).

PROPOSITION 4. Let E be a strict inductive limit of FM-spaces E_n. If E has a continuous norm, E' is a holomorphically bornological space.

PROOF. From Proposition 54 of [1] it is enough to show that $\mathcal{H}_M(U) = \mathcal{H}(U)$ for every $U \subset E'$ and E' is a holomorphically infrabarreled space.

1) Let us show $\mathcal{H}_M(U) = \mathcal{H}(U)$: If p is a continuous norm on E, then p is the Minkoswki functional of $V = \{\varphi \in E : p(\varphi) \leq 1\}$. Since E' is Montel, $K = V^o$ is a compact subset of E' and from Lemma 3 K is determining for every $f \in \mathcal{H}_M(W)$ where W is a convex balanced open subset of E', that is, if $f \in \mathcal{H}_M(W)$ and $f|K \cap W = 0$ then $f \equiv 0$. As $E' = \varprojlim_{\pi_n} E'_n$ is an open and compact surjective limit (see Example 2.10 [5]) we may assume without loss of generality that $U = \pi_m^{-1}(W)$ for some m (where W is an open subset of E'_m). Now let $f \in \mathcal{H}_M(U)$. As E' is an open and compact surjective representation of E' by DFM-spaces, in order to show that $f \in \mathcal{H}(U)$ it suffices to show that f factors through some E'_n i.e. there exists an $n \in \mathbb{N}$ such that $f(x+y) = f(x)$ for all x, $x+y \in U$ whenever $\pi_n(y) = 0$. If f does not factor through some n, then for each n there exists z_n, $z_n+y_n \in U$ where $\pi_n(y_n) = 0$, $y_n \neq 0$ and $f(z_n+y_n) \neq f(z_n)$. Note that (y_n) is very strongly convergent to zero in E' (since $\pi_r(y_n) = 0$ whenever $n \geq r$). For fixed n, $z \mapsto f(z+y_n) - f(z)$ defines an element of $\mathcal{H}_M(V)$ where V is some convex balanced neighbourhood of zero

in E', and hence there exists $x_n \in K \cap (1-\alpha)U$ $(\alpha < 1)$ such that $f(x_n+y_n) \neq f(x_n)$. For $n \geq m$, $g_n(\lambda) = f(x_n+\lambda y_n) - f(x_n)$ is a non constant entire function on $\mathbb{C}$, and hence there exists $\lambda_n \in \mathbb{C}$ such that $|g_n(\lambda_n)| > n + |f(x_n)|$, i.e. $|f(x_n+\lambda_n y_n)| > n$. However $(x_n+\lambda_n y_n)_{n\geq m}$ is a relatively compact subset of U whose closure is contained in U and f is unbounded on $(x_n+\lambda_n y_n)_{n\geq m}$ contradicting the fact that $f \in \mathcal{H}_M(U)$. Hence f must factor through some E'_n, i.e. $f \in \mathcal{H}(U)$.

2) Let us prove that E' is a holomorphically infrabarreled space: Suppose $(f_\alpha)_{\alpha\in A}$ is a τ_o-bounded subset of $\mathcal{H}(U)$, U open in E'. We want to show $(f_\alpha)_{\alpha\in A}$ is locally bounded. As in part 1) of this proof, we may assume without loss of generality that $U = \pi_m^{-1}(W)$ for some m where W is an open subset of E'_m. We claim that $(f_\alpha)_{\alpha\in A}$ factors uniformly through E'_n for some n i.e., there exists $n \in \mathbb{N}$ such that $f_\alpha = \tilde{f}_\alpha \circ \pi_n$ for all $\alpha \in A$ where $\tilde{f}_\alpha \in \mathcal{H}(\pi_n(U))$. If not, we can, as in part 1) of this proof, find a sequence $(x_n+\lambda_n y_n) \subset (1-\alpha)U$ which is relatively compact and a sequence $(f_{\alpha_n})_{n\geq m}$ such that $|f_{\alpha_n}(x_n+\lambda_n y_n)| \geq n$. This contradicts the fact that $(f_\alpha)_{\alpha\in A}$ is τ_o-bounded, and hence $(f_\alpha)_{\alpha\in A}$ factors uniformly through some E'_n. Now E' is a compact surjective limit and hence $(\tilde{f}_\alpha)_{\alpha\in A}$ is τ_o-bounded in $\mathcal{H}(\pi_n(U))$. Therefore from Proposition 6 [4] we conclude that $(\tilde{f}_\alpha)_{\alpha\in A}$ is locally bounded and hence $(f_\alpha)_{\alpha\in A}$ is locally bounded on U (since E' is an open surjective limit). This completes the proof of Proposition 4.

We would like to prove that if E is a strict inductive limit of FM-spaces E_n such that E' is a holomorphically bornological space, then E has a continuous norm. As we know that under such hypothesis E_n has a continuous norm for every n (see Proposition 2 [8]), the first idea to do this was to prove that if

E is a strict inductive limit of FM-spaces E_n such that every E_n has a continuous norm, then E has a continuous norm. Unfortunatelly Floret proved, giving a counterexample, that this statement is not true. Actually it is false even in the case when the E_n are FN-spaces. These counterexamples can be found in [6].

The original problem remains an open problem. We know now that if E is a strict inductive limit of FM-spaces E_n, at least one of the following two statements must be false:

(1) E' is a holomorphically bornological space $\Rightarrow$ E has a continuous norm.

(2) $P_M(^kE') = P(^kE')$ for every $k \Rightarrow H_M(U) = H(U)$ for every $U \subset E'$.

If (2) is correct, the following assertions are equivalent:

(a) E_n has a continuous norm for every n.

(b) E' is a holomorphically bornological space.

In [3] Dierolf and Floret prove the following:

LEMMA 5. Let F be a linear subspace of E which is closed with respect to a continuous norm p on E. Then every continuous norm q on F can be extended to E as a continuous norm.

This lemma allows us to state the

PROPOSITION 6. Let E be a strict inductive limit of FM-spaces E_n such that each E_n is norm-closed in E_{n+1}. Then E' is a holomorphically bornological space.

PROOF. It is a consequence of Lemma 5 and Proposition 3 [8].

LEMMA 7. If E is the strict inductive limit of Fréchet-spaces E_n such that each E_n has a continuous norm. If E has an unconditional basis then there exists a continuous norm on E.

PROOF. See Floret [6].

PROPOSITION 8. If E is the strict inductive limit of FM-spaces E_n and E has an unconditional basis, then the following are equivalent:

(a) E_n has a continuous norm for every n.

(b) E' is a holomorphically bornological space.

PROOF. (a) $\Rightarrow$ (b): It is a consequence of Lemma 7 and Proposition 3 [8].

(b) $\Rightarrow$ (a): It is a consequence of Proposition 2 [8].

SECTION 2. HOLOMORPHIC INDUCTIVE LIMITS

PROPOSITION 9. Let E be the holomorphic inductive limit of the locally convex spaces $(E_i)_{i\in I}$ by $(\rho_i)_{i\in I}$. The following statements are true:

(a) If E_i is a holomorphically bornological space for every $i \in I$, then E is a holomorphically bornological space.

(b) If E_i is a holomorphically barreled space for every $i \in I$, then E is a holomorphically barreled space.

(c) If E_i is a holomorphically infrabarreled space for every $i \in I$, then E is a holomorphically infrabarreled space.

(d) If E_i is a holomorphically Mackey space for every $i \in I$, then E is a holomorphically Mackey space.

PROOF. (a) Let $f: U \to F$ be a algebraically holomorphic mapping such that f is bounded on every compact subset of U. Let $i \in I$ arbitrary. Then $f\circ\rho_i$ is algebraically holomorphic as ρ_i is linear, and $f\circ\rho_i$ is bounded on every compact subset of U_i as ρ_i is continuous. Since E_i is a holomorphically bornological space, $f\circ\rho_i \in \mathcal{H}(U_i;F)$ for every $i \in I$ and for every F and therefore $f \in \mathcal{H}(U;F)$ (as E is a holomorphic inductive limit). This completes the proof of (a).

(b) From Proposition 35 of [1] it follows that it suffices to show that for every $U \subset E$ we have that each collection $\mathfrak{X} \subset \mathcal{H}(U)$ is locally bounded if $\mathfrak{X}$ is bounded on every finite dimensional compact subset of U, i.e., $\mathfrak{X}$ is τ_{of}-bounded. Let $\mathfrak{X} \subset \mathcal{H}(U)$ be τ_{of}-bounded. As $\rho_i: E_i \to E$ is linear and continuous, the image of every finite dimensional compact subset of $U_i = \rho_i^{-1}(U)$ is a finite dimensional compact subset of U. So, for every $i \in I$ $\mathfrak{X}_i = \{f \circ \rho_i : f \in \mathfrak{X}\} \subset \mathcal{H}(U_i)$ is τ_{of}-bounded and therefore $\mathfrak{X}_i$ is locally bounded (since E_i is a holomorphically barreled space for every $i \in I$). Consider $g: U \to l^\infty(\mathfrak{X})$ defined by $g(x)(f) = f(x)$ for $x \in U$ and $f \in \mathfrak{X}$. We claim that $g \circ \rho_i: U_i \to l^\infty(\mathfrak{X})$ is algebraically holomorphic and locally bounded; hence $g \circ \rho_i \in$ $\in \mathcal{H}(U_i; l^\infty(\mathfrak{X}))$: If S is a finite dimensional vector subspace of E meeting U_i and K is a compact subset of $S \cap U_i$, $g \circ \rho_i(K)$ is bounded as $\rho_i(K)$ is a finite dimensional subset of U. So, $g \circ \rho_i$ is G-holomorphic. On the other hand, since $\mathfrak{X}_i \subset \mathcal{H}(U_i)$ is locally bounded, it is clear that $g \circ \rho_i$ is locally bounded. So, we have $g \circ \rho_i \in \mathcal{H}(U_i; l^\infty(\mathfrak{X}))$ for every $i \in I$. As E is a holomorphic inductive limit, $g \in \mathcal{H}(U; l^\infty(\mathfrak{X}))$ and consequently g is locally bounded, that is, $\mathfrak{X}$ is locally bounded. This completes the proof of (b).

(c) Let $\mathfrak{X} \subset \mathcal{H}(U)$ be τ_o-bounded. As $\rho_i: E_i \to E$ is continuous, the image of every compact subset of $U_i = \rho_i^{-1}(U)$ is a compact subset of U and consequently, for every $i \in I$ $\mathfrak{X}_i = \{f \circ \rho_i : f \in \mathfrak{X}\}$ $\subset \mathcal{H}(U_i)$ is τ_o-bounded; as E_i is a holomorphically infrabarreled space, $\mathfrak{X}_i$ is locally bounded for every $i \in I$. Now we prove as in (b) that $g: U \to l^\infty(\mathfrak{X})$ defined by $g(x)(f) = f(x)$ for $x \in U$ and $f \in \mathfrak{X}$ belongs to $\mathcal{H}(U; l^\infty(\mathfrak{X}))$ and therefore $\mathfrak{X}$ is locally bounded. This completes the proof of (c).

(d) Let U be an open subset of E. Consider $f: U \to F$ such that $\psi \circ f \in \mathcal{H}(U)$ for every $\psi \in F'$. Since $\rho_i: E_i \to E$ is linear and continuous and $\psi \circ f \in \mathcal{H}(U)$ we have that $\psi \circ f \circ \rho_i \in \mathcal{H}(U_i)$ where

$U_i = \rho_i^{-1}(U)$ is an open subset of E_i. As E_i is a holomorphically Mackey space for every $i \in I$ and $\psi \circ f \circ \rho_i \in \mathcal{H}(U_i)$ for every $\psi \in F'$, we conclude that $f \circ \rho_i \in H(U_i;F)$ for every $i \in I$ that is, $f \circ \rho_i \in F^{U_i} \cap \mathcal{H}(U_i;\hat{F})$ for every $i \in I$. Now, E is a holomorphic inductive limit and so we infer that $f \in F^U \cap \mathcal{H}(U;\hat{F}) = H(U;F)$. This completes the proof of (d).

REMARK. If E is a DFM-space, E has a fundamental sequence of compact sets, $(B_n)_{n=1}^{\infty}$, which we may suppose are convex, balanced and increasing. For each n let E_{B_n} denote the vector space spaned by B_n and endowed with the norm generated by the Minkowski functional of B_n. For every $n \in \mathbb{N}$, E_{B_n} is a Banach space and so, it is a holomorphically barreled space. The space E is isomorphic to the inductive limit $\varinjlim_n E_{B_n}$ in the category of locally convex spaces and continuous linear mappings (see §2 of [4]). We conjecture if $E = \varinjlim_n E_{B_n}$ is a holomorphic inductive limit. This conjecture is equivalent to that of Dineen in the p. 163 of [4].

Matos introduced, in [7], an other notion of holomorphically bornological space. We will call S-holomorphically bornological the spaces that are holomorphically bornological in the sense of Matos [7] and holomorphically bornological the spaces that are holomorphically bornological in the sense of Barroso, Matos and Nachbin. Every S-holomorphically bornological space is holomorphically bornological. In [4] Dineen proves that DFM-spaces are holomorphically bornological and asks if they are S-holomorphically bornological (see conjecture, p. 163 of [4]). We are going to prove that the holomorphic inductive limit of S-holomorphically bornological spaces is a S-holomorphically bornological space.

Let $\mathcal{B}_E$ denote the family of all closed absolutely convex bounded subsets of E. For each $B \in \mathcal{B}_E$, let E_B denote the vector space spanned by B and endowed with the norm generated by the Minkowski functional of B.

DEFINITION 10. A mapping f from U into F is called S-holomorphic in U if f is finitely holomorphic in U and for every $B \in \mathcal{B}_E$ the mapping $f|U \cap E_B$ is continuous (or, equivalently, holomorphic) relative to the normed topology.

Let $\mathcal{H}_S(U;F)$ denote the vector space of all S-holomorphic mappings from U into F.

DEFINITION 11. The space E is a S-holomorphically bornological space if for every U and F it is true that $\mathcal{H}(U;F) = \mathcal{H}_S(U;F)$.

For basic properties, see [7].

PROPOSITION 12. Let E be the holomorphic inductive limit of the locally convex spaces $(E_i)_{i\in I}$ by $(\rho_i)_{i\in I}$. If E_i is a S-holomorphically bornological space for every $i \in I$, then E is a S-holomorphically bornological space.

PROOF. Let $f: U \to F$ be a G-holomorphic mapping which is bounded on the strict compact subsets of U.

1) If $E = \varinjlim_{\rho_i} E_i$ is a inductive limit and K_i is a strict compact subset of E_i, then $\rho_i(K_i)$ is a strict compact subset of E: by definition, K_i is a strict compact subset of E_i iff there exists $B_i \in \mathcal{B}_{E_i}$ such that $K_i \subset (E_i)_{B_i}$ is compact in $(E_i)_{B_i}$. Since ρ_i is linear and continuous, $\overline{\rho_i(B_i)} \in \mathcal{B}_E$ and we are going to prove that $\rho_i(K_i)$ is compact in $E_{\overline{\rho_i(B_i)}}$ in the topology of the norm. As $E_{\overline{\rho_i(B_i)}}$ is a normed space, $\rho_i(K_i)$ is compact iff every sequence $(x_n) \subset \rho_i(K_i)$ admits subsequence which converges (in the norm topology) to a point of $\rho_i(K_i)$. Let $y_n = \rho_i^{-1}(x_n)$ for every $n \in \mathbb{N}$. Then (y_n) is a sequence in K_i and, as K_i is

a compact subset of $(E_i)_{B_i}$, there exists (y_{n_k}) = subsequence of (y_n) which converges in $(E_i)_{B_i}$ to $y_o \in K_i$, i.e., $\|y_{n_k}-y_o\|_{B_i} \to 0$ as $k \to \infty$. It follows that given $\varepsilon > 0$, there exists N_o such that for every $k > N_o$ we have $\|y_{n_k}-y_o\|_{B_i} < \varepsilon$ i.e., $\inf\{\lambda > 0 : y_{n_k}-y_o \in \lambda B_i\} < \varepsilon$. So, given $\varepsilon > 0$, for each $k > N_o$ there exists $\lambda_{\varepsilon,k}$ such that $0 < \lambda_{\varepsilon,k} < \varepsilon$ and $y_{n_k}-y_o \in \lambda_{\varepsilon,k}B_i$. But this implies $\rho_i(y_{n_k}-y_o) \in \lambda_{\varepsilon,k}\rho_i(B_i)$ and consequently $\rho_i(y_{n_k}) - \rho_i(y_o) \in \lambda_{\varepsilon,k}\rho_i(B_i) \subset \lambda_{\varepsilon,k}\overline{\rho_i(B_i)}$. It follows $\inf\{\lambda > 0 : \rho_i(y_{n_k}) - \rho_i(y_o) \in \lambda\overline{\rho_i(B_i)}\} < \varepsilon$ and this is true for every $k > N_o$. So, $(x_{n_k}) = (\rho_i(y_{n_k}))$ is a subsequence of (x_n) which converges to $\rho_i(y_o) \in \rho_i(K_i)$ (as $y_o \in K_i$) in $E_{\overline{\rho_i(B_i)}}$. So, $\rho_i(K_i) \subset E_{\overline{\rho_i(B_i)}}$ and is compact, i.e., $\rho_i(K_i)$ is a strict compact subset of E.

2) We consider now $f: U \to F$ G-holomorphic and bounded on the compact subsets of U. It is clear that $\rho_i^{-1}(U) = U_i$ is an open subset of E_i (as ρ_i is continuous). If $K_i \subset \rho_i^{-1}(U)$ is a strict compact subset of E_i, $\rho_i(K_i) \subset U$ is a strict compact subset of E, by 1). So, $f_i = f\circ\rho_i$ is bounded on every $K_i \subset \rho_i^{-1}(U)$ such that K_i is a strict compact subset of E_i (as f is bounded on the strict compact subsets of U, by hypothesis). So, $f_i \in \mathcal{H}_S(U_i;F)$ and, as E_i is a S-holomorphically bornological space by hypothesis, f_i is continuous for every $i \in I$. But the inductive limit is holomorphic and so, $f \in \mathcal{H}(U_i;F)$ for every $i \in I$ implies $f \in \mathcal{H}(U;F)$. This completes the proof of Proposition 8.

ACKNOWLEDGEMENTS. I would like to express my thanks to Professor Leopoldo Nachbin and to Professor Mario Matos for some useful discussions concerning this paper. This research was supported in part by FINEP, to which I express may gratitude.

REFERENCES

1. J.A. BARROSO, M.C. MATOS and L. NACHBIN, On holomorphy versus linearity in classifying locally convex spaces. Infinite Dimensional Holomorphy and Applications, Ed. M.C. Matos, North Holland Math. Studies, 12, 1977, p. 31-74.

2. P.J. BOLAND and S. DINEEN, Duality theory for spaces of germs and holomorphic functions on nuclear spaces. Advances in Holomorphy. Ed. J.A. Barroso, North Holland Math. Studies, 34, 1979, p. 179-207.

3. S. DIEROLF and K. FLORET, Über die Fortsetzbarkeit stetiger Normen. Archiv. der Math., 35, 1980, p. 149-154.

4. S. DINEEN, Holomorphic functions on strong duals of Fréchet-Montel spaces. Infinite Dimensional Holomorphy and Applications, Ed. M.C. Matos, North Holland Math. Studies, 12, 1977, p. 147-166.

5. S. DINEEN, Surjective limits of locally convex spaces and their application to infinite dimensional holomorphy. Bull. Soc. Math. France, 103, 1975, p. 441-509.

6. K. FLORET, Continuous norms on locally convex strict inductive limit spaces. Preprint.

7. M. MATOS, Holomorphically bornological spaces and infinite dimensional versions of Hartogs' theorem. J. London Math. Soc., 2, 17, 1978, p. 363-368.

8. L.A. MORAES, Holomorphic functions on strict inductive limits. Resultate der Math., 4, 1981, p. 201-212.

Universidade Federal do Rio de Janeiro
Instituto de Matemática
Caixa Postal 68.530
21.944 - Rio de Janeiro, RJ, Brasil

Functional Analysis, Holomorphy and Approximation Theory II, G.I. Zapata (ed.)
©Elsevier Science Publishers B.V. (North-Holland), 1984

NUCLEAR KÖTHE QUOTIENTS OF FRÉCHET SPACES

V.B. Moscatelli(*)

The structure theory of Fréchet spaces is, at present, the object of an intensive study not only because of its intrinsic interest, but also because of its applications to approximation theory and to concrete function spaces. Within this framework, one is led to problems concerned with the determination of what kinds of subspaces and quotients can be found in arbitrary Fréchet spaces, and here I shall attempt to sketch briefly the history of one of these problems up to its present state. In order to introduce the problem, let us first explain the title. Of course, all our spaces will be infinite-dimensional. Background references are [4], [8], [9] and [11].

We recall that a <u>Fréchet space</u> E is a projective limit of operators $u_k: E_{k+1} \to E_k$ $(k \in \mathbb{N})$ on Banach spaces, that is, E is the set of all sequences (x_k) such that $x_k = u_k(x_{k+1})$ $(k \in \mathbb{N})$ with the product topology. E is said to be <u>nuclear</u> if we can choose Banach spaces E_k and linking maps $u_k: E_{k+1} \to E_k$ such that each u_k can be represented as

$$u_k(x) = \sum_n \langle x, x'_{kn} \rangle y_{kn} \qquad (x \in E_{k+1}),$$

where $(x'_{kn}) \subset E'_{k+1}$, $(y_{kn}) \subset E_k$ and $\sum_n \|x'_{kn}\|_{k+1} \|y_{kn}\|_k < \infty$.

(*) The author gratefully acknowledges partial support from the Italian CNR through a travel grant.

Given a set P of non-negative sequences $a = (a_n)$, the <u>Köthe space</u> $\lambda(P)$ is defined as

$$\lambda(P) = \{(\xi_n) : \sum_n a_n|\xi_n| < \infty \quad \text{for each} \quad (a_n) \in P\}$$

with the locally convex topology generated by the semi-norms

$$p_a(\xi_n) = \sum_n a_n|\xi_n|.$$

Here we assume, to have a Hausdorff topology, that for each $m \in \mathbb{N}$ there exists $(a_n) \in P$ with $a_m > 0$.

A sequence (x_n) in a topological vector space E is a <u>basis</u> if for each $x \in E$ there exists a unique scalar sequence (ξ_n) such that $x = \sum_n \xi_n x_n$ in E. A <u>basic sequence</u> is a sequence which is a basis for the closed subspace it generates.

Now let E be a nuclear Fréchet space with a basis (x_n). Then, by the fundamental Basis Theorem of Dynin-Mitiagin [6], (x_n) is an <u>absolute basis</u> in the sense that the above series converges absolutely for each of the semi-norms defining the topology of E. From this it follows, putting $a_n^k = p_k(x_n)$, that E is isomorphic to the Köthe space $\lambda(P)$, where $P = ((a_n^k))$. The matrix P is a representation of the basis (x_n) and it can always be taken to satisfy $0 \leq a_n^k \leq a_n^{k+1}$ and the following condition, known as the Grothendiek-Pietsch criterion:

(*) for each k there is a j such that $\sum_n \frac{a_n^k}{a_n^j} < \infty$.

Thus we see that the collection of all nuclear Fréchet spaces with basis is the same as the collection of Köthe spaces $\lambda(P)$ with P countable and satisfying (*).

Finally, if there exists a continuous norm on E we say that E <u>admits a continuous norm</u>. In the case of a nuclear Köthe space $\lambda(P)$, this is the same as assuming that $a_n^1 > 0$ for all n.

With a slight abuse of language, from now on by a nuclear Köthe space I will mean a nuclear Fréchet space which has a basis and admits a continuous norm, and the problem under consideration is:

<u>Which Fréchet spaces have nuclear Köthe quotients</u>?

REMARK 1. The continuous norm business is crucial here. Already in 1936 Eidelheit [7] showed that any non-normable Fréchet space has a quotient isomorphic to ω (the topological product of countably many copies of the real line) and, of course, ω does not have a continuous norm. The proof is simple: represent the dual E' as the union of an increasing sequence (E'_n) of Banach spaces, where we can assume $E'_n \subsetneqq E'_{n+1}$. Pick elements $x'_n \in E'_{n+1} \sim E'_n$; the required quotient is then $E/[\text{span}(x'_n)]^\circ$.

REMARK 2. Nuclearity is also crucial in the sense that the problem is likely to be much more difficult without it. Indeed, the answer is unknown even in the Banach space case and it is a celebrated open problem to know whether every Banach space has a separable quotient.

Thus, nuclearity rules out Banach spaces, but the above remark points at the difficulty that might lie at the heart of the problem and, indeed, this has been solved so far only for separable Fréchet spaces (see (4) below).

The problem may be raised, of course, for subspaces as well as quotients and it is instructive to look at the subspace case. Again, nuclearity rules out Banach spaces (but it is an old and classical result that every Banach space has a subspace with a basis) and the subspace problem for Fréchet spaces was ultimately solved about twenty years ago by Bessaga, Pełczyński and Rolewicz [2], [3], who showed that

(1) <u>a non-normable Fréchet space</u> E <u>has a nuclear Köthe subspace if and only if</u> E <u>is not isomorphic to a product of the form</u> $X \times \omega$, <u>with</u> X <u>a Banach space</u> (<u>possibly</u> $\{0\}$).

Note that all closed subspaces of $X \times \omega$ (X Banach) are either of the same form, or Banach or isomorphic to ω. A method of proof can quickly be summarized as follows. If (p_n) is an increasing sequence of semi-norms defining the topology of E, first one finds a separable subspace $F \subset E$ on which the semi-norms p_n are mutually non-equivalent norms. Next, one chooses inductively basic sequences $(x_k^n : k \in N)$ in each Banach space $(F,p_{n+1})^{\sim}$ such that, denoting by X_n their closed linear spans in $(F,p_{n+1})^{\sim}$, all embeddings $X_n \to X_{n-1}$ are nuclear. Finally, one takes suitable linear combinations of elements from the set $(x_k^n : n,k \in N)$ to construct a basic sequence (x_k) in F (hence in E) whose closed linear span is the required nuclear Köthe subspace.

Now let us go back to our problem. To work directly with quotients is generally difficult and so one is tempted to work with subspaces in the dual E' of a Fréchet space E and then conclude by duality. What I mean is that one is led to represent E' as the union of an increasing sequence of Banach spaces E_n', then to look for a subspace F of E' on which the E_n'-norms form a decreasing sequence of mutually non-equivalent norms and, finally, to try to construct a basic sequence (x_k) in F as above. Well, this approach does not work because the sequence (x_k) thus obtained is basic in each Banach space E_n' but might not be basic in E' (strong topology). Indeed, there are examples to the contrary due to Dubinsky and we refer to [5] for this as well as for related pathologies. Of course, this is not surprising, for it occurs all the time when one deals with inductive limits. We note however that the approach through the dual space works, but with entirely different methods, if the original space is already nuclear (cf.[5]), leading to the positive result that

(2) Every nuclear Fréchet space not isomorphic to ω has a nuclear Köthe quotient.

Now let us see what can be said on the negative side. There is a class of Fréchet spaces which can be ruled out without any assumption whatsoever: it is the class of those Fréchet spaces that are now called quojections. I introduced this class in [10] as the class of Fréchet spaces E that are projective limits of a sequence of surjective operators on Banach spaces. Obviously, a countable product of Banach spaces is a quojection, but there are a lot of quojections which are not products (these were called "twisted" in [10]). Quojections fail to have nuclear Köthe quotients in a very strong way, for it is not difficult to show that

(3) If E is a quojection and F is a quotient of E, then:

(a) F is nuclear if and only if it is isomorphic to ω;

(b) F admits a continuous norm if and only if it is Banach.

By (3)(a), ω is the only nuclear quojection, so in the light of (2) and (3) we can ask:

Are quojections the only Fréchet spaces without nuclear Köthe quotients?

The answer is unknown and in general no more can be said at present (remember Remark 2). However, there is an important positive result obtained only recently by Bellenot and Dubinsky [1] under the assumption of separability. It is the following, which generalizes (2):

(4) <u>A separable Fréchet space</u> E <u>has a nuclear Köthe quotient if and only if</u> E' <u>is not the union of an increasing sequence of Banach spaces</u> E'_n <u>with each</u> E'_n <u>being a closed subspace of</u> E'_{n+1}.

What we are saying here is that there is a subspace F of E' and an increasing sequence (p_n) of semi-norms on E such that the dual norms (p'_n) form a (clearly decreasing) sequence of mutually non-equivalent norms on F. Unfortunately, the proof of (4) is quite technical and its heavy use of separability points at the difficulty that may be encountered in trying to solve our problem in general. In proving (4), first one goes over to a quotient with a continuous norm p and whose dual has the property stated in (4). Then separability comes in, and we may choose a sequence (d_n) which is dense in this quotient. Calling E_o the linear span of (d_n), we then use the Bessaga-Pełczyński-Rolewicz method mentioned above to construct a biorthogonal system (x_n, f_n) such that:

(i) $(x_n) \subset E_o$, $(f_n) \subset (E_o, p)'$;

(ii) $f_m(d_n) = 0$ for $m > n$;

(iii) $\dfrac{\|f_n\|_{k+1}}{\|f_n\|_k} \leq n^{-2}$ for $k < n$.

Condition (ii) enables us to extract a subsequence $(f_{n_j}) \subset (f_n)$ such that, if $N = \bigcap_j f_{n_j}^{-1}(0)$ and $\phi\colon E_o \to E_o/N$ is the quotient map, then $\phi(x_{n_j})$ is a basis in the completion $(E_o/N)^{\sim}$. By construction, the latter space is a Köthe quotient of E and condition (iii), when reflected on the $\phi(x_{n_j})$, ensures nuclearity.

Let us remark that the condition of (4) is on the dual E' of E. We know that the dual of every quojection satisfies the condition. Is the converse true? This is at present unknown and an answer to it would settle our original problem in the separable case. What we can say is that if the condition of (4) holds, then E'^{x} (the space of bounded linear functionals on E') is a quojection (easy to prove) and therefore we can conclude that

(5) Within the class of separable, reflexive Fréchet spaces, quojections are exactly those spaces without nuclear Köthe quotients.

REFERENCES

1. S.F. BELLENOT and E. DUBINSKY, Fréchet spaces with nuclear Köthe quotients, Trans. Amer. Math. Soc. (to appear).
2. C. BESSAGA and A. PEŁCZYŃSKI, On a class of B_0-spaces, Bull. Acad. Polon. Sci., V.4 (1957) 375-377.
3. C. BESSAGA, A. PEŁCZYŃSKI and S. ROLEWICZ, On diametral approximative dimension and linear homogeneity of F-spaces, Bull. Acad. Polon. Sci., IX, 9 (1961) 677-683.
4. E. DUBINSKY, The structure of nuclear Fréchet spaces, Lecture Notes in Mathematics 720, Springer 1979.
5. E. DUBINSKY, On (LB)-spaces and quotients of Fréchet spaces, Proc. Sem. Funct. Anal., Holomorphy and Approx. Theory, Rio de Janeiro 1979, Marcel Dekker Lecture Notes (to appear).
6. A.S. DYNIN and B.S. MITIAGIN, Criterion for nuclearity in terms of approximative dimension, Bull. Acad. Polon. Sci., III, 8 (1960) 535-540.
7. M. EIDELHEIT, Zur Theorie der systeme linearer Gleichungen, Studia Math., 6 (1936) 139-148.
8. H. JARCHOW, Locally convex spaces, Teubner 1981.
9. G. KÖTHE, Topological vector spaces I, Springer 1969.
10. V.B. MOSCATELLI, Fréchet spaces without continuous norms and without bases, Bull. London Math. Soc., 12 (1980) 63-66.

11. A. PIETSCH, Nuclear locally convex spaces, Springer 1969.

Dipartimento di Matematica

Università - C.P. 193

73100 Lecce - Italy

Functional Analysis, Holomorphy and Approximation Theory II, G.I. Zapata (ed.)

A COMPLETENESS CRITERION FOR INDUCTIVE LIMITS OF BANACH SPACES

Jorge Mujica(*)

INTRODUCTION

By an (LB)-space $X = \varinjlim X_j$ we mean the locally convex inductive limit of an increasing sequence of Banach spaces X_j, where $X = \bigcup_{j=0}^{\infty} X_j$ and where each inclusion mapping $X_j \to X_{j+1}$ is continuous. It is often of crucial importance to know whether a given (LB)-space is complete or not, but there are very few criteria to establish completeness for (LB)-spaces, and in many situations these criteria do not apply. In Theorem 1 we show that if there exists a Hausdorff locally convex topology τ on the (LB)-space $X = \varinjlim X_j$ with the property that the closed unit ball of each X_j is τ-compact, then X is complete. As an easy consequence of Theorem 1 we prove that the space $H(K)$ of all germs of holomorphic functions on a compact subset K of a complex Fréchet space is always complete. This result had already been established by Dineen [4] in a much more complicated way. In Theorem 1 we also give a sufficient condition for an (LB)-space to be the strong dual of a quasi-normable Fréchet space. As an application of this criterion we show that if K is a compact subset of a complex quasi-normable Fréchet space, then H(K) is the strong dual of a quasi-normable Fréchet space. This improves a previous result of Avilés and the author [1].

(*) This research, partially supported by FAPESP, Brazil, was performed when the author was a visiting lecturer at the University College Dublin, Ireland, during the academic year 1980-1981.

I would like to thank Richard Aron and Séan Dineen for many helpful discussions that we had during the preparation of this paper. I would also like to thank Klaus-Dieter Bierstedt, for a problem we had discussed some time ago was one of the principal motivations for this research. That problem, that up to my knowledge still remains open, is the following: Is every regular (LB)-space complete?

1. THE MAIN RESULT

A basic tool in this paper is Berezanskii's inductive topology on the dual of a locally convex space. If Y is a locally convex space then we will denote by Y'_i the dual Y' of Y, endowed with the locally convex inductive topology defined by $Y'_i = \varinjlim (Y')_{V^o}$, where V varies among all neighborhoods of zero in Y. This inductive topology is stronger than the strong topology. See Berezanskii [2] or Floret [7]. Köthe has also studied this inductive topology on Y' in the case where Y is a metrizable locally convex space, and he has proved that in that case the space Y'_i is always complete. See Köthe [9, p.400]. This result will be used without further reference. We should remark also that the topology τ_ω introduced by Nachbin on the space $P(^mY)$ of all continuous m-homogeneous polynomials on Y reduces to the Berezanskii topology in the case $m = 1$. See Dineen [5, p.51].

THEOREM 1. Let $X = \varinjlim X_j$ be an (LB)-space.

(a) If there exists a Hausdorff locally convex toplogy τ on X with the property that the closed unit ball K_j of each X_j is τ-compact, then $X = Y'_i$ for a suitable Fréchet space Y. In particular X is complete.

(b) If in addition X has a base of τ-closed, convex, balanced neighborhood of zero, then X is actually the strong dual of Y, and Y is a distinguished Fréchet space.

PROOF. The statement and proof of part (a) are nothing but an adaptation of a characterization of dual Banach spaces, due to Ng [11, Th.1]. Ng's result is, on the other hand, a variant of an old result of Dixmier [6, Th. 19].

First of all we observe that the identity mapping $X \to (X,\tau)$ is continuous. Now, let Y denote the vector space of all linear forms on X whose restrictions to each set K_j are τ-continuous. If we endow Y with the topology of uniform convergence on the sets K_j, then it is clear that Y is a Fréchet space (it is actually a closed subspace of the strong dual X'_b of X). Let $J: X \to Y^*$ (Y^* = algebraic dual of Y) denote the evaluation mapping. Since $Y \supset (X,\tau)'$ and since τ is Hausdorff, we see that Y separates the points of X and hence the mapping J is injective. Let $^\circ$ and $^\bullet$ denote the polars with respect to the dual pairs (X,Y) and (Y,Y^*), respectively. Since clearly $J(K_j) \subset K_j^{\circ\bullet}$ for every j, we see that J maps X continuously into Y'_i. Now, the mapping

$$J: (K_j,\tau) \to (Y',\sigma(Y',Y))$$

is clearly continuous, by the definition of Y.

Hence $J(K_j)$ is $\sigma(Y',Y)$-compact. Since $J(K_j)$ is $\sigma(Y',Y)$-dense in $(J(K_j))^{\bullet\bullet}$, by the Bipolar Theorem, we conclude that $J(K_j) = (J(K_j))^{\bullet\bullet}$. Thus, since clearly $(J(K_j))^\bullet = K_j^\circ$, we conclude that

$$J(K_j) = (J(K_j)^{\bullet\bullet} = K_j^{\circ\bullet}$$

and hence J is a topological isomorphism between X and Y'_i.

To show (b) let U be a τ-closed, convex, balanced neighborhood of zero in X. Then $U = U^{\circ\circ}$, by the Bipolar Theorem. Now, since J maps X onto Y' we see that $J(\Phi)^{\circ} = \Phi^{\bullet}$ for each $\Phi \subset Y$ and hence

$$J(U) = J(U^{\circ\circ}) = U^{\circ\bullet}.$$

Since U° is clearly bounded in Y we conclude that $J(U)$ is a 0-neighborhood in Y'_b and hence the mapping $J^{-1}: Y'_b \to X$ is continuous. Since the identity mapping $Y'_i \to Y'_b$ is always continuous, the proof is complete.

To verify the second condition in Theorem 1, the following lemma will be useful.

LEMMA 1. Let $X = \varinjlim X_j$ be an (LB)-space, and assume that there exists a Hausdorff locally convex topology τ on X with the following properties:

(i) The clsed unit ball K_j of each X_j is τ-compact;

(ii) for each 0-neighborhood U in X there exists a sequence of τ-closed, convex, balanced 0-neighborhoods V_j in X such that $V_j \cap K_j \subset U$.

Then X has a base of τ-closed, convex, balanced neighborhoods of zero.

PROOF. Let U be a 0-neighborhood in X. We choose a sequence (ϵ_j) of positive numbers with $\sum_{j=0}^{\infty} \epsilon_j \leq 1$ such that

$$\sum_{j=0}^{\infty} \epsilon_j K_j \subset \frac{1}{2} U \tag{1}$$

where $\sum_{j=0}^{\infty} \epsilon_j K_j$ denotes the set $\bigcup_{n=0}^{\infty} \sum_{j=0}^{n} \epsilon_j K_j$.

By (ii) we can find a sequence of τ-closed, convex, balanced 0-neighborhoods V_j in X such that

$$V_j \cap (j+1)K_j \subset \frac{1}{2} U. \tag{2}$$

Define

$$(3) \qquad V = \bigcap_{n=0}^{\infty} \left(\sum_{j=0}^{n} \varepsilon_j K_j + V_n \right).$$

Then V absorbs every K_j, it is convex and balanced, and by (i) it is τ-closed. Since X is barrelled we conclude that V is a 0-neighborhood in X. To conclude the proof is suffices to show that $V \subset U$. Let $z \in V$ and choose n such that $z \in nK_n$. By (3) we can write

$$(4) \qquad z = x+y, \quad \text{with} \quad x \in \sum_{j=0}^{n} \varepsilon_j K_j \quad \text{and} \quad y \in V_n .$$

Hence

$$(5) \qquad y = z-x \in nK_n + \sum_{j=0}^{n} \varepsilon_j K_j \subset (n+1)K_n$$

since $\sum_{j=0}^{\infty} \varepsilon_j \leq 1$ and since we may assume, without loss of generality, that the sequence (K_j) is increasing. Thus from (4), (5) and (2) we conclude that

$$(6) \qquad y \in V_n \cap (n+1)K_n \subset \tfrac{1}{2} U$$

and therefore

$$z = x+y \in \sum_{j=0}^{n} \varepsilon_j K_j + \tfrac{1}{2} U \subset U$$

by (4), (6) and (1). Thus $V \subset U$ and the proof is complete.

2. APPLICATIONS TO COMPLEX ANALYSIS

If K is a compact subset of a complex Fréchet space E then the space $\mathcal{H}(K)$ of all germs of holomorphic functions on K is defined as the locally convex inductive limit

$$\mathcal{H}(K) = \lim_{\rightarrow} \mathcal{H}^{\infty}(U_j),$$

where (U_j) is a decreasing fundamental sequence of open neighborhoods of K and where $\mathcal{H}^{\infty}(U_j)$ denotes the Banach space of all bounded holomorphic functions on U_j, with the norm of the supremum. This (LB)-space has received a good deal of attention in recent years and we refer to the survey article of Bierstedt and Meise [3] or to the recent book of Dineen [5] for background information and open problems concerning $\mathcal{H}(K)$. The problem of completeness of $\mathcal{H}(K)$ remained open for several years until it was finally solved by Dineen [4, Th. 8], who proved that the space $\mathcal{H}(K)$ is always complete. Dineen's proof is quite complicated, but we can now obtain Dineen's result as an easy consequence of Theorem 1.

THEOREM 2. Let K be a compact subset of a complex Fréchet space E. Then $\mathcal{H}(K) = Y'_i$ for a suitable Fréchet space Y. In particular, $\mathcal{H}(K)$ is complete.

Before proving Theorem 2 we fix some notation. Let $\mathcal{H}(U)$ denote the space of all holomorphic functions on an open subset U of a complex locally convex space E. If $f \in \mathcal{H}(U)$ and $x \in U$ then we let $f^{(n)}(x)$ denote the nth term in the Taylor series expansion of f at x. If $f \in \mathcal{H}(U)$, $A \subset U$ and $B \subset E$ then we set $\|f^{(n)}\|_{A,B} = \sup_{\substack{x \in A \\ s \in B}} |f^{(n)}(x)(s)|$.

PROOF OF THEOREM 2. Let (U_j) be a decreasing fundamental sequence of open neighborhoods of K. Let τ_o denote the compact-open topology on $\mathcal{H}(U_j)$, and by abuse of notation let τ_o also denote the locally convex inductive topology on $\mathcal{H}(K)$ which is defined by

$$(\mathcal{H}(K),\tau_o) = \lim_{\rightarrow} (\mathcal{H}(U_j),\tau_o).$$

As Nicodemi [12] has remarked, $(\mathcal{H}(K),\tau_o)$ is a Hausdorff space, for the seminorms $f \to |f^{(n)}(x)(s)|$ are well-defined and continuous on $(\mathcal{H}(K),\tau_o)$ for all $n \in \mathbb{N}$, $x \in K$ and $s \in E$. On the other hand, by Ascoli theorem, the closed unit ball of $\mathcal{H}^\infty(U_j)$ is compact in $(\mathcal{H}(U_j),\tau_o)$, and hence in $(\mathcal{H}(K),\tau_o)$. An application of Theorem 1 completes the proof.

REMARK. The locally convex space $(\mathcal{H}(K),\tau_o)$ has recently been studied by the author [10] in great detail and it turns out that the Fréchet space Y that appears in Theorem 2 is the strong dual of $(\mathcal{H}(K),\tau_o)$.

Avilés and the author [1, Th.2] have shown that $\mathcal{H}(K)$ satisfies the strict Mackey convergence condition whenever K is a compact subset of a complex quasi-normable Fréchet space. This result can be improved as follows:

THEOREM 3. Let K be a compact subset of a complex quasi-normable Fréchet space E. Then $\mathcal{H}(K)$ is the strong dual of a quasi-normable Fréchet space.

We refer to Grothendieck [8] for information concerning quasi-normable spaces and the strict Mackey convergence condition. To prove Theorem 3 we need the following lemma, which is essentially a reformulation of the proof of [1, Th.2].

LEMMA 2. Let K be a compact subset of a complex quasi-normable Fréchet space E. Then there exists a decreasing fundamental sequence of open, convex, balanced 0-neighborhoods U_j in E such that, if we let $\mathcal{K}_j$ denote the closed unit ball of $\mathcal{H}^{\infty}(K+U_j)$, then for each 0-neighborhood $\mathcal{U}$ in $\mathcal{H}(K)$ there exists a sequence of convex, balanced 0-neighborhoods $\mathcal{V}_j$ in $\mathcal{H}(K)$ with the following properties:

(i) each $\mathcal{V}_j$ is closed in $(\mathcal{H}(K),\tau_o)$;

(ii) $\mathcal{V}_j \cap \mathcal{K}_j \subset \mathcal{U}$ for every j.

PROOF. Since E is metrizable and quasi-normable, we can inductively find a fundamental sequence of open, convex, balanced 0-neighborhoods U_j in E such that:

(a) $2U_{j+1} \subset U_j$ for every j;

(b) for every j and for every $\delta > 0$ there exists a bounded set B in E such that $2U_{j+1} \subset B + \delta U_j$.

Let $\mathcal{K}_j$ denote the closed unit ball of $\mathcal{H}^\infty(K+2U_j)$. Since $\mathcal{U}$ is a 0-neighborhood in $\mathcal{H}(K)$ we can find a sequence of positive numbers ε_j such that $3\varepsilon_j \mathcal{K}_j \subset \mathcal{U}$ for every j. Fix j and fix $f \in \mathcal{K}_j$. Then using (a) and the Cauchy integral formulas, we can write, for each $N \in \mathbb{N}$:

$$\|f\|_{K+2U_{j+1}} \leq \sum_{n=0}^{N-1} \|f^{(n)}\|_{K,2U_{j+1}} + \sum_{n=N}^{\infty} \|f^{(n)}\|_{K,U_j}$$

(1) $$\|f\|_{K+2U_{j+1}} \leq \sum_{n=0}^{N-1} \|f^{(n)}\|_{K,2U_{j+1}} + \sum_{n=N}^{\infty} 2^{-n} .$$

Now, since $f \in \mathcal{K}_j$ the Cauchy integral formulas imply that

$$|f^{(n)}(x)(s)| \leq 1 \quad \text{for all} \quad x \in K \quad \text{and} \quad s \in 2U_j$$

and hence that

(2) $$|(f^{(n)}(x))^{(k)}(s)(t)| \leq 1 \quad \text{for all} \quad x \in K \quad \text{and} \quad s,t \in U_j .$$

Next we note that by (b), given $\delta > 0$ there exists $B \subset E$ bounded such that each $s \in 2U_{j+1}$ can be written in the form

$$s = b + \delta t, \quad \text{with} \quad b \in B \quad \text{and} \quad t \in U_j .$$

Hence, for each $x \in K$, (2) implies that

$$|f^{(n)}(x)(b) - f^{(n)}(x)(s)| \leq \sum_{k=1}^{n} \delta^k |(f^{(n)}(x))^{(k)}(s)(t)| \leq \sum_{k=1}^{n} \delta^k$$

and we conclude that

(3) $$\|f^{(n)}\|_{K,2U_{j+1}} \leq \|f^{(n)}\|_{K,B} + \sum_{k=1}^{n} \delta^k .$$

First we choose $N \in \mathbb{N}$ such that $\sum_{n=N}^{\infty} 2^{-n} \leq \varepsilon_{j+1}$ and next we choose $0 < \delta < 1$ such that $\sum_{k=1}^{\infty} \delta^n \leq \frac{\varepsilon_{j+1}}{N}$. If B is the bounded set associated with δ in (b) then from (1) and (3) we conclude that

$$(4) \qquad \|f\|_{K+2U_{j+1}} \leq \sum_{n=0}^{N-1} \|f^{(n)}\|_{K,B} + 2\varepsilon_{j+1} .$$

If we define

$$\mathcal{V}_j = \{f \in \mathcal{H}(K) : \sum_{n=0}^{N-1} \|f^{(n)}\|_{K,B} \leq \varepsilon_{j+1}\}$$

then $\mathcal{V}_j$ is a convex, balanced 0-neighborhood in $\mathcal{H}(K)$, $\mathcal{V}_j$ is closed in $(\mathcal{H}(K),\tau_o)$ and by (4)

$$\mathcal{V}_j \cap \mathcal{K}_j \subset 3\varepsilon_{j+1}\mathcal{K}_{j+1} \subset \mathcal{U}.$$

The proof of Lemma 2 is now complete.

PROOF OF THEOREM 3. Let (U_j) be the fundamental sequence of neighborhoods of zero in E given by Lemma 2. Since we already know that the closed unit ball of each $\mathcal{H}^\infty(K+U_j)$ is compact in $(\mathcal{H}(K),\tau_o)$, then from Lemma 1 and Lemma 2 we conclude that $\mathcal{H}(K)$ has a base of convex, balanced neighborhoods of zero, each of which is closed in $(\mathcal{H}(K),\tau_o)$. Then we conclude from Theorem 1 that $\mathcal{H}(K)$ is the strong dual of a Fréchet space Y. But since by [1, Th.2] $\mathcal{H}(K)$ satisfies the strict Mackey convergence condition, we conclude that Y must be quasi-normable. The proof is now complete.

REFERENCES

1. P. AVILES and J. MUJICA, Holomorphic germs and homogeneous polynomials on quasi-normable metrizable spaces, Rend. Mat. 10 (1977), 117-127.

2. J.A. BEREZANSKII, Inductively reflexive locally convex spaces, Soviet Math. Dokl. 9 (1968), 1080-1082.

3. K.-D. BIERSTEDT and R. MEISE, Aspects of inductive limits in spaces of germs of holomorphic functions on locally convex spaces and applications to a study of $(H(U),\tau_\omega)$, in Advances in Holomorphy (J.A. Barroso, ed.), North-Holland, Amsterdam, 1979, p. 111-178.

4. S. DINEEN, Holomorphic germs on compact subsets of locally convex spaces, in Functional Analysis, Holomorphy and Approximation Theory (S. Machado, ed.), Lecture Notes in Math. 843, Springer, Berlin, 1981, p. 247-263.

5. S. DINEEN, Complex Analysis in Locally Convex Spaces, North-Holland, Amsterdam, 1981.

6. J. DIXMIER, Sur un Théorème de Banach, Duke Math. J. 15 (1948), 1057-1071.

7. K. FLORET, Über den Dualraum eines lokalkonvexen Unterraumes, Arch. Math. (Basel) 25 (1974), 646-648.

8. A. GROTHENDIECK, Sur les espaces (F) et (DF), Summa Brasil. Math. 3 (1954), 57-123.

9. G. KÖTHE, Topological Vector Spaces I, Springer, Berlin, 1969.

10. J. MUJICA, A new topology on the space of germs of holomorphic functions (preprint).

11. K.F. NG, On a theorem of Dixmier, Math. Scand. 29 (1971), 279-280.

12. O. NICODEMI, Homomorphisms of algebras of germs of holomorphic functions, in Functional Analysis, Holomorphy and Approximation Theory (S. Machado, ed.), Lecture Notes in Math. 843, Springer, Berlin, 1981, p. 534-546.

Department of Mathematics
University College Dublin
Belfield, Dublin 4
Ireland

and

Instituto de Matemática
Universidade Estadual de Campinas
Caixa Postal 6155
13100 Campinas, SP - Brazil
(current address)

Functional Analysis, Holomorphy and
Approximation Theory II, G.I. Zapata (ed.)

ABOUT THE CARATHEODORY COMPLETENESS OF ALL REINHARDT DOMAINS

Peter Pflug

It is well known that in the theory of complex analysis there are different notions of distances on a bounded domain G in $\mathbb{C}^n$, for example, the Caratheodory-distance dealing with bounded holomorphic functions, the Bergmann-metric measuring how many L^2-holomorphic functions do exist or the Kobayashi-distance describing the sizes of analytic discs in G. A survey on these notions, also generalized to infinite dimensional holomorphy, can be found in the book of Franzoni-Vesentini [3].

The main problem working with these distances is to decide which domain G is complete w.r.t. one of these distances. There is a fairly general result for the Bergmann-metric due to T. Ohsawa and P. Pflug [6,7] which states that any pseudoconvex domain with C^1-boundary is complete w.r.t. the Bergmann-metric. On the other hand it is well known that the Caratheodory-distance can be compared with the other two, in fact, it is the smallest one, but there is no relation between the Bergmann-metric and the Kobayashi-metric [2]. Thus the question remains which domains are complete w.r.t. the Caratheodory-distance or, at least, w.r.t. the Kobayashi-distance.

In this short note it will be shown that any bounded complete Reinhardt domain G which is pseudoconvex is complete in the sense of the Caratheodory-distance; in fact, it will be proved that any Caratheodory ball is a relatively compact subset of G. Using the above remark on the comparability of the distances it is clear that those domains are also complete w.r.t. the two other distances.

First, some definitions should be repeated.

DEFINITION 1. A domain $G \subset \mathbb{C}^n$ is called a complete Reinhardt domain if for any $z^o \in G$ the polycylinder $\{z \in \mathbb{C}^n: |z_i| \le |z_i^o|$ for $1 \le i \le n\}$ has to be contained in the domain G.

It is well known that a complete Reinhardt domain G is pseudoconvex iff G is logarithmically convex which means the set

$$\log|G| := \{x \in R^n: \text{for } x \ \exists\ z \in G: x = (\log|z_1|,\ldots,\log|z_n|)\}$$

is convex in the usual sense.

DEFINITION 2. Let G be a domain in $\mathbb{C}^n$ then, for points z', z'' in G,

$$C_G(z',z'') := \sup\left\{\frac{1}{2}\log\frac{1+|f(z'')|}{1-|f(z'')|} : f:G\to E \text{ holomorphic with } f(z')=0\right\}$$

is called the Caratheodory distance between z' and z''; here and in the future E denotes the unit disc $E = \{\lambda \in \mathbb{C}: |\lambda| < 1\}$.

It is easy to check that $C_G(\ ,\ )$ is, in fact, a distance on a bounded G; hence $(G, C_G(\ ,\ ))$ is a metric space. Asking whether this space is complete it suffices to establish that any Caratheodory ball $\{z \in G: C_G(z,z^o) < M\}$ around $z^o \in G$ is a relatively compact subset of G; this is called the strong Caratheodory-completeness of G.

It is well known that a Caratheodory-complete domain (i.e. $(G, C_G(\ ,\ ))$ is complete) has to be $H^\infty(G)$-convex and a domain of bounded holomorphy [8]; the converse, in general, is false. In fact there exists a $H^\infty(G)$-convex domain of bounded holomorphy which is not Caratheodory-complete. On the other hand it should be repeated that any pseudoconvex domain with a smooth boundary is $H^{\infty}(G)$-convex and also a domain of holomorphy [1]. This remark may induce the following problem: is any pseudoconvex domain with smooth boundary Caratheodory-complete or, at least, complete w.r.t. the Kobayashi-distance? Here only a simple partial result can be presented.

THEOREM. Any bounded complete Reinhardt domain G, which is pseudoconvex, is strongly Caratheodory-complete.

PROOF. Without loss of generality we can assume that G is contained in the unit-polycylinder. Then assuming the proposition is false there exists a sequence $\{z^v\} \subset G$ with $z^v \to z^o \in G$ such that $C_G(z^v,0) \leq M < \infty$ for all $v \in \mathbb{N}$. First, $\prod_{v=1}^{n} z_v^o \neq 0$ is assumed.
Then $x^o := (\log|z_1^o|,\dots,\log|z_n^o|)$ belongs to the boundary of the convex set $\log|G|$. Hence a linear functional $L\colon R^n \to R$ can be found such that

$$L(x) = \sum_{i=1}^{n} \xi_i x_i < L(x^o) =: C$$

for all $x \in \log|G|$ with $C \leq 0$. Using the completeness of the Reinhardt domain G it's clear that the numbers ξ_i are nonnegative.

Assume that $\xi_{i_1},\dots,\xi_{i_k}$ are positive - the remaining ones should be zero - then, compare [4], it is possible, for any $N \in \mathbb{N}^*$, to find integers $\beta_{1,N},\dots,\beta_{k,N}$ with $1 \leq k_N \leq N^k$ and $|\xi_{j_v} - \frac{\beta_{v,N}}{k_N}| \leq \frac{1}{Nk_N} \leq \frac{1}{N}$. Provided N large enough it follows: $\beta_{v,N} > 0$.

Defining

$$f_N(z) = e^{-Ck_N} \cdot z_{i_1}^{\beta_{1,N}} \cdots z_{i_k}^{\beta_{k,N}} \qquad \text{and}$$

$$g_N(z) := \frac{f_N(z)}{\|f_N\|_{\bar{G}}}$$

we have obtained holomorphic functions $g_N: G \to E$ for which the following inequalities for $z \in G$, z near z^o, can be proved:

$$\log|f_N(z)| = -Ck_N + \sum_{v=1}^{k} \beta_{v,N} \log|z_{i_v}|$$

$$= -k_N(C-L(\log|z_1|,\ldots,\log|z_n|))+k_N \sum_{v=1}^{k} (\frac{\beta_{v,N}}{k_N} - \xi_{i_v})\log|z_{i_v}|$$

$$\geq -N^k(C-L(\log|z_1|,\ldots,\log|z_n|)) - k\cdot k_N \cdot \frac{M^*}{Nk_N}$$

or

$$|f_N(z)| \geq \exp[-N^k(C-L(\log|z_1|,\ldots,\log|z_n|)) - \frac{kM^*}{N}].$$

To estimate $\|f_N\|_{\bar{G}}$ it is enough to look for those $z \in G$ with: $\prod_{v=1}^{n} z_v \neq 0$ and $\sum_{v=1}^{k} \beta_{v,N} \cdot \log|z_{i_v}| \geq C\cdot k_N$ which implies

$$|\log|z_{i_v}|| \leq |C| \frac{k_N}{\beta_{v,N}} \quad \text{for} \quad 1 \leq v \leq k.$$

Using the choice of the β's and N large enough the following inequality can be found:

$$|\log|z_{i_v}|| \leq |C|k_N \frac{2}{\min_{1\leq\lambda\leq k} \xi_{i_\lambda}} \frac{1}{k_N} =: \alpha .$$

Hence one has received for all $z \in G$:

$$|f_N(z)| \leq \exp \frac{k\alpha k_N}{k_N N} = \exp \frac{k\cdot\alpha}{N} .$$

Combining the above estimates one ends up (N and v large enough) with:

$$|g_N(z^v)| \geq \exp[-N^k(C-L(\log|z_1^v|,\ldots,\log|z_n^v|)) - \frac{kM^*}{N} - \frac{k\alpha}{N}]$$

from which follows

$$1 > \frac{e^{2M}-1}{e^{2M}+1} \geq |g_N(z^v)| \xrightarrow[N,v\to\infty]{} 1.$$

In the remaining case z^o can be assumed as $z^o = (z_1^o,\ldots,z_\ell^o,0,\ldots 0)$ with $\prod_{v=1}^{n} z_v^o \neq 0$.

Then $G' := \{z' \in \mathbb{C}^{\ell}: z' \in \pi(G)\}$, where $\pi: G \to \mathbb{C}^{\ell}$ denotes the usual projection, can be easily recognized as a bounded pseudoconvex complete Reinhardt domain in $\mathbb{C}^{\ell}$ with boundary point $z^{o'} := \pi(z^o)$. Hence it follows:

$$c_{G'}(0,\pi(z^v)) \leq c_G(0,z^v) \leq M$$

which contradicts the case discussed before.

Therefore the proof of the theorem is complete.

It should be mentioned that by a lemma due to E. Low (reported by K. Diederich) that the strong Caratheodory-completeness implies the sequential H^{∞}-convexity one has the following consequence:

COROLLARY. Any sequence of points $\{z^v\}$ in a bounded pseudoconvex complete Reinhardt domain G with $z^v \to z^o \in \partial G$ contains a subsequence $\{z^{1v}\}$ such that there exists a holomorphic function $f: G \to E$ with $\lim|f(z^{1v})| = 1$ and $f(0) = 0$.

For the convenience of the reader a proof will be presented.

PROOF. It is easy using the strong Caratheodory completeness of G to find a subsequence $z^v \to z^o$ and holomorphic functions $f_v: G \to E$ with $f_v(0) = 0$ and $f_v(z^v) > 1 - \varepsilon_v$ where $\frac{1}{\varepsilon_v} > 2^{2v}$.
Setting $\tilde{f}_v(z) := -\frac{f_v(z)+1}{f_v(z)-1}$ one can define

$$F(z) := \sum_{k=1}^{\infty} \frac{1}{2^k} \tilde{f}_k(z).$$

Then an easy exercise proves that for any $K \subset G$ - K compact - there exists a number $0 < R < 1$ such that, for all $z \in K$ and all $v \in N$, $|f_v(z)| \leq R$ is valid. This remark implies that the above series converges uniformly on compact subsets of G. Hence F gives a holomorphic map $F: G \to \{z \in \mathbb{C}: \mathrm{Re}\, z > 0\} =: H$, for which

$$F(0) = 1 \quad \text{and} \quad |F(z^\nu)| \geq \frac{1}{2^\nu} \tilde{f}_\nu(z^\nu) > 2^\nu \xrightarrow[\nu\to\infty]{} \infty$$

hold.

By $\hat{f}(z) = \frac{F(z)-1}{F(z)+1}$ a holomorphic function $\hat{f}: G \to E$ is constructed with:

$$\hat{f}(0) = 0 \quad \text{and} \quad |\hat{f}(z^\nu)| \geq \frac{1-\frac{1}{|F(z^\nu)|}}{1+\frac{1}{|F(z^\nu)|}} \xrightarrow[\nu\to\infty]{} 1$$

which ends the proof.

Another application is concerned with the Serre-problem:

COROLLARY. A locally trivial holomorphic fibre bundle with Stein base, whose fiber is a bounded pseudoconvex complete Reinhardt domain, is already a Stein space.

REFERENCES

1. CATLIN, D.: Boundary behaviour of holomorphic functions on pseudoconvex domains; Journal Diff. Geometry 15, 605-625 (1980).

2. DIEDERICH, K. and E. FORNAESS: Comparison of the Bergmann and the Kobayashi metric; Math. Annalen 254, 257-262 (1980).

3. FRANZONI, T. and E. VESENTINI: Holomorphic and invariant distances; Notas de Matemática 69 (1980).

4. GAMELIN. T.W.: Peak points for algebras on circled sets; Math. Annalen 238, 131-139 (1978).

5. KOBAYASHI, S.: Geometry of bounded domains; Trans. Amer. Math. Soc. 92, 267-289 (1959).

6. OHSAWA, T.: A remark on the completeness of the Bergmann metric; Proc. of the Japan Academy 57, 238-240 (1981).

7. PFLUG, P.: Various applications of the existence of well growing holomorphic functions; in Functional Analysis, Holomorphy and Approximation Theory ed. by J.A. Barroso (1982).

8. SIBONY, N.: Prolongement analytique des fonctions holomorphes bornés; Séminaire Lelong, Année 1972-1973, 44-66 (1974).

ADDED IN PROOF. For dimension $n=2$ our result has been proved independently also by J.-P. Vigué in an article "La distance de Caratheodory n'est pas intérieure" which will appear in Resultate der Mathematik.

Universität Osnabrück
-Abteilung Vechta-
Fachbereich
Naturwissenschaften/Mathematik
Postfach 1349 - D-2848 Vechta

Functional Analysis, Holomorphy and
Approximation Theory II, G.I. Zapata (ed.)

BEST SIMULTANEOUS APPROXIMATION

João B. Prolla

Throughout this paper $(F,|\cdot|)$ is a non-trivial non-archimedean valued division ring, and $(E,\|\cdot\|)$ is a non-zero non-archimedean normed space over $(F,|\cdot|)$.

DEFINITION 1. Let $M \subset E$ be a closed linear subspace, and $x \in E$. A <u>best approximation of</u> x <u>in</u> M is any element $y \in M$ such that

$$\|x-y\| = \inf\{\|x-z\| ; z \in M\}$$
$$= \mathrm{dist}(x;M).$$

We denote by $\mathcal{P}_M(x)$ the set of all best approximations of x in M. There are two problems to be considered, once x and M are given:

(1) When is $\mathcal{P}_M(x) \neq \emptyset$, i.e., the <u>existence</u> of best approximations;

(2) When $\mathcal{P}_M(x)$ contains no more than one element, i.e. the <u>uniqueness</u> of best approximations.

Let us start with the second problem, which in the non-archimedean case has a very simple solution: when $M \neq \{0\}$, there is no uniqueness. More exactly, we have the following result (Monna [3]):

THEOREM 2. Let $M \subset E$ be a closed linear subspace. For every $x \in E$, $x \notin M$, if $y \in \mathcal{P}_M(x)$ and $t \in M$ with $\|t-y\| < r$, where $r = \|x-y\|$, then $t \in \mathcal{P}_M(x)$.

PROOF. Since $x \notin M$, and $y \in M$, the distance $r = \|x-t\|$ is >0. By the strong triangle inequality,

$$\|x-t\| = \|x-y\| = r$$

for all $t \in E$ such that $\|t-y\| < r$. Hence $t \in \rho_M(x)$.

It remains the problem of existence of best approximations.

DEFINITION 3. A closed linear subspace $M \subset E$ is called <u>proximinal</u> if $\rho_M(x)$ contains at least one element for all $x \in E$.

The definition above poses two problems:

(i) Let $M \subset E$ be a closed linear subspace. Give sufficient conditions on M so that M is proximinal, i.e., every $x \in E$ has at least one best approximation in M.

(ii) Give sufficient conditions on E so that every closed linear subspace is proximinal.

Concerning problem (i), one has the following result of Monna [3], who introduced the notion of orthogonal projection for non-archimedean spaces.

DEFINITION 4. A continuous linear map $P: E \to E$ such that $P^2 = P$ is called a <u>continuous linear projection</u> from E onto $P(E)$.

It follows that, for any non-zero continuous linear projection P one has $1 \leq \|P\|$.

A continuous linear projection P is called an <u>orthogonal projection onto</u> $M = P(E)$, if for all $x \in E$, Px is a best approximation of x in M.

It follows that, for any orthogonal projection P one has $\|P\| \leq 1$. Indeed, for any $x \in E$:

$$\|Px\| = \|Px-x+x\| \leq \max(\|Px-x\|, \|x\|) \leq \|x\|,$$

because $\|Px-x\| \leq \|0-x\| = \|x\|$, since $0 \in M$ and Px is a best approximation of x in M. Hence for any non-zero orthogonal projection P one has $\|P\| = 1$. Let us prove the converse. (See Monna [4], p.478.)

THEOREM 5. Every continuous linear projection of norm one is an orthogonal projection.

PROOF. Clearly $M = P(E)$ is the set of all $y \in E$ such that $y = Py$. Hence M is closed. Let $x \in E$. If $x \in M$, then $Px = x$ is a best approximation of x in M. If $x \notin M$, then for every $y \in M$ we have

$$\|x-Px\| = \|x-y-P(x-y)\| \leq$$
$$\leq \max(\|x-y\|, \|P(x-y)\|) \leq \|x-y\|;$$

that is $\|x-Px\| \leq \operatorname{dist}(x;M)$. Since $Px \in M$, $\|x-Px\| = \operatorname{dist}(x;M)$, and Px is a best approximation of x in M.

COROLLARY 6. Let M be a closed linear subspace of E such that $M = P(E)$ for some continuous linear projection P with $\|P\| = 1$. Then M is proximinal.

COROLLARY 7. Every spherically complete linear subspace is proximinal.

PROOF. By [6], 6.10, given a spherically complete subspace $M \neq \{0\}$, there is a continuous linear projection P onto M with $\|P\| = 1$.

COROLLARY 8. Assume that $(F,|\cdot|)$ is spherically complete. Then every finite-dimensional subspace of E is proximinal.

PROOF. Apply Theorem 6.12, [6].

Let us now consider the problem of <u>best simultaneous approximation</u>.

DEFINITION 7. Let $(E,\|\cdot\|)$ be a normed space over $(F,|\cdot|)$, $G \subset E$, and B be a bounded subset of E. Define the <u>relative Chebyshev</u>

radius of B (with respect to G)

$$\mathrm{rad}_G(B) = \inf_{g\in G} \sup_{f\in B} \|g-f\| .$$

If $G = E$, then we write

$$\mathrm{rad}_E(B) = \mathrm{rad}(B)$$

and call it the Chebyshev radius of B.

The elements $g_o \in G$ where the infimum is attained are called relative Chebyshev centers of B (with respect to G), and we denote by $\mathrm{cent}_G(B)$ the set of all such $g_o \in G$.

If $G = E$, then we write

$$\mathrm{cent}_E(B) = \mathrm{cent}(B)$$

and call it the set of Chebyshev centers of B.

We say that G has the relative Chebyshev center property in E if $\mathrm{cent}_G(B) \neq \emptyset$ for all non-empty bounded sets $B \subset E$. Since $B = \{f\}$ is bounded, any subspace G which has the relative Chebyshev center property in E is proximinal in E.

When $G = E$, and $\mathrm{cent}(B) \neq \emptyset$ for every non-empty bounded subset $B \subset E$, i.e. if E has the relative Chebyshev center property in E, we say that E admits Chebyshev centers.

The following result generalizes Corollary 7, and will be given a direct proof.

THEOREM 8. Every spherically complete linear subspace of a non-archimedean normed space has the Chebyshev center property.

PROOF. Let $(E,\|\cdot\|)$ be a non-archimedean normed space and let G be a spherically complete subspace. Let $B \subset E$ be any non-empty bounded subset. For each $g \in G$, put

$$\rho(g) = \sup_{f\in B} \|g-f\| .$$

Consider the family $\mathcal{C}$ of closed balls on G given by

$$\mathcal{C} = \{B(g;\rho(g));\ g \in G\}.$$

The family $\mathcal{C}$ has the binary intersection property. Indeed, if g and g' belong to G, then for all $f \in B$,

$$\|g-g'\| \leq \mathrm{Max}(\|g-f\|,\ \|f-g'\|).$$

Hence $\|g-g'\| \leq \max(\rho(g),\rho(g'))$.

Since G is spherically complete, there is some $g_o \in G$ such that $g_o \in B(g;\rho(g))$ for all $g \in G$. This is equivalent to say that $\|g_o-g\| \leq \rho(g)$, for all $g \in G$. Now

$$\begin{aligned}\sup_{f\in B} \|g_o-f\| &= \sup_{f\in B} \|g_o-g+g-f\| \\ &\leq \sup_{f\in B} (\max(\|g_o-g\|,\ \|g-f\|)) \\ &\leq \sup_{f\in B} (\max(\rho(g),\ \|g-f\|) = \rho(g)\end{aligned}$$

for all $g \in G$. Hence

$$\sup_{f\in B} \|g_o-f\| \leq \inf_{g\in G} \rho(g) = \inf_{g\in G} \sup_{f\in B} \|g-f\|.$$

This proves that $g_o \in \mathrm{cent}_G(B)$.

COROLLARY 9. Every finite dimensional subspace of a non-archimedean normed space over a spherically complete valued division ring $(F,|\cdot|)$ has the Chebyshev center property.

COROLLARY 10. Every spherically complete n.a. normed space admits Chebyshev centers.

Let X be a compact Hausdorff space. It is well known that closed subalgebras of $C(X;\mathbb{R})$ are proximinal. This result was extended by Smith and Ward [7], who proved that every closed subalgebra of $C(X;\mathbb{R})$ has in fact the Chebyshev center property in $C(X;\mathbb{R})$: if $A \subset C(X;\mathbb{R})$ is a closed subalgebra, and $B \subset C(X;\mathbb{R})$ is any non-empty bounded subset, then $\mathrm{cent}_A(B) \neq \emptyset$. (See Theorem 1, [7]). On the other hand, if one considers vector-valued continuous functions, i.e., if X is as before a compact Hausdorff space and E is a normed space, then for suitable E (over the reals), the space $\mathcal{C}(X;E)$ admits centers. For example, if E an arbitrary real Hilbert space, then $\mathcal{C}(X;E)$ admits centers. (See Theorem 2, Ward [8].) Another result true for vector-valued functions is the following: every Stone-Weierstrass subspace of $\mathcal{C}(X;E)$ is proximinal, for suitable E. For example, this is true if E is a Lindenstrauss space over $\mathbb{R}$ (see Blatter [1]), or if E is a uniformly convex Banach space over $\mathbb{R}$ or $\mathbb{C}$ (see Olech [5], Theorem 2).

DEFINITION 11. Let X be a compact Hausdorff space and let $(E,\|\cdot\|)$ be a normed space over a non-archimedean non-trivially valued division ring $(F,|\cdot|)$. A closed vector subspace $W \subset \mathcal{C}(X;E)$ is called a Weierstrass-Stone subspace if there exists a compact Hausdorff space Y and a continuous surjection $\pi: X \to Y$ such that

$$W = \{g \circ \pi : g \in \mathcal{C}(Y;E)\}.$$

Clearly, W contains the constants.

The results of [6], §5, allow a characterization of the Weierstrass-Stone subspaces, since $(F,|\cdot|)$ is a non-archimedean non-trivially valued division ring. Indeed, let $W \subset \mathcal{C}(X;E)$ be a Weierstrass-Stone subspace and let

$$A = \{b \circ \pi ;\ b \in \mathcal{C}(Y;F)\}.$$

Then A is a subalgebra of $\mathcal{C}(X;F)$, containing the constants and such that $\{\pi^{-1}(y);\ y \in Y\}$ is the set of equivalence classes modulo X/A. Moreover, W is an A-module. By [6], Th. 5.5, and the fact that W is closed, if follows that $f \in \mathcal{C}(X;E)$ belongs to W

if, and only if, f is constant on each equivalence class $\pi^{-1}(y)$, $y \in Y$.

Conversely, let $W \subset \mathcal{C}(X;E)$ be the vector subspace of all $f \in \mathcal{C}(X;E)$ which are constant on each equivalence class modulo X/A, where $A \subset \mathcal{C}(X;F)$ is some subalgebra containing the constants. Let Y be the quotient space of X modulo X/A and let $\pi: X \to Y$ be the quotient map. Then Y is a compact Hausdorff space. Clearly, each $f \in W$ factors through π, i.e., $f = g \circ \pi$. By the definition of the quotient topology, $g \in \mathcal{C}(Y;E)$. Hence $W \subset \{g \circ \pi, g \in \mathcal{C}(Y;E)\}$. Conversely, each function of the form $g \circ \pi$ is clearly continuous and constant on each equivalence class modulo X/A.

PROPOSITION 12. Let X, E and $(F,|\cdot|)$ be as in Definition 11, and let $W \subset \mathcal{C}(X;E)$ be a Weierstrass-Stone subspace. Then W is the closure of $A \otimes E$ in $\mathcal{C}(X;E)$, where $A = \{b \circ \pi;\ b \in \mathcal{C}(Y;F)\}$, and π and Y are given by Definition 11.

PROOF. Clearly, $A \otimes E$ is an A-module, and by 5.5, [6], each $f \in W$ belongs to the closure of $A \otimes E$. Indeed, given any equivalence class $\pi^{-1}(y)$, $y \in Y$, and any $\varepsilon > 0$, consider the constant value v of on $\pi^{-1}(y)$, and the constant function $b = 1$ on Y. Then $g = (b \circ \pi) \otimes v$ belongs to $A \otimes E$ and is equal to f throughout $\pi^{-1}(y)$:

$$\|g(x)-f(x)\| = 0 < \varepsilon$$

for all $x \in \pi^{-1}(y)$.

DEFINITION 13. Let X and Z be two topological spaces. A mapping φ from X into the non-empty subsets of Z is called a carrier.

A carrier φ from X into the non-empty subsets of Z is said to be lower semicontinuous if $\{x \in X;\ \varphi(x) \cap G \neq \emptyset\}$ is open in X for every open subset $G \subset Z$.

A continuous mapping $f: X \to Z$ is called a <u>continuous selection for</u> a carrier φ if $f(x) \in \varphi(x)$ for all $x \in X$.

The following result is an obvious consequence of Michael [2], Theorem 2, page 233.

THEOREM 14. Let X be a 0-dimensional compact T_1-space and let $(E, \|\cdot\|)$ be a Banach space over a non-trivially valued division ring $(F, |\cdot|)$. Every lower semicontinuous carrier φ from X into the non-empty, closed subsets of E admits a continuous selection.

THEOREM 15. Let X be a 0-dimensional compact T_1-space and let $(E, \|\cdot\|)$ be a non-archimedean Banach space over a non-trivially valued division ring $(F, |\cdot|)$ such that $\mathrm{cent}(K) \neq \emptyset$ for every non-empty compact subset $K \subset E$.

Then every Weierstrass-Stone subspace $W \subset \mathcal{C}(X;E)$ is proximinal.

PROOF. Let $\pi: X \to Y$ be a continuous surjection of X onto a compact Hausdorff space Y such that

$$W = \{g \circ \pi;\ g \in \mathcal{C}(Y;E)\}.$$

Let $f \in \mathcal{C}(X;E)$ be given with $f \notin W$. Then $\delta = \mathrm{dist}(f;W) > 0$, since W is a closed subspace.

Let us define a carrier φ from Y into the non-empty closed subsets of E. For each $y \in Y$, define

$$\varphi(y) = \{s \in E;\ \sup_{x \in \pi^{-1}(y)} \|f(x) - s\| \leq \delta\}.$$

It is clear that $\varphi(y)$ is closed. Now $\pi^{-1}(y)$ is a compact subset of X and therefore $K = \{f(x);\ x \in \pi^{-1}(y)\}$ is compact. By hypothesis, there exists some $s_o \in E$ such that

$$(*) \qquad \sup_{x\in\pi^{-1}(y)} \|f(x)-s_o\| = \mathrm{rad}(K).$$

We claim that

$$(**) \qquad \mathrm{rad}(K) \le \delta .$$

Indeed, let $g \in W$ be given. Then

$$\mathrm{rad}(K) = \inf_{z\in E} \sup_{x\in\pi^{-1}(y)} \|f(x)-z\| \le$$

$$\le \sup_{x\in\pi^{-1}(y)} \|f(x)-g(x)\|$$

$$\le \|f-g\| .$$

Since g was arbitrary,

$$\mathrm{rad}(K) \le \inf_{g\in W} \|f-g\| ,$$

and (**) is true.

Now, from (*) and (**) it follows that $s_o \in \varphi(y)$. Hence, $\varphi(y) \neq \emptyset$, for all $y \in Y$.

We claim that φ is lower semi-continuous, i.e. that

$$\{y \in Y;\ \varphi(y) \cap G \neq \emptyset\}$$

is open in Y for each open subset $G \subset E$.

Let $y_o \in Y$ be such that $\varphi(y_o) \cap G \neq \emptyset$. Let $s_o \in \varphi(y_o) \cap G$. Then

$$\sup_{x\in\pi^{-1}(y_o)} \|f(x)-s_o\| \le \delta .$$

This means that $f(\pi^{-1}(y_o)) \subset B(s_o;\delta)$, where $B(s_o;\delta) = \{s \in E;\ \|s-s_o\| \le \delta\}$. Notice that $B(s_o;\delta)$ is open, and that $\pi^{-1}(y_o) \subset \subset f^{-1}(B(s_o;\delta))$. Since X is compact and Y is Hausdorff, the map π is closed. Hence there is some saturated open set V in X with

$$\pi^{-1}(y_o) \subset V \subset f^{-1}(B(s_o;\delta)).$$

Then $U = \pi(V)$ is open in Y (because $\pi^{-1}(U) = V$) and for any $y \in U$, $\pi^{-1}(y) \subset V \subset f^{-1}(B(s_o;\delta))$. Hence $f(\pi^{-1}(y)) \subset B(s_o;\delta)$ for all $y \in U$; that is

$$\|f(t)-s_o\| \leq \delta$$

for all $t \in \pi^{-1}(y)$, $y \in U$. But this means that $s_o \in \varphi(y)$ for all $y \in U$, and φ is lower semi-continuous.

By Theorem 14, there is a continuous selection $g \in C(Y;E)$ for φ, i.e., $g(y) \in \varphi(y)$ for all $y \in Y$. Let $w = g \circ \pi$. Then $w \in W$ and, for any $x \in X$ let $y = \pi(x)$. Then

$$\|f(x)-w(x)\| = \|f(x)-g(y)\| \leq \delta.$$

Hence $\|f-w\| \leq \operatorname{dist}(f;W)$, and therefore $w \in \mathcal{P}_W(f)$. This ends the proof that W is proximinal.

When the Banach space $(E,\|\cdot\|)$ admits Chebyshev centers, a better result can be proved, namely that $\operatorname{cent}_W(B) \neq \emptyset$ not only a for $B = \{f\}$, but for equicontinuous bounded sets $B \subset C(X;E)$.

THEOREM 16. Let X be a 0-dimensional compact T_1-space and let $(E,\|\cdot\|)$ be a non-archimedean Banach space over a non-trivially valued division ring $(F,|\cdot|)$. If E admits Chebyshev centers, and $W \subset C(X;E)$ is a Weierstrass-Stone subspace, then $\operatorname{cent}_W(B) \neq \emptyset$ for every non-empty bounded subset $B \subset C(X;E)$ which is equicontinuous at every point of X.

PROOF. Let $\pi: X \to Y$ be a continuous surjection of X onto a compact Hausdorff space Y such that

$$W = \{g \circ \pi;\ g \in C(Y;E)\}.$$

Let $B \subset C(X;E)$ be a non-empty bounded subset which is equicontinuous at every point of X.

Let $\delta = \mathrm{rad}_W(B)$.

CASE I: $\delta > 0$.

Define a carrier φ from Y into the non-empty closed subsets of E by

$$\varphi(y) = \{s \in E;\ \sup_{f\in B}\ \sup_{x\in\pi^{-1}(y)} \|f(x)-s\| \le \delta\}.$$

It is clear that $\varphi(y)$ is closed, for each $y \in Y$. Since $B \subset \mathcal{C}(X;E)$ is bounded,

$$B(y) = \{f(x);\ x \in \pi^{-1}(y),\ f \in B\}$$

is bounded in E, and by hypothesis $\mathrm{cent}(B(y)) \ne \emptyset$, i.e., there exists $s_o \in E$ such that

$$\sup_{f\in B}\ \sup_{x\in\pi^{-1}(y)} \|f(x)-s_o\| = \mathrm{rad}(B(y)).$$

We claim that

$$(*) \qquad \mathrm{rad}(B(y)) \le \delta.$$

Indeed, for any $g \in W$, we have

$$\mathrm{rad}(B(y)) \le \sup_{f\in B}\ \sup_{x\in\pi^{-1}(y)} \|f(x)-g(x)\|$$

because g is constant on $\pi^{-1}(y)$. Hence

$$\mathrm{rad}(B(y)) \le \sup_{y\in Y}\ \sup_{f\in B}\ \sup_{x\in\pi^{-1}(y)} \|f(x)-g(x)\|$$

$$= \sup_{f\in B}\ \sup_{y\in Y}\ \sup_{x\in\pi^{-1}(y)} \|f(x)-g(x)\|$$

$$= \sup_{f\in B} \|f-g\|.$$

Now g was arbitrary, so

$$\mathrm{rad}(B(y)) \le \inf_{g\in W}\ \sup_{f\in B} \|f-g\| = \mathrm{rad}_W(B)$$

and so (*) is true, as claimed.

Therefore $s_o \in \varphi(y)$, and $\varphi(y)$ is non-empty.

We claim that φ is lower semicontinuous, i.e., that

$$\{y \in Y;\ \varphi(y) \cap G \neq \emptyset\}$$

is open in Y, for each open subset $G \subset E$. Let $y_o \in Y$ be such that $\varphi(y_o) \cap G \neq \emptyset$. Choose $s_o \in \varphi(y_o) \cap G$. Then

$$\sup_{f \in B} \sup_{x \in \pi^{-1}(y_o)} \|f(x)-s_o\| \leq \delta .$$

Since $\pi^{-1}(y_o)$ is a compact subset of X, there exists a finite open covering $V_1, V_2, \ldots, V_n$ of $\pi^{-1}(y_o)$, with $V_i \cap \pi^{-1}(y_o) \neq \emptyset$, $1 \leq i \leq n$, such that

$$x, x' \in V_i \Rightarrow \|f(x)-f(x')\| < \delta$$

for all $f \in B$. This is possible because the set $B \subset \mathcal{C}(X;E)$ is equicontinuous at every point of X.

Let $V_o = V_1 \cup V_2 \cup \ldots \cup V_n$. We claim that $\|f(x)-s_o\| \leq \delta$ for all $x \in V_o$ and $f \in B$. Indeed, given $x \in V_o$ choose V_i such that $x \in V_i$, and choose $t \in V_i \cap \pi^{-1}(y_o)$. Then, for all $f \in B$

$$\|f(x)-s_o\| \leq \max(\|f(x)-f(t)\|, \|f(t)-s_o\|) \leq \delta .$$

Therefore

$$\pi^{-1}(y_o) \subset V_o \subset \bigcap_{f \in B} f^{-1}(B(s_o;\delta)).$$

Choose a saturated open set V in X with $\pi^{-1}(y_o) \subset V \subset V_o$. (This is possible because π is a closed map). Then $U = \pi(V)$ is an open neighborhood of y_o in Y, and for every $y \in U$, $\pi^{-1}(y) \in$ $\in V \subset f^{-1}(B(s_o;\delta))$ for all $f \in B$. Hence

$$\|f(x)-s_o\| \leq \delta$$

for all $x \in \pi^{-1}(y)$ and $f \in B$. This means that $s_o \in \varphi(y)$ for all $y \in U$, and so φ is lower semicontinuous.

By Theorem 14 there exists a continuous selection $g: Y \to E$ such that $g(y) \in \varphi(y)$ for all $y \in Y$. Let $w = g \circ \pi$. Then $w \in W$ and for any $x \in X$, let $y = \pi(x)$. Then for any $f \in B$ we have

$$\begin{aligned}\|f(x)-w(x)\| &= \|f(x)-g(y)\| \\ &\leq \sup_{t \in \pi^{-1}(y)} \|f(t)-g(y)\| \leq \delta .\end{aligned}$$

Hence $\sup_{f \in B} \|f-w\| \leq \delta$, and so $w \in \mathrm{cent}_W(B)$.

CASE II: $\delta = 0$.

Now $\mathrm{rad}_W(B) = 0$ implies $B = \{f\}$ and $\mathrm{dist}(f;W) = \mathrm{rad}_W(B) = 0$. Therefore $f \in W$ and there is nothing to prove.

REMARK. In Olech [5] the formula

$$(*) \qquad \mathrm{dist}(f;W) = \sup_{y \in Y} \mathrm{rad}(f(\pi^{-1}(y)))$$

was proved for Weierstrass-Stone subspaces $W \subset \mathcal{C}(X;E)$, where X is compact and E is a uniformly convex Banach space (over R or C). We will show that (*) is a consequence of the Stone-Weierstrass Theorem.

THEOREM 17. Let X be a compact Hausdorff space, and $(E,\|\cdot\|)$ be a normed space over a non-archimedean non-trivially valued division ring $(F,|\cdot|)$. For every $f \in \mathcal{C}(X;E)$ and every Weierstrass-Stone subspace $W \subset \mathcal{C}(X;E)$ we have

$$\mathrm{dist}(f;W) = \sup_{y \in Y} \mathrm{rad}(f(\pi^{-1}(y))),$$

where $\pi: X \to Y$ is the continuous surjection of X onto a compact Hausdorff space Y such that

$$W = \{g \circ \pi;\ g \in \mathcal{C}(Y;E)\}.$$

PROOF. Let $y \in Y$. Then, for every $w \in W$ we have

$$\inf_{z \in E} \sup_{x \in \pi^{-1}(y)} \|f(x)-z\| \leq$$

$$\leq \sup_{x \in \pi^{-1}(y)} \|f(x)-w(x)\| \leq \|f-w\|$$

because w is constant on $\pi^{-1}(y)$. Since w was arbitrary,

$$\mathrm{rad}(f(\pi^{-1}(y))) \leq \mathrm{dist}(f;W)$$

and then

$$\sup_{y \in Y} \mathrm{rad}(f(\pi^{-1}(y))) \leq \mathrm{dist}(f;W).$$

Conversely, by Theorem 6.4, [6], we have

$$\mathrm{dist}(f;W) = \sup_{y \in Y} \inf_{w \in W} \sup_{x \in \pi^{-1}(y)} \|f(x)-w(x)\|.$$

Let $y \in Y$. For each $z \in E$, the constant function $g_z(x) = z$, $x \in X$, belongs to W. Hence, for each $z \in E$

$$\inf_{w \in W} \sup_{x \in \pi^{-1}(y)} \|f(x)-w(x)\| \leq$$

$$\leq \sup_{x \in \pi^{-1}(y)} \|f(x)-z\|.$$

Since z was arbitrary, we have

$$\inf_{w \in W} \sup_{x \in \pi^{-1}(y)} \|f(x)-w(x)\| \leq \mathrm{rad}(f(\pi^{-1}(y)))$$

and from this it clearly follows that

$$\mathrm{dist}(f;W) \leq \sup_{y \in Y} \mathrm{rad}(f(\pi^{-1}(y))).$$

REMARK. In the proof given above we used the following properties of W:

(1) every $w \in W$ is constant on each $\pi^{-1}(y)$, $y \in Y$;

(2) for each $y \in Y$, and $z \in E$, there is some $w \in W$ such

that $w(x) = z$ for all $x \in \pi^{-1}(y)$;

(3) W is an A-module, where A is a subalgebra of $\mathcal{C}(X;F)$ such that $\pi^{-1}(\pi(x))$ is the equivalence class of x modulo X/A, for each $x \in X$.

Hence the following result is true:

THEOREM 18. Let X be a compact Hausdorff space and let $(E,\|\cdot\|)$ be a normed space over a non-archimedean non-trivially valued division ring $(F,|\cdot|)$. Let $A \subset \mathcal{C}(X;F)$ be a subalgebra and let $\pi: X \to Y$ be the quotient map of X onto the quotient space Y of all equivalence classes modulo X/A. Let $W \subset \mathcal{C}(X;E)$ be an A-module such that

(1) every $w \in W$ is constant on each equivalence class $\pi^{-1}(y)$, $y \in Y$;

(2) for each $y \in Y$ and $z \in E$ there is some $w \in W$ such that $w(x) = z$ for all $x \in \pi^{-1}(y)$.

For each $f \in \mathcal{C}(X;E)$ we have

$$\mathrm{dist}(f;W) = \sup_{y\in Y} \mathrm{rad}(f(\pi^{-1}(y))).$$

REMARK. Under the hypothesis of Theorems 17 and 18 it is natural to ask the following question: given a subspace $W \subset \mathcal{C}(W;E)$ which is an A-module, where $A \subset \mathcal{C}(X;F)$ is a separating subalgebra, and W is such that $W(x) = \{w(x);\ w \in W\} \subset E$ is proximinal in E, for each $x \in X$, does it follow that W is proximinal in $\mathcal{C}(W;E)$?

Our next result shows that the answer is yes if X and $(E,\|\cdot\|)$ satisfy the hypothesis of the selection Theorem 14.

THEOREM 19. Let X be a 0-dimensional compact T_1-space and let $(E,\|\cdot\|)$ be a non-archimedean Banach space over a non-trivially valued division ring $(F,|\cdot|)$. Let $A \subset \mathcal{C}(X;F)$ be a separating subalgebra and let $W \subset \mathcal{C}(X;E)$ be a closed A-module such that $W(x)$ is proximinal in E, for every $x \in X$.

Then W is proximinal in $\mathcal{C}(X;E)$.

PROOF. Let $f \in \mathcal{C}(X;E)$ be given with $f \notin W$. Then $\delta = \mathrm{dist}(f;W) > 0$, because W is closed. Let us define a carrier φ from X into the non-empty closed subsets of E. For each $x \in X$, define

$$\varphi(x) = W(x) \cap \{s \in E; \|f(x)-s\| \leq \delta\}.$$

For each $x \in X$, there is some $w \in W$ such that

$$\|w(x)-f(x)\| \leq \mathrm{dist}(f(x);W(x)) \leq \delta$$

and so $\varphi(x) \neq \emptyset$. Clearly, $\varphi(x)$ is closed. We claim that φ is lower semicontinuous. Let $G \subset E$ be open and $\varphi(x_o) \cap G \neq \emptyset$. Choose $s_o \in \varphi(x_o) \cap G$. There is some $w \in W$ such that $s_o = w(x_o)$ and $\|f(x_o)-w(x_o)\| \leq \delta$. Hence $x_o \in (f-w)^{-1}(B(0;\delta))$ where $B(0;\delta) = \{s \in E; \|s\| \leq \delta\}$. By continuity, there is some neighborhood U of x_o such that $w(x) \in G$ and $x \in (f-w)^{-1}(B(0;\delta))$ for all $x \in U$, because $B(0;\delta)$ is open. Then $w(x) \in \varphi(x) \cap G$, for all $x \in U$, and φ is lower semicontinuous.

By Theorem 14 there is a continuous selection $g \in \mathcal{C}(X;E)$ for φ. Then $g(x) \in W(x)$ for all $x \in X$, and so $g \in W$, by [6], Th. 6.4. On the other hand

$$\|f(x)-g(x)\| \leq \delta = \mathrm{dist}(f;W)$$

for all $x \in X$, and therefore

$$\|f-g\| \leq \mathrm{dist}(f;W),$$

i.e., $g \in \mathcal{P}_W(f)$ and W is proximinal.

REMARK. When $W(x)$ is not only proximinal in E but has the Chebyshev center property in E, a better result can be proved, namely that $\mathrm{cent}_W(B) \neq \emptyset$ for every equicontinuous bounded subset

$B \subset \mathcal{C}(X;E)$.

THEOREM 20. Let X and $(E,\|\cdot\|)$ be as in Theorem 19. Let $A \subset \mathcal{C}(X;F)$ be a separating subalgebra and let $W \subset \mathcal{C}(X;E)$ be a closed A-module such that $W(x)$ has the relative Chebyshev center property in E, for every $x \in X$.

Then $\text{cent}_W(B) \neq \emptyset$, for every non-empty equicontinuous bounded subset $B \subset \mathcal{C}(X;E)$.

PROOF. Let $B \subset \mathcal{C}(X;E)$ be a non-empty bounded subset which is equicontinuous at every point of X. Let $\delta = \text{rad}_W(B)$. If $\delta = 0$, then B is a singleton $\{f\}$ with $f \in W$ and there is nothing to prove. Hence we may assume that $\delta > 0$.

Define a carrier φ from X into the non-empty closed subsets of E by

$$\varphi(x) = W(x) \cap \{s \in E;\ \sup_{f\in B} \|f(x)-s\| \leq \delta\}.$$

For each $x \in X$, $B(x) = \{f(x);\ f \in B\}$ is bounded in E, and there is some $w \in W$ such that

$$\sup_{f\in B} \|f(x)-w(x)\| \leq \text{rad}_{W(x)}(B(x)).$$

Now

$$\text{rad}_{W(x)}(B(x)) = \inf_{w\in W} \sup_{f\in B} \|f(x)-w(x)\|$$
$$\leq \inf_{w\in W} \sup_{f\in B} \|f-w\| = \delta.$$

Hence $\varphi(x) \neq \emptyset$. Clearly, $\varphi(x)$ is closed. We claim that φ is lower semicontinuous, i.e. that $\{x \in X;\ \varphi(x) \cap G \neq \emptyset\}$ is open, for every open subset $G \subset E$. Let $x_o \in X$ be such that $\varphi(x_o) \cap G \neq$ $\neq \emptyset$ and choose $s_o \in \varphi(x_o) \cap G$. There is some $w \in W$ such that $s_o = w(x_o)$ and $\sup_{f\in B} \|f(x_o)-w(x_o)\| \leq \delta$.

Hence $x_o \in (f-w)^{-1}(B(0;\delta))$ for every $f \in B$, where $B(0;\delta) = \{s \in E;\ \|s\| \leq \delta\}$. By continuity of w and equicontinuity of $\{f-w;\ f \in B\}$, there is some neighborhood U of x_o in X such that $w(x) \in G$ and $x \in (f-w)^{-1}(B(0;\delta))$ for all $f \in B$ and $x \in U$. Then $w(x) \in \varphi(x) \cap G$, for all $x \in U$, and the carrier φ is lower semicontinuous.

By Theorem 14 there is a continuous selection $g \in \mathcal{C}(X;E)$ for the carrier φ. Then $g(x) \in W(x)$ for all $x \in X$, and so $g \in W$, by [6], 6.4. On the other hand

$$\sup_{f\in B} \|f(x)-g(x)\| \leq \delta$$

for all $x \in X$, and therefore

$$\sup_{f\in B} \|f-g\| \leq \delta = \mathrm{rad}_W(B).$$

COROLLARY 21. Let X and $(E,\|\cdot\|)$ be as in Theorem 19. Assume that E admits Chebyshev centers. For each closed subset $Z \subset X$, the closed vector subspace $W \subset \mathcal{C}(X;E)$, given by

$$W = \{g \in \mathcal{C}(X;E);\ g(x) = 0,\quad x \in Z\},$$

is such that $\mathrm{cent}_W(B) \neq \emptyset$ for every non-empty equicontinuous bounded subset $B \subset \mathcal{C}(X;E)$.

PROOF. W is a $\mathcal{C}(X;F)$-module, and X being a 0-dimensional T_1-space, $\mathcal{C}(X;F)$ is separating over X. On the other hand, for every $x \in X$, $W(x) = 0$, if $x \in Z$; and $W(x) = E$ if $x \notin Z$.

COROLLARY 22. Let X and $(E,\|\cdot\|)$ be as in Theorem 19. If the space E admits Chebyshev centers, then $\mathrm{cent}(B) \neq \emptyset$ for every non-empty equicontinuous bounded subset $B \subset \mathcal{C}(X;E)$.

PROOF. $W = \mathcal{C}(X;E)$ is a $\mathcal{C}(X;F)$-module, and $W(x) = E$ for every $x \in X$. Since $\mathcal{C}(X;F)$ is separating, the result follows from Theorem 20.

REFERENCES

1. BLATTER, J., Grothendieck spaces in approximation theory, Memoirs Amer. Math. Soc. 120 (1972).

2. MICHAEL, E., Selected selection theorems, Amer. Math. Monthly 63 (1956), 233-238.

3. MONNA, A.F., Sur les espaces linéaires normés non-archimédiens, I. Indagationes Mathematicae 18 (1956), 475-483.

4. MONNA, A.F., Remarks on some problems in linear topological spaces over fields with non-archimedean valuation, Indagationes Mathematicae 30 (1968), 484-496.

5. OLECH, C., Approximation of set-valued functions by continuous functions, Colloquium Mathematicum 19 (1968), 285-293.

6. PROLLA, J.B., Topics in Functional Analysis over Valued Division Rings, North-Holland Publ. Co., Amsterdam, 1982.

7. SMITH, P.W. and J.D. WARD, Restricted centers in subalgebras of $C(X)$, Journal of Approximation Theory 15 (1975), 54-59.

8. WARD, J.D., Chebyshev centers in spaces of continuous functions, Pacific Journal of Mathematics 52 (1974), 283-287.

Departamento de Matemática
UNICAMP - IMECC
Campinas, SP - Brazil

Functional Analysis, Holomorphy and
Approximation Theory II, G.I. Zapata (ed.)

ABSTRACT FROBENIUS THEOREM - GLOBAL FORMULATION APPLICATIONS TO LIE GROUPS

Reinaldo Salvitti

The main goal of this work is to give the Global Formulation of the Abstract Frobenius Theorem in the context of Scales of Banach Spaces and to applie it in the construction of Lie Subgroups. The motivation of this work was the study of germs of analytic transformations of C^n that vanish at the origin, $gh(n,C)$, as it was studied by Pisanelli in [1]. The applications of this work, in the scale $gh(n,C)$, will appear in another paper.

1. INTRODUCTION

1.1 DEFINITION. A Scale X of Complex Banach Spaces is a topological vector space, obtained from the union of a family of Complex Banach Spaces X_s, with norm $\| \|_s$ and $0 < s \leq 1$, and such that:

(a) $X_s \subset X_{s'}$, $\| \|_{s'} \leq \| \|_s$ for all pair s, s', such that $0 < s' < s \leq 1$;

(b) $X = \varinjlim X_s$, Hausdorff and sequentially complete.

1.2 EXAMPLES

1.2.1 $X_s = X$ for all s, $0 < s \leq 1$, then X is a Banach Space.

1.2.2 If $X = \bigcup_{n\geq 1} Y_n$ is a Silva Space, letting $X_s = Y_n$ for $\frac{1}{n+1} < s \leq \frac{1}{n}$, we get that every Silva Space are Scale of Banach Spaces.

1.3 DEFINITION. Let X and Y be locally convex sequentially complete spaces, $\Omega \subset X$ open. A map f

$$f\colon \Omega \to Y$$

is LF-analytic if for each map g

$$g\colon W \to \Omega$$

$W \subset C$ open, the map $f \circ g$ is analytic.

1.4 LIE GROUP [2]

Let X be a Hausdorff, sequentially complete locally complex space, $\mathcal{G} \subset X$ open and endowed of a group structure such that the maps

$$\begin{array}{llll} \emptyset\colon \mathcal{G}\times\mathcal{G} & \to \mathcal{G} & \qquad & \mathcal{G} \to \mathcal{G} \\ (x,y) & \to \emptyset(x,y) = xy & & x \to x^{-1} \end{array}$$

are LF-analytics. We then call $\mathcal{G}$ a Lie group.

The linear maps

$$\begin{array}{ll} L(x)\colon & X \to X \\ & h \to \emptyset'_y(x,e)h \end{array}$$

$$\begin{array}{ll} \mathcal{L}(x)\colon & X \to X \\ & h \to \emptyset'_y(x^{-1},x)h \end{array}$$

are the inverses of one another, for each $x \in \mathcal{G}$, where $\underline{e}$ is the group's unit.

From the associative law and differentiating with respect to the second variable we get the Lie equations for the group:

$$\begin{cases} \emptyset'_y(x,y)h = L(\emptyset)\mathcal{L}(y)h \\ \emptyset(x,e) = x \end{cases}$$

$$\begin{cases} \psi'_y(x,y)h = L(\psi)\mathcal{L}(y)h \\ \psi(x,x) = e \end{cases} \qquad \psi(x,y) = \emptyset(x^{-1},y).$$

The LF-analytic map Lh defined by

$$\mathcal{G} \to X$$
$$x \to L(x)h$$

is called infinitesimal transformation of the group $\mathcal{G}$ at vector h. We also have $L(e)h = h$, for each h, $h \in X$.

1.5 LIE ALGEBRA OF THE INFINITESIMAL TRANSFORMATIONS. [1]

Let X be a Hausdorff, sequentially complete locally convex space and $\mathcal{G} \subset X$ open. The vector space $\mathcal{H}(\mathcal{G}:X)$ of LF-analytic maps is a Complex Lie Algebra with the bracket

$$[\xi,\eta] = \xi'\eta - \eta'\xi.$$

When $\mathcal{G}$ is a Lie Group the set of infinitesimal transformations is a Lie Subalgebra of $\mathcal{H}(\mathcal{G}:X)$. This Subalgebra is called the Lie Algebra of the group $\mathcal{G}$. It is possible to endow X with a Lie algebra structure, isomorphic to that infinitesimal transformation by defining

$$[h,k] = L'(e)kh - L'(e)hk.$$

1.6 Let X and Y be complex, Hausdorff, sequentially complete locally convex space. Let $A \subset X$ and $B \subset Y$ be open subsets and

$$f\colon A \times B \times X \to Y$$

a LF-analytic map. The map f satisfies the integrability condition if

$$f'_x(x,y)kh + f'_y(x,y)(f(x,y)k)h$$

is symmetric for all $(h,k) \in X \times Y$ and each pair $(x,y) \in A \times B$.

1.7 LOCAL FROBENIUS THEOREM. [3]

Let $X = \bigcup_{0<s\leq 1} X_s$ and $Y = \bigcup_{0<s\leq 1} Y_s$ be Scale of Banach Spaces. Consider the system (E)

$$(E)\quad \begin{cases} y'(x)h = f(x,y)h \\ y(x_o) = y_o \end{cases}$$

where $x_o \in X$ and $y_o \in Y$ and

$$f:\ \bigcup_{0<s\leq 1} B_s(0,R) \times \bigcup_{0<s\leq 1} D_s(0,R) \times X \to Y$$

with $B_s(0,R)$: ball of center 0 and radius R of X_s,

$D_s(0,R)$: ball of center 0 and radius R of Y_s.

Suppose that f restricted to $B_s(0,R) \times D_s(0,R) \times X_s$ takes its values inside $Y_{s'}$, $0 < s' < s \leq 1$, is G-analytic and

$$\|f(x,y)h\|_{s'} \leq C \frac{s}{s-s'} \|h\|_s$$

with $f(x,y)h$ satisfying the integrability condition. Then there is a unique LF-analytic map defined on $\bigcup_{0<s\leq 1} B_s(0,\rho) \times \bigcup_{0<s\leq 1} B_s(0,\rho) \times \bigcup_{0<s\leq 1} D_s(0,\rho)$ taking values inside Y, where $\rho = \dfrac{R}{192e^2C}$.

REMARK. With the same system, in Frobenius Theorem, taking f

$$f:\ \Big(\bigcup_{0<s\leq 1} B_s(w_o,R)\Big) \times \Big(\bigcup_{0<s\leq 1} D_s(z_o,R)\Big) \times X \to Y$$

such that f restricted to $B_s(w_o,R) \times D_s(z_o,R) \times X_s$ taking its values in $Y_{s'}$, is G-analytic with $\|f(x,y)h\|_{s'} \leq C \frac{s}{s-s'} \|h\|_s$ and satisfying the integrability condition we get the existence of a unique LF-analytic map $y(x,x_o,y_o)$ defined in

$$\Big(\bigcup_{0<s\leq 1} B_s(w_o,\rho)\Big) \times \Big(\bigcup_{0<s\leq 1} B_s(w_o,\rho)\Big) \times \Big(\bigcup_{0<s\leq 1} D_s(z_o,\rho)\Big)$$

taking its values in Y, with $\rho = \dfrac{R}{192e^2C}$.

2. INTEGRAL MANIFOLD

We always denote X, Y, Z, etc, complex, Hausdorff, sequentially complete, locally convex spaces.

2.1 ANALYTIC MANIFOLD

Let M be a subset of X, with $M = \bigcup_{i\in I} O_i$, where:

(a) there is a bijection ϕ_i of O_i onto $\phi_i(O_i)$, and open subset of space Y_i ;

(b) $\phi_i(O_i\cap O_j)$ is open in $\phi_i(O_i)$ for all i, j of I;

(c) $\phi_i\circ\phi_j^{-1}: \phi_j(O_i\cap O_j) \to \phi_i(O_i\cap O_j)$ is analytic for all $i,j \in I$;

(d) $\phi_i^{-1}: \phi_i(O_i) \to X$ is analytic;

(e) $(\phi_i^{-1})'(z_i): Y_i \to X$ is injective, for all $z \in O_i$ and $i \in I$, where $z_i = \phi_i(z)$.

Each pair (ϕ_i,O_i) is <u>a chart</u> of the collection $\mathfrak{A} = (\phi_i,O_i)_{i\in I}$ and $\mathfrak{A}$ is <u>an atlas</u> of M.

There is a unique topology in M which makes the maps ϕ_i homeomorphisms and the subsets O_i open [4].

A couple of atlases, $\mathfrak{A}$ and $\tilde{\mathfrak{A}}$ in M are compatible if:

(a) the identities $(M,\mathfrak{A}) \to (M,\tilde{\mathfrak{A}})$ and $(M,\tilde{\mathfrak{A}}) \to (M,\mathfrak{A})$ are continuous;

(b) the maps $\tilde{\phi}_j\circ\phi_i^{-1}$ and $\phi_i\circ\tilde{\phi}_j^{-1}$ are analytic in $\phi_i(O_i\cap\tilde{O}_j)$ and $\tilde{\phi}_j(\tilde{O}_j\cap O_i)$ respectively for all i, j.

A <u>analytic manifold</u> M in X is a subset of X together with a equivalent class of atlas of giving by the compatibility relation.

2.2 TANGENT SPACE

Let M be a analytic manifold in X and $z \in M$. The

tangent space to M at z is the vector space

$$T_z(M) = (\phi_i^{-1})'(z_i)Y_i$$

where $z_i = \phi_i(z)$.

The vector space $T_z(M)$ does not depend on the chart (ϕ_i, O_i) that was chosen [2].

2.3 INTEGRABLE DISTRIBUTION

Let U be an open subset of X and H a closed vector subspace of X with topological supplement.

2.3.1 DEFINITION. A distribution $\mathcal{D}$ in U is a map f,

$$\begin{aligned} f:\ U\times H &\to X \\ (a,h) &\to f(z)h \end{aligned}$$

analytic and such that for each $z \in U$ the map

$$\begin{aligned} f(z):\ H &\to X \\ h &\to f(z)h \end{aligned}$$

is linear and the vector spaces H and $f(z)H$ are top-linear isomorphics.

2.3.2 DEFINITION. Let M contained in U be an analytic manifold. We call M an integral manifold of a distribution $\mathcal{D}$ in U if for all $z \in M$

$$T_z(M) = f(z)H.$$

Whenever it is convenient we will denote by M_z an integral manifold M that contains z and that we will say that M passes through z.

2.3.3 DEFINITION. Let $\mathcal{D}$ be a distribution in U.

(a) We call $\mathcal{D}$ integrable if for each z, $z \in U$, there is an integrate manifold of $\mathcal{D}$ passing through z.

(b) A <u>integrable distribution</u> $\mathcal{D}$ has the (P) property if for each $z_o \in U$ there is a continuous projection p onto $f(z_o)H$ and a neighborhood of z_o, $V(z_o) \subset U$, such that for each $\underline{a}$, $\underline{a} \in V(z_o)$, the restriction of p to a neighborhood of $\underline{a}$, $W_a \subset M_a$, is a chart.

Furthermore there is an analytic map $z = z(x,a)$,

$$z\colon V(p(z_o)) \times V(z_o) \to X$$
$$(x,a) \to z(x,a)$$

where $V(p(z_o))$ is an open neighborhood of $p(z_o)$ in $f(z_o)H$ and for each $a \in V(z_o)$, the restriction of $z(x,a)$ to $p(W_a) \times \{a\}$ is the inverse of p restrict on W_a.

REMARK. We always can consider $p(V(z_o)) \subset V(p(z_o))$ because p is continuous.

2.4 THEOREM. Let $\mathcal{D}$ be an integrable distribution in U with the (P) property. Let z_o be contained in U and suppose that for each z, $z \in V(z_o)$, the map

$$p\circ f(z)\colon H \to f(z_o)H$$

is invertible and for each h, $h \in f(z_o)H$, the map

$$V(z_o) \to H$$
$$z \to [p\circ f(z)]^{-1}h$$

is LF-analytic. Then for $\underline{a}$ in a neighborhood of z_o and $\underline{x}$ in a neighborhood of $p(z_o)$ in $f(z_o)H$ the system (S),

$$(S)\quad \begin{cases} z'(x,a) = f(z) \circ [p\circ f(z)]^{-1} \\ z(p(a),a) = a \end{cases}$$

is satisfied, where $z'(x,a)$ is the differential of z with respect to x.

2.5 THEOREM. Let $\mathcal{D}$ be a distribution in U and for each $z_o \in U$ suppose that there is a continuous projection p onto $f(z_o)H$ such that the map

$$p\circ f(z) : H \to f(z_o)H$$

is invertible for all z, $z \in W(z_o)$, where $W(z_o)$ is a neighborhood of z_o, and for each $h \in H$ the map

$$\begin{array}{ccc} W(z_o) & \to & H \\ z & & [p\circ f(z)]^{-1}h \end{array}$$

is LF-analytic. If the system (S)

$$(S)\quad \begin{cases} z'(x,a) = f(z)\,[p\circ f(z)]^{-1} \\ z(p(a),a) = a \end{cases}$$

has analytic solution $z(x,a)$

$$z(x,a) : V(p(z_o)) \times V(z_o) \to W(z_o)$$

then $\mathcal{D}$ is an integrable distribution with the (P) property, where

$V(p(z_o))$ is a neighborhood of $p(z_o)$ in $f(z_o)H$ and

$V(z_o)$ is a neighborhood of z_o in U.

2.6 COROLLARY 1. Under the same hypotheses (2.5) we have

$$z(y,z(x,a)) = z(y,a)$$

for $\underline{a}$ in a neighborhood of z_o in U and x and y in a neighborhood of $p(z_o)$ in $f(z_o)H$.

2.7 COROLLARY 2. Let S_o be the topological supplementary of $f(z_o)H$, parallel to p. If $\underline{a}$ belongs to $z_o + S_o$ then $z(p(z_o),a) = a$, where $\underline{a}$ is in a neighborhood of z_o.

2.8 Under the hypotheses of the Theorem (2.5) we define two analytic maps, F and G. We take $V(z_o)$, without lost of generality, like a product of neighborhoods of $p(z_o)$ in $f(z_o)H$ and

$q(z_o)$ in S_o, $q = I-p$. We define

$$F: V(z_o) \to X \quad F(a) = p(a) + z(p(z_o),a) - p(z_o)$$

$$G: V(z_o) \to X \quad G(a) = z(p(A), A-p(A) + p(z_o)).$$

We see that $F(z_o) = z_o$ and $G(z_o) = z_o$.

2.9 THEOREM. In convenient neighborhoods of z_o we have

(i) $(G\circ F)(a) = a$ (ii) $(F\circ G)(A) = A$

2.10 REMARK. Let X be a complex Banach space. Then

$$F'(G(A))G'(A) = I_X \quad \text{and} \quad G'(F(a))F'(a) = I_X.$$

If $A = a = z_o$ we have

$$F'(z_o)G'(z_o) = I_X \quad \text{and} \quad G'(z_o)F'(z_o) = I_X.$$

By the Inverse Function Theorem there is a neighborhood of z_o where F and G are inverses to each other.

2.11 THEOREM. In a convenient neighborhood of z_o we have

$$F'(a)f(a)H = f(z_o)H.$$

2.12 DEFINITION. A distribution $\mathcal{D}$ is <u>involutive</u> if $[f(z)h,f(z)k] \in f(z)H$ for all $h,k \in H$ and for all $z \in U$, where the bracket $[\ ,\ ]$ is

$$[f(z)h,f(z)k] = f'(z)(f(z)k)h - f'(z)(f(z)h)k$$

and $f'(z)h$ is differential of f with respect to z for h fixed.

2.13 THEOREM. Let $\mathcal{D}$ be a distribution and for all $z_o \in U$ we have a continuous projection p onto $f(z_o)H$. If for all z_o the map

$$p\circ f(z): H \to f(z_o)H$$

is invertible, where z belongs to a neighborhood $V(z_o)$ and for each $h \in f(z_o)H$ we have that the map

$$\begin{aligned} V(z_o) &\to H \\ z &\to [p \circ f(z)]^{-1}h \end{aligned}$$

is analytic then $\mathcal{D}$ is an involutive distribution if only if the bracket $[\phi(z)h, \phi(z)k]$ belongs to $f(z)H$ for all $h,k \in f(z_o)H$, where $\phi(z)h = f(z)[p \circ f(z)]^{-1}h$.

2.14 THEOREM. Let $\mathcal{D}$ be a distribution with the same hypotheses of Theorem (2.4). Then the second term of

$$z'(x,a)h = f(z)[p \circ f(z)]^{-1}h$$

satisfies the integrability condition.

2.15 COROLLARY. On integrable distribution with the property (P) is involutive.

2.16 THEOREM. Let $\mathcal{D}$ be an involutive distribution and $z_o \in U$ such that p is a continuous projection onto $f(z_o)H$. Suppose as well that

$$p \circ f(z) \colon H \to f(z_o)H$$

is invertible for each z belonging to $V(z_o)$ and for each $h \in f(z_o)H$ fixed, the map

$$\begin{aligned} V(z_o) &\to H \\ z &\to [p \circ f(z)]^{-1}h \end{aligned}$$

is analytic. Then the second term of the equation

$$z'(x,a)h = f(z)[p \circ f(z)]^{-1}h$$

satisfies the integrability condition for all $z \in V(z_o)$.

2.17 LOCAL FROBENIUS THEOREM FOR A DISTRIBUTION IN A SCALE OF BANACH SCALES

We can now apply the Theorem (1.7) to an involutive distribution. We assume that if $X = \bigcup_{0<s\leq 1} X_s$ is a Scale of Banach Spaces and H is closed vectorial subspace then H is also a Scale of Banach Spaces, in fact

$$H = \lim_{\rightarrow} (H \cap X_s).$$

THEOREM. Let $\mathfrak{D}$ be an involutive distribution in U, $U \subset X$ open and $X = \bigcup_{0<s\leq 1} X_s$ a Scale of Banach Spaces. We suppose that for each $z_o \in U$ there is a continuous projection p onto $f(z_o)H$ such that the map

$$p \circ f(z)\colon H \rightarrow f(z_o)H$$

is invertible for $z \in V(z_o) = \bigcup_{0<s\leq 1} B_s(z_o,R)$ and for each $h \in f(z_o)H$ the map

$$\begin{aligned} v(z_o) &\rightarrow H \\ z &\rightarrow [p\circ f(z)]^{-1}h \end{aligned}$$

is LF-analytic. Consider the system

$$(S)\quad \begin{cases} z'(x,a)h = f(z)\,[p\circ f(z)]^{-1}h \\[2ex] z(p(a),a) = a \end{cases}$$

where $\|f(z)[p\circ f(z)]^{-1}h\|_{s'} \leq C \frac{s}{s-s'}\|h\|_s$ $z \in V(z_o)$, $h \in f(z_o)H$. Then the distribution $\mathfrak{D}$ is integrable.

PROOF. By (2.15) we have the integrability condition for $z'(x,a) = f(z)[p\circ f(z)]^{-1}h$. Applying (1.7) and (2.5) we have that $\mathfrak{D}$ is a integrable distribution.

2.18 THEOREM. Let $\mathcal{G}$, $\mathcal{G} \subset X$, be a Lie Group and H a Lei Subalgebra and as well as a closed subset of X. Let $L(z)H$ be the distribution in $\mathcal{G}$ where L is an infinitesimal transformation of the group $\mathcal{G}$. If ξ and η are analytic maps

$$\xi,\eta\colon \Omega \rightarrow X, \quad \Omega \subset \mathcal{G}$$

such that $\xi(z)$ and $\eta(z)$ belongs to $L(z)H$ then $[\xi,\eta]_z \in L(z)H$ where $[\xi,\eta]_z = \xi'(z)\eta(z) - \eta'(z)\xi(z)$.

2.19 COROLLARY 1. The distribution $L(z)H$ is involutive.

2.20 COROLLARY 2. We suppose that for each $z_o \in \mathcal{G}$ there is a continuous projection onto $f(z_o)H$ such that the map

$$p \circ L(z): H \to L(z_o)H$$

is invertible for each $z \in V(z_o)$. Then the second term of the equation

$$z'(x,a)h = L(z)[p \circ L(z)]^{-1}h$$

satisfies the integrability condition.

3. FROBENIUS THEOREM - GLOBAL FORMULATION

First of all we suppose that we are always with the (2.5) hypotheses. Then we have an integrable distribution $\mathcal{D}$ in U with the (P) property. Then for a neighborhood of z_o, $z_o \in U$, the integrable manifolds are solutions of

$$(S) \begin{cases} z'(x,a) = f(z)[p \circ f(z)]^{-1} \\ z(p(a),a) = a \end{cases}$$

We will prove that for each $z_o \in U$ there is a <u>connected maximal integrable manifold</u> $\mathfrak{F}_{z_o}$, i.e., every connected integrable manifold of $\mathcal{D}$ passing through z_o is contained in $\mathfrak{F}_{z_o}$.

3.1 A NEW TOPOLOGY

We define for U a new topology τ_N by a complete system of neighborhoods at each point $z \in U$ that will be denote by $\mathcal{V}_z$.

A neighborhood of z_o is a subset of U that contains the image by $z = z(x,z_o)$ of a neighborhood of $p(z_o)$ in $f(z_o)H$, $z(x,z_o)$ being the solution of the system (S) with the initial condition $z(p(z_o),z_o) = z_o$.

(1) For each z_o, we have $\mathcal{V}_{z_o} \neq \emptyset$ because the system (S) has solution.

(2) For each z_o and $V \in \mathcal{V}_{z_o}$ we have $z_o \in V$ because $z(p(z_o),z_o) = z_o$.

(3) If $V \in \mathcal{V}_{z_o}$ and $V' \supset V$ then $V' \in \mathcal{V}_{z_o}$.

(4) If V and W belongs to $\mathcal{V}_{z_o}$ then $V \cap W$ belongs to $\mathcal{V}_{z_o}$ because the solution $z(x,z_o)$ is injective.

(5) We will prove that if $V \in \mathcal{V}_{z_o}$ there is $O \in \mathcal{V}_{z_o}$ such that $O \subset V$ and $O \in \mathcal{V}_z$ for all $z \in O$. The proof of this will require two lemmas.

3.2 LEMMA. Since z_1 belongs to a convenient neighborhood of z_o in M_{z_o} we have that $(F \circ \phi_1)(x) - F(z_1)$ has values in $f(z_o)H$ when x is taken in a convenient neighborhood of $p_1(z_1)$ of $f(z_1)H$. Here

F is the map defined in (2.8),

p_1 is a continuous projection onto $f(z_1)H$,

ϕ_1 is the solution of (S) replacing p and z_o by p_1 and z_1 respectively.

We denote $\phi_1(x,z_1) = \phi_1(x)$ and $z(x,z_o) = z(x)$.

PROOF. By the Theorem (2.11) we have

$$F'(a)f(a)H = f(z_o)H$$

where $\underline{a}$ belongs to a convenient neighborhood $V(z_o)$.

Let $V(p(z_o))$ be a convenient neighborhood in $f(z_o)h$ such that $z(V(p(z_o))) \subset V(z_o)$.

Let $z_1 \in z(V(p(z_o)))$. There is a connected neighborhood $W(p_1(z_1))$ in $f(z_1)H$ such that

$$\phi_1(W(p_1(z_1))) \subset V(z_o).$$

We have

$$[(F\circ\phi_1)(x) - F(z_1)]'f(z_1)H = F'(\phi_1(x))\phi_1'(x)f(z_1)H =$$

$$= F'(\phi_1(x))f(\phi_1(x))[p_1\circ f(\phi_1(x))]^{-1}f(z_1)H =$$

$$= F'(\phi_1(x))f(\phi_1(x))H.$$

But $\phi_1(x) \in V(z_o)$ hence

$$[F\circ\phi_1(x) - F(z_1)]'f(z_1)H = f(z_o)H.$$

Applying $q = I-p$ to the two terms of the last equation we have

$$[(q\circ F\circ\phi_1)(x) - (q\circ F)(z_1)]'f(z_1)H = 0.$$

Since $(q\circ F\circ\phi_1)(x_1) - (q\circ F)(z_1) =$ constant, because $W(p_1(z_1))$ is connected, we have, replacing x_1 by $p_1(z_1)$,

$$(q\circ F\circ\phi_1)(x_1) - (q\circ F)(z_1) = 0$$

and therefore

$$[(F\circ\phi_1)(x) - F(z_1)] \in f(z_o)H, \qquad x \in W(p_1(z_1)).$$

3.3 LEMMA. There is a neighborhood $V(p(z_o))$ in $f(z_o)H$ such that for all z_1, $z_1 \in z(V(p(z_o)))$, there exists a neighborhood $W(p_1(z_1))$ in $f(z_1)H$ such that $\phi_1(W(p_1(z_1)) \subset z(V(p(z_o)))$.

PROOF. By the Theorems (2.9) and (2.11) we have a neighborhood $V(z_o)$ such that for $a \in V(z_o)$

$$F(a)f(a)H = f(z_o)H$$

and

$$(G \circ F)(a) = a.$$

As in (3.2) let $V(p(z_o))$ such that $z(V(p(z_o))) \subset V(z_o)$ and $z_1 \in z(V(p(z_o)))$. We denote $x_1 = p(z_1)$.

Let $V(0)$ be a neighborhood in $f(z_o)H$ such that

$$x_1 + V(0) \subset V(p(z_o)) \quad \text{and} \quad F(z_1) + V(0) \subset V(z_o).$$

The maps ϕ_1 and $F \circ \phi_1$ are continuous hence there is a neighborhood $W = W(p_1(z_1))$ in $f(z_1)H$ such that $\phi_1(W) \subset V(z_o)$ and

$$(F \circ \phi_1)(W) - F(z_1) \subset V(0).$$

Then

$$(F \circ \phi_1)(W) \subset F(z_1) + V(0).$$

Applying G to both terms we have

$$\phi_1(W) \subset G(F(z_1) + V(0)).$$

Let us see what is $G(F(z_1) + V(0))$

$$F(z_1) = p(z_1) + z(p(z_o),z_1) - p(z_o) = x_1 + z_o - p(z_o).$$

Taking $t \in V(0)$ we have

$$G(F(z_1)+t) = G(x_1+z_o-p(z_o)+t) =$$

$$= z(p(x_1+z_o-p(z_o)+t), x_1+z_o-p(z_o)+t-p(x_1+z_o-p(z_o)+t)-p(z_o)) =$$

$$= z(x_1+t, x_1+z_o-p(z_o)+t-x_1-t+p(z_o)) = z(x_1+t,z_o) = z(x_1+t).$$

Then $G(F(z_1) + V(0) = z(x_1 + V(0),z_o) = z(x_1 + V(0))$ hence

$$\phi_1(W) \subset z(x_1+V(0)) \subset z(V(p(z_o))).$$

Now it is simple to prove the property 5.

Let $V \in \mathcal{V}_{z_o}$, by the (3.3) there is $O = z(V(p(z_o))) \subset V$ such that for each $z_1 \in O$ there is $W = W(p_1(z_1))$ with $\phi_1(W) \subset z(V(p(z_o)))$, (recall $W \in \mathcal{V}_{z_1}$).

With those 5 properties it is defined a new topology in U that we denote τ_N.

3.4 THEOREM. The identity $(U,\tau_N) \to U$ is continuous.

PROOF. Let $V(z_o)$ be a neighborhood of z_o. The solution of the system (S) is continuous hence there is a neighborhood $V(p(z_o))$ such that $z(V(p(z_o))) \subset V(z_o)$. The set $z(V(p(z_o)))$ is a neighborhood of z_o in (U,τ_N) hence the inclusion above is continuous.

3.5 THEOREM. Let M be an integrable manifold of the distribution $\mathfrak{D}$ in U. Then the inclusion

$$M \to (U,\tau_N)$$

is continuous.

To prove this theorem we use two lemmas like (3.2) and (3.3).

3.6 COROLLARY. For each $z_o \in U$ let z be the solution of the system (S). The map z, restrict to a neighborhood of $p(z_o)$, is continuous one taking values in (U,τ_N).

As a consequence of the last corollary there is a fundamental system of connected neighborhood in (U,τ_N) because H is a locally convex space. Then the connected components of (U,τ_N) are open.

Now we consider in (U,τ_N) the connected component $\mathfrak{F}_{z_o}$ for each z_o. It is easy to prove that $\mathfrak{F}_{z_o}$ is an integrable manifold of the distribution $\mathfrak{D}$ in U. From the last theorem if M_{z_o} is a connected integrable manifold which pass through z_o then M_{z_o} is connected in (U,τ_N), so $M_{z_o} \subset \mathfrak{F}_{z_o}$. Then $\mathfrak{F}_{z_o}$ is the maximal integrable manifold of the distribution $\mathfrak{D}$.

3.7 THE EXISTENCE OF A MAXIMAL INTEGRABLE MANIFOLD IN SCALE OF BANACH SCALES.

The existence of a maximal integrable manifold came from the solution of the system (S) in the neighborhood of each point z_o

of an open set U. Then under, the hypotheses of (2.17), an involutive distribution $\mathcal{D}$ in a Scale of Banach Spaces is integrable and for each point $z_o \in U$ there is a connected maximal integral manifold which contains z_o.

3.8 REMARK. 1. We constructed the maximal integral manifold based on the existence of solutions of the system (S) for each $z_o \in U$ where the distribution is defined. It was necessary the existence of a family of projections p_z, $z \in U$. The Theorem (3.5) shows that the existence of a connected maximal integral manifold does not depend on the family of projections used, i.e., if p_z, $z \in V$ and p'_z, $z \in U$ are two families of projections satisfying the hypothesis of Theorem (2.5) then they produce the same connected maximal integral manifolds.

2. If X is a complex Banach Space any family of projections satisfies the hypothesis of Theorem (2.5). We will see this proof in another paper.

4. APPLICATION TO A LIE GROUP

4.1 THEOREM. Let X be a complex, Hausdorff, sequentially complete, locally convex sapce and $\mathcal{G}$, $\mathcal{G} \subset X$, a Lie Group. Let H be a Lie subalgebra of X with topological supplementary. We consider the distribution $\mathcal{D}$ of $\mathcal{G}$ giving by L(z)H, where L is the infinitesimal transformation of the group $\mathcal{G}$. We also suppose that for each $z_o \in \mathcal{G}$ there is a projection p onto $L(z_o)H$ such that

$$p \circ L(z) : H \to L(z_o)H$$

is invertible for all $z \in W(z_o)$ and the system

$$(S) \begin{cases} \phi(x,a)h = L(\phi)[p \circ L(\phi)]^{-1}h & h \in L(z_o)H \\ \phi(p(a),a) = a \end{cases}$$

has holomorphic solution $\phi(x,a)$

$$\phi(x,a) : V(p(z_o)) \times V(z_o) \to W(z_o)$$

where $V(p(z_o))$ is a neighborhood in $L(z_o)H$.

Then for each point $z \in \mathcal{G}$ there is a connected maximal integral manifold and that one which $\underline{e}$, the unit of the group, belongs to is a Lie subgroup of $\mathcal{G}$.

PROOF. We have to prove only that the connected maximal integral manifold who passes through $\underline{e}$ is a Lie Subgroup of $\mathcal{G}$. We need two lemmas.

4.2 LEMMA. Let $\mathcal{G}$ be endowed with the new topology τ_N defined in (3.1). Then for each a, $a \in \mathcal{G}$, the translation

$$\begin{aligned} (\mathcal{G},\tau_N) &\to (\mathcal{G},\tau_N) \\ z &\to az = \phi(a,z) \end{aligned}$$

is continuous, where ϕ is the operation of the group $\mathcal{G}$.

PROOF. First of all we will see that if N is a integral manifold of the distribution $\mathcal{D}$ then aN is also a integral manifold of $\mathcal{D}$. In fact, if $z(x)$ is a parametrization of points of N, where $x \in Y$, Y parameter space. Consider $\phi(a,z(x))$,

$$[\phi(a,z(x))]'_x\, Y = \phi'_2(a,z(x)) \cdot z'(x)Y.$$

From the Lie equations of the group $\mathcal{G}$ and the fact that N is an integral manifold we have

$$[\phi(a,z(x))]'_x\ Y = L(\phi(a,z(x)))\mathcal{L}(z(x))L(z(x))H$$

$$[\phi(a,z(x))]'_x\ Y = L(\phi(a,z(x))H.$$

Then the tangent space at az is $L(\phi(a,z)H)$ and aN is an integral manifold of $\mathcal{D}$.

Let z_o belong to $\mathcal{G}$ and W be a neighborhood of a z_o in $(\mathcal{G},\tau_N)$. From (3.5) we know that the inclusion

$$aN \to (\mathcal{G},\tau_N)$$

is continuous. Hence there is a neighborhood $V(p(z_o))$ in $L(z_o)H$ such that $\phi(a,z(V(p(z_o)),z_o)) \subset W$.

Since $z(V(p(z_o)),z_o)$ is a neighborhood of z_o in $(\mathcal{G},\tau_N)$ the translation

$$(\mathcal{G},\tau_N) \to (\mathcal{G},\tau_N)$$
$$a \to az$$

is continuous.

4.3 LEMMA. The connected maximal integral manifold Me is a group with the operation of $\mathcal{G}$.

PROOF. Let x, y belong to Me. From (4.2) xMe is a connected integral manifold which passes through x. Since $x \in Me$, $xMe \subset Me$ and $xy \in Me$.

Analogously $x^{-1}Me$ contains $\underline{e}$ so $x^{-1}Me \subset Me$ therefore $x^{-1} \in Me$.

PROOF OF THE (4.1). We may, without lost of generality, suppose $e = 0$. If p is a projection onto $L(e)H = H$ we have $p(e) = e$. We denote $\Omega = p^{-1}(V(p(e)))$ and we define two functions f and g, inverse to each other,

$$f(z) = z - \phi(p(z)) - p(z)$$

$$g(w) = \phi(p(w) + w - p(w))$$

where $\phi(x) = \phi(x,0)$.

$$f \circ g = I_\Omega \quad \text{and} \quad g \circ f = I_\Omega .$$

Because f and g are continuous there are $U,V,W,W = \Omega \cap G$, (ϕ_1,U,V,W) a local group [2] around 0, where

$$\phi_1(z,w) = f[\phi(g(z),g(w))].$$

We have

$$L_1(z)h = (\phi_1)'_w(z,0)h = f'(g(z))\phi'_y(g(z),0)\cdot g'(0)h$$

but for $z \in V(p(e))$ we have $g(z) = \phi(z)$.

Taking $h \in H$

$$g'(0)h = \phi'(0)h = L(\phi(0))(p\circ L(\phi(0)))^{-1}h = h$$

therefore

$$L_1(z)h = f'(g(z))L(g(z))h.$$

As $(f\circ g)(z) = z$, $z \in \Omega$, we have

$$f'(g(z))g'(z)h = h, \qquad h \in X,$$

and so
$$f'(g(z))g'(z)H = H.$$

Taking $z \in V(p(e))$ we have $g(z) = \phi(z)$ and $g'(z) = \phi'(z)$ hence

$$\phi'(z)H = L(\phi(z)H$$

$$g'(z)H = L(g(z))H$$

so

$$f'(g(z))L(g(z))H = H.$$

Then

$$L_1(z)H = H$$

and so
$$\mathcal{L}_1(z)H = H,$$

where $z \in V(p(e))$.

So if z and w belong to a neighborhood of $\underline{e}$ in H we have $L_1(z)\mathcal{L}_1(w)H \subset H$ (*).

Take z and w in a neighborhood of $\underline{e}$ in H, $h \in H$ and consider the systems

$$(S_1)\quad \begin{cases} (\phi_1)'_w(z,w)h = L_1(\phi_1)\mathcal{L}_1(w)h \\ \phi_1(z,e) = z \end{cases}$$

$$(S_2)\quad \begin{cases} (\psi_1)'_w(z,w)h = L_1(\psi_1)\mathcal{L}_1(w)h \\ \psi_1(z,z) = e. \end{cases}$$

The solution of (S_1) is

$$zw = \phi_1(z,w) = \sum_{m\geq 0} \left[\delta^{(w-e,L_1\mathcal{L}_1(w-e)}{}_{\pi}\right]^m (e,z),\ [5],$$

where π is the projection,

$$X \times X \to X$$
$$(z,w) \to w$$

and

$$[\delta^{(k,fk)}{}_{\pi}](x,y) = \pi'_x(x,y)k + \pi'_y(x,y)(f(x,y)k).$$

The solution of (S_2) is

$$z^{-1} = \psi_1(z,e) = \sum_{m\geq 0} \left[\delta^{(e-z,L_1\mathcal{L}_1(e-z))}{}_{\pi}\right]^m (z,e),\quad [5].$$

From (*) each term of the series above is in H and since H is closed zw and z^{-1} belong to H.

We then have a local group $G_1 \subset H$,

$$G_1 = (\phi_1,U_1,V_1,W_1).$$

Also the map g is a local homomorphism between $\mathcal{G}_1$ and $\mathcal{G}$. Consider the local group in $\mathcal{G}$, [6],

$$(\phi,g(U_1),g(V_1),g(W_1))$$

and the topological group that passes through $\underline{e}$

$$N = \bigcup_{n\geq 1} [g(U_1)]^n = \bigcup_{n\geq 1} \bigcup_{a\in[g(U_1)]^{n-1}} ag(U_1).$$

We can take U_1 conncected then follows that $g(U_1)$ is connected so N is connected too. Since the set $g(U_1)$ is a connected integral manifold which passes through $\underline{e}$, N is a connected integral manifold which contains e. Hence $N \subset \mathfrak{F}_e$.

The set N is open in $\mathfrak{F}_e$ because $g(U_1) = \emptyset(V(p(e)))$ is open in $(\mathcal{G},\tau_N)$. The set $\bigcup_{a\in\mathfrak{F}_e/N} aN$ is open therefore N is closed in $\mathfrak{F}_e$, since $\mathfrak{F}_e$ is connected $N = \mathfrak{F}_e$. So $\mathfrak{F}_e$ is a topological group.

Now we prove that the operations

$$\begin{aligned} \mathfrak{F}_e\times\mathfrak{F}_e &\to \mathfrak{F}_e \\ (z,w) &\to zw \end{aligned}$$

$$\begin{aligned} \mathfrak{F}_e &\to \mathfrak{F}_e \\ z &\to z^{-1} \end{aligned}$$

are analytics.

Let $z_o \in \mathfrak{F}_e$, $w_o \in \mathfrak{F}_e$ and $z(t)$, $w(\tau)$, parametrizations with $z(t_o) = z_o$, $w(\tau_o) = w_o$.

The map

$$(t,\tau) \to p_o(\emptyset(z(t),w(t))$$

is analytic, where p_o is a projection onto $L(z_ow_o)H$.

Therefore the map $(z,w) \to zw$ is analytic.

Analogously the map $z \to z^{-1}$ is analytic.

REFERENCES

1. PISANELLI, D., An example of a infinitive Lie group. Proc. Am. Math. Soc. 62 (1976) nº 1, 156-160 (1977).

2. PISANELLI, D., Grupos Analíticos Finitos de Transformações, Publicação da Sociedade Brasileira de Matemática - Escola de Análise de 1977.

3. PISANELLI, D., Théorème d'Ovcyannicov, Frobenius et groupes de Lie locaux dans une échelle d'espaces de Banach, C.R.A. S. Paris, 277 (1973).

4. LANG, S., Differential Manifolds, Addison, Wesley D.C., 1972.

5. PISANELLI, D., Linear Connected Subgroup of a Lie Group in a locally convex space. Anais da Academia Brasileira de Ciências (1979), 51(4).

6. PISANELLI, D., Sull'integrazione di un sistema di differenziali totali in uno spazio di Banach. Academia Nationale dei Lincei, Rendiconti della Classe di Scienze Fisiche, Matematiche e Naturali Serie VII, Vol. XLVI, fase 6, giugno 1969.

Instituto de Matemática e Estatística
Universidade de São Paulo
CX 20570 - Agência Iguatemi
São Paulo, SP - Brasil

Functional Analysis, Holomorphy and Approximation Theory II, G.I. Zapata (ed.)
© Elsevier Science Publishers B.V. (North-Holland), 1984

OPTIMIZATION BY LEVEL SET METHODS. II: FURTHER DUALITY FORMULAE IN THE CASE OF ESSENTIAL CONSTRAINTS

Ivan Singer

ABSTRACT

Let G be a non-empty subset of a real locally convex space F and $h: F \to \bar{R}$ a functional. We show that if $h(y') < \inf h(G)$ for some $y' \in F$ (this happens e.g. in the theory of best approximation of y' by the elements of G), then, in the duality formulae of [4] for $\inf h(G)$, one can replace the set $G^s = \{\psi \in F^* \mid \psi \neq 0, \sup \psi(G) < +\infty\}$ by its subset $\{\psi \in G^s \mid \sup \psi(G) < \psi(y')\}$ or other similar subsets, and one can obtain new duality theorems, which reduce the computation of $\inf h(G)$ to the computation of the infima of h over some support hyperplanes of G.

§0. INTRODUCTION

In the present paper we shall continue the study of the following general optimization problem, considered in [4]: Given a locally convex space F (which will be assumed <u>real</u>, without any special mention), a subset G of F (which will be assumed <u>non-empty</u>), called <u>the constraint set</u>, and a functional $h: F \to \bar{R} = [-\infty, +\infty]$, find convenient formulae for the number

$$a = \inf h(G) = \inf_{g \in G} h(g) (\in \bar{R}). \tag{0.1}$$

In [4] we have shown that the existence of functionals $\psi \in F^*$ (the set of all continuous linear functionals on F), which support G (i.e., $\psi \neq 0$ and $\sup \psi(G) < +\infty$) and which separate G from certain level sets of h, closedly or openly or nicely (in the sense of V. Klee [1]), imply duality formulae which reduce the computation of $\inf h(G)$ to the computation of the infima of h on some closed half-spaces or strips or closed strips containing G and that, conversely, under some additional connectedness assumptions on G and on certain level sets of h, these support and separation conditions are also necessary for our duality formulae to hold. The duality formulae of [4] involved functionals $\psi \in G^s$, or $\psi \in -G^s$, where

$$G^s = \{\psi \in F^* \mid \psi \neq 0, \sup \psi(G) < +\infty\}, \tag{0.2}$$

(i.e., <u>support functionals</u> ψ of G), which have the advantage that they can be easily computed for certain classes of sets G (see [4], formulae (1.19) and (1.21)).

In [4], remark 1.2 c), we have observed that the duality formulae of [4] present interest only in the case when the constraint set G is essential, i.e., when

$$\inf h(F) < \inf h(G); \tag{0.3}$$

clearly, this inequality is equivalent to the condition that there should exist an element $y' \in F$ satisfying

$$h(y') < \inf h(G). \tag{0.4}$$

The assumption (0.3) (or (0.4)) implies, obviously, that

$$a = \inf h(G) \in (-\infty, +\infty]. \tag{0.5}$$

In the present paper we shall study further the optimization problem (0.1), under the assumption (0.3) (hence (0.5) will hold throughout the paper). We shall show that if $y' \in F$ is an arbitrary fixed element satisfying (0.4), one can replace in the duality formulae of [4] the set G^s by its subset $\{\psi \in G^s \mid \sup \psi(G) < \psi(y')\}$

or other similar subsets, and one can obtain new duality theorems, which reduce the computation of $\inf h(G)$ to the computation of the infima of h over some support hyperplanes of G. Thus, the division of this paper into sections and subsections will be similar to that of [4], with two additional subsections on results of weak and strong duality in terms of support hyperplanes of G.

Let us observe that a fixed element $y' \in F$ satisfying (0.4) arises quite naturally in some concrete optimization problems (0.1). For example, in the theory of best approximation, there are given a normed linear space F, a subset G of F and an element $y' \in F$, and one wants to compute

$$\operatorname{dist}(y',G) = \inf_{g \in G} \|y'-g\|, \tag{0.6}$$

which is a problem of type (0.1), for the functional $h: F \to R_+ = [0,+\infty)$ defined by

$$h(y) = \|y'-y\| \qquad (y \in F); \tag{0.7}$$

clearly, we have $0 < \operatorname{dist}(y',G)$ if and only if (0.4) holds. Since the balls with center y', used in the theory of best approximation to obtain duality formulae for (0.6), are nothing else than the level sets of the functional h defined by (0.7), it is natural to attempt to extend the methods of the theory of best approximation to level set methods for the general problem (0.1), under the assumptions (0.3), (0.4). In [2], [3], we have shown that this leads to useful duality formulae for (0.1) when G and some level sets of h are convex and satisfy certain topological assumptions. In the present paper these convexity and topological assumptions are weakened so as to become both necessary and sufficient for our duality formulae to hold.

§1. THEOREMS OF WEAK DUALITY

1.0 PRELIMINARIES

For convenience, let us recall and complement some of the main terminology and notations of [4].

By "hyperplane" we shall always mean: closed hyperplane. In addition to the non-symmetric separation properties used in [4], we shall say that a hyperplane

$$H = \{y \in F \mid \psi(y) = c\}, \tag{1.1}$$

where $\psi \in F^*$, $\psi \neq 0$, $c \in R$, <u>strictly separates</u> G_1 <u>from</u> G_2, if $G_1 \subset \{y \in F \mid \psi(y) < c\}$ and $G_2 \subset \{y \in F \mid \psi(y) > c\}$; correspondingly, a functional $\psi \in F^*\setminus\{0\}$ <u>strictly separates</u> G_1 <u>from</u> G_2, if there exists $c \in R$ such that the hyperplane (1.1) strictly separates G_1 from G_2.

Given a set $G \subset F$ and a functional $h: F \to \bar{R}$, in the sequel an important role will be played by the level sets

$$A_c = \{y \in F \mid h(y) < c\} \qquad (c \in R), \tag{1.2}$$

$$S_c = \{y \in F \mid h(y) \leq c\} \qquad (c \in R); \tag{1.3}$$

the assumptions (0.3) and (0.4) mean that $A_a \neq \emptyset$, $y' \in A_a$, where $a = \inf h(G)$.

As in [4], we shall say that a subset G of F is F^*-<u>connected</u>, if $\psi(G) \subset R$ is connected (i.e., an interval $\langle\alpha,\beta\rangle$, finite or infinite, closed or open from the right or from the left), for each $\psi \in F^*$. We shall denote by $\bar{G}$ the closure of the set G.

Finally, for $\alpha < \beta$ in $\bar{R}$, as usual, $(\alpha,\beta) = \{\gamma \in R \mid \alpha < \gamma < \beta\}$, $[\alpha,\beta) = (\alpha,\beta) \cup \{\alpha\}$, $(\alpha,\beta] = (\alpha,\beta) \cup \{\beta\}$, $[\alpha,\beta] = [\alpha,\beta) \cup \{\beta\}$.

1.1 RESULTS OF WEAK DUALITY IN TERMS OF CLOSED HALF-SPACES CONTAINING G

THEOREM 1.1. Let F be a locally convex space, G a subset of F and $h: F \to \bar{R}$ a functional with $A_a \neq \emptyset$, where $a = \inf h(G)$, and let $y' \in A_a$. The following statements are equivalent:

1°. For each $c \in (h(y'),a)$ there exists $\psi_c \in G^s$ satisfying

$$\sup \psi_c(G) < \psi_c(y) \qquad (y \in A_c). \tag{1.4}$$

2°. For each $c \in [h(y'),a)$ there exists $\psi_c \in G^s$ satisfying

$$\sup \psi_c(G) < \psi_c(y) \qquad (y \in S_c). \tag{1.5}$$

3°. There holds

$$\inf h(G) = \sup_{\psi \in G^s} \inf_{\substack{y \in F \\ \psi(y) \le \sup \psi(G)}} h(y). \tag{1.6}$$

4°. There holds

$$\inf h(G) = \sup_{\substack{\psi \in G^s \\ \sup \psi(G) < \psi(y')}} \inf_{\substack{y \in F \\ \psi(y) \le \sup \psi(G)}} h(y). \tag{1.7}$$

$1'$. For each $c \in (h(y'),a)$ there exists $\psi_c \in -G^s$ satisfying

$$\psi_c(y) < \inf \psi_c(G) \qquad (y \in A_c). \tag{1.4'}$$

$2'$. For each $c \in [h(y'),a)$ there exists $\psi_c \in -G^s$ satisfying

$$\psi_c(y) < \inf \psi_c(G) \qquad (y \in S_c). \tag{1.5'}$$

$3'$. There holds

$$\inf h(G) = \sup_{\psi \in -G^s} \inf_{\substack{y \in F \\ \psi(y) \ge \inf \psi(G)}} h(y). \tag{1.6'}$$

$4'$. There holds

$$\inf h(G) = \sup_{\substack{\psi \in -G^s \\ \psi(y') < \inf \psi(G)}} \inf_{\substack{y \in F \\ \psi(y) \ge \inf \psi(G)}} h(y). \tag{1.7'}$$

PROOF. The equivalences $1^{\circ} \Leftrightarrow 2^{\circ} \Leftrightarrow 3^{\circ} \Leftrightarrow 1' \Leftrightarrow 2' \Leftrightarrow 3'$ hold by [4], Theorem 1.1 and Remark 2.10 c).

$3^{\circ} \Leftrightarrow 4^{\circ}$. Since $y' \in A_a$, we have

$$\sup_{\substack{\psi \in G^s \\ \psi(y') \leq \sup \psi(G)}} \inf_{\substack{y \in F \\ \psi(y) \leq \sup \psi(G)}} h(y) \leq h(y') < a = \inf h(G), \tag{1.8}$$

and hence, if 3° holds, we obtain 4°. The implication $4^{\circ} \Rightarrow 3^{\circ}$ is obvious (by the obvious inequality $\geq$ in (1.6)).

Finally, considering $\psi' = -\psi$, we obtain the equivalence $4^{\circ} \Leftrightarrow 4'$. This completes the proof of Theorem 1.1.

REMARK 1.1. a) Since $\{\psi \in G^s \mid \sup \psi(G) < \psi(y')\} \subset \subset \{\psi \in G^s \mid \sup \psi(G) \leq \psi(y')\} \subset G^s$, one can obviously add in Theorem 1.1 the equivalent condition

5°. There holds

$$\inf h(G) = \sup_{\substack{\psi \in G^s \\ \sup \psi(G) \leq \psi(y')}} \inf_{\substack{y \in F \\ \psi(y) \leq \sup \psi(G)}} h(y). \tag{1.9}$$

However, in this paper we shall consider conditions of this type (i.e., involving $\{\psi \in G^s \mid \sup \psi(G) \leq \psi(y')\}$) only when they are not equivalent to the corresponding conditions of type (1.7) (i.e., involving $\{\psi \in G^s \mid \sup \psi(G) < \psi(y')\}$); see Theorems 1.3 and 2.3 below.

b) If $\varepsilon > 0$, one can add in Theorem 1.1 the following equivalent conditions, with the convention that if $A_c = \emptyset$, then (1.4) holds for any $\psi_c \neq 0$:

6°. $G^s \neq \emptyset$ and for each $c < a$ there exists $\psi_c \in G^s$ satisfying (1.4).

7°. $G^s \neq \emptyset$ and for each $c \in (a-\varepsilon, a)$ there exists $\psi_c \in G^s$ satisfying (1.4).

Indeed, the equivalences $7^{\circ} \Leftrightarrow 6^{\circ} \Leftrightarrow 3^{\circ}$ have been shown in [4],

Remark 2.10 c) and Theorem 1.1. Similar remarks can be also made for (1.5), (1.15), (1.16), (1.4'), etc. in the theorems of the present section, but we shall omit them. The advantage of considering $(h(y'),a)$ instead of $(-\infty,a)$ or $(a-\varepsilon,a)$ consists in the fact that for each $c \in (h(y'),a)$ we have $y' \in A_c \neq \emptyset$ (and, similarly, for $c \in [h(y'),a)$ we have $y' \in S_c \neq \emptyset$), so we do not need the above convention for $A_c = \emptyset$ (or $S_c = \emptyset$); on the other hand, for $\varepsilon' = a-h(y')$ we have $(h(y'),a) = (a-\varepsilon',a)$.

c) By the obvious inequalities $\geq$ in (1.6) and (1.7), one can express 3° and 4° of Theorem 1.1 in the following equivalent forms, respectively:

8°. For each $c \in (h(y'),a)$ there exists $\psi_c \in G^s$ satisfying

$$\inf_{\substack{y \in F \\ \psi_c(y) \leq \sup \psi_c(G)}} h(y) \geq c. \tag{1.10}$$

9°. For each $c \in (h(y'),a)$ there exists $\psi_c \in G^s$ satisfying (1.10) and

$$\sup \psi_c(G) < \psi_c(y'). \tag{1.11}$$

Similar remarks can be also made for 3', 4' and for the other results of §1.

REMARK 1.2. Geometrically, formulae (1.9) and (1.7) mean that in (1.6) it is enough to take the sup over all $\psi \in G^s$ which separate (respectively, which separate strictly) G from y'. In these cases, the hyperplane

$$H_{\psi,G} = \{y \in F \mid \psi(y) = \sup \psi(G)\} \tag{1.12}$$

supports G and separates (respectively, has a translate separating strictly) G from y'. One can also write (1.9) and (1.7) in the equivalent form

$$\inf h(G) = \sup_{\substack{\psi \in G^s \\ y' \notin \operatorname{Int} D_{\psi,G}}} \inf h(D_{\psi,G}) \tag{1.13}$$

and, respectively,

$$\inf h(G) = \sup_{\substack{\psi \in G^s \\ y' \notin D_{\psi,G}}} \inf h(D_{\psi,G}) \tag{1.14}$$

where, as in [4], Remark 1.1, $D_{\psi,G}$ is the smallest closed half-space in the direction ψ, containing G. Since the geometric interpretations of most of the other formulae in this paper can be obtained similarly from the corresponding ones of [4], we shall omit them in the sequel (with the exception of Remark 1.5 below).

It will be worth while to state separately the following sufficient condition in order to have (1.6), (1.7), (1.9), (1.13), (1.14):

THEOREM 1.2. Let F be a locally convex space, G a subset of F and $h: F \to \bar{R}$ a functional with $A_a \neq \emptyset$, and let $y' \in A_a$. If G and the sets S_c with $c \in [h(y'),a)$ are convex and closed for a locally convex topology τ on F, weaker than or equal to the initial topology on F, and if either G or the sets S_c with $c \in [h(y'),a)$ are compact for τ, then we have 2° of Theorem 1.1, whence also (1.6), (1.7), (1.9), (1.13), (1.14).

The proof is similar to the second part of the proof of [4], Theorem 1.2.

1.2 RESULTS OF WEAK DUALITY IN TERMS OF STRIPS CONTAINING G

THEOREM 1.3. Let F be a locally convex space, G a subset of F and $h: F \to \bar{R}$ a functional with $A_a \neq \emptyset$, and let $y' \in A_a$. Consider the following statements:

1°. For each $c \in (h(y'),a)$ there exists $\psi_c \in G^s$ satisfying (1.4).

2°. For each $c \in [h(y'),a)$ there exists $\psi_c \in G^s$ satisfying (1.5).

3°. For each $c \in (h(y'),a)$ there exists $\psi_c \in G^s$ satisfying either (1.4) or

$$\psi_c(g) < \inf \psi_c(A_c) \qquad (g \in G). \tag{1.15}$$

4°. For each $c \in [h(y'),a)$ there exists $\psi_c \in G^s$ satisfying either (1.5) or

$$\psi_c(g) < \inf \psi_c(S_c) \qquad (g \in G). \tag{1.16}$$

5°. There holds

$$\inf h(G) = \sup_{\psi \in G^s} \inf_{\substack{y \in F \\ \psi(y) \in \psi(G)}} h(y). \tag{1.17}$$

6°. There holds

$$\inf h(G) = \sup_{\substack{\psi \in G^s \\ \psi(y') \notin \psi(G)}} \inf_{\substack{y \in F \\ \psi(y) \in \psi(G)}} h(y). \tag{1.18}$$

7°. There holds

$$\inf h(G) = \sup_{\substack{\psi \in G^s \\ \sup \psi(G) \leq \psi(y')}} \inf_{\substack{y \in F \\ \psi(y) \in \psi(G)}} h(y). \tag{1.19}$$

8°. There holds

$$\inf h(G) = \sup_{\substack{\psi \in G^s \\ \sup \psi(G) < \psi(y')}} \inf_{\substack{y \in F \\ \psi(y) \in \psi(G)}} h(y). \tag{1.20}$$

1'-8', obtained from 1°-8° replacing $\psi_c \in G^s$ by $\psi_c \in -G^s$ and (1.4), (1.5), (1.15)-(1.20) respectively by (1.4'), (1.5'),

$$\sup \psi_c(A_c) < \psi_c(g) \qquad (g \in G), \tag{1.15'}$$

$$\sup \psi_c(S_c) < \psi_c(g) \qquad (g \in G), \tag{1.16'}$$

$$\inf h(G) = \sup_{\psi \in -G^s} \inf_{\substack{y \in F \\ \psi(y) \in \psi(G)}} h(y), \tag{1.17'}$$

$$\inf h(G) = \sup_{\substack{\psi \in -G^s \\ \psi(y') \notin \psi(G)}} \inf_{\substack{y \in F \\ \psi(y) \in \psi(G)}} h(y), \tag{1.18'}$$

$$\inf h(G) = \sup_{\substack{\psi \in -G^s \\ \psi(y') \leq \inf \psi(G)}} \inf_{\substack{y \in F \\ \psi(y) \in \psi(G)}} h(y), \tag{1.19'}$$

$$\inf h(G) = \sup_{\substack{\psi \in -G^s \\ \psi(y') < \inf \psi(G)}} \inf_{\substack{y \in F \\ \psi(y) \in \psi(G)}} h(y). \tag{1.20'}$$

i) We have the implications $2^o \Rightarrow 4^o \Rightarrow 3^o \Rightarrow 7^o \Rightarrow 5^o$ and $2^o \Rightarrow 1^o \Rightarrow \Rightarrow 3^o$, as well as $1^o \Rightarrow 8^o \Rightarrow 7^o$, and the equivalence $5^o \Leftrightarrow 6^o$.

ii) If G is F^*-connected, then $5^o \Leftrightarrow 6^o \Leftrightarrow 7^o$.

iii) If G and the sets A_c with $c \in (h(y'),a)$ are F^*-connected, then $3^o \Leftrightarrow 5^o \Leftrightarrow 6^o \Leftrightarrow 7^o$.

iv) If G and the sets S_c with $c \in [h(y'),a)$ are F^*-connected, then $3^o \Leftrightarrow \dots \Leftrightarrow 7^o$.

v) We have $n^o \Leftrightarrow n'$ $(n = 1,\dots,8)$.

PROOF. i) The implications $2^o \Rightarrow 4^o \Rightarrow 3^o$, $2^o \Rightarrow 1^o \Rightarrow 3^o$, $8^o \Rightarrow 7^o \Rightarrow \Rightarrow 5^o$ and $6^o \Rightarrow 5^o$ are obvious (since $A_c \subset S_c$ and since the inequalities $\geq$ in (1.19) and (1.17) are obvious).

$1^o \Rightarrow 8^o$. Since $y' \in A_a$, there exists $c \in (h(y'),a)$. Then, by 1°, for any such c there exists $\psi_c \in G^s$ satisfying (1.4) and hence, by [4], Proposition 1.2 i), we have

$$\inf_{\substack{y \in F \\ \psi_c(y) \in \psi_c(G)}} h(y) \geq c. \tag{1.21}$$

Furthermore, by (1.4) and $y' \in A_c$ we have, in particular, (1.11), whence, by (1.21),

$$b' = \sup_{\substack{\psi \in G^s \\ \sup \psi(G) < \psi(y')}} \inf_{\substack{y \in F \\ \psi(y) \in \psi(G)}} h(y) \geq c.$$

Consequently,

$$b' \geq \sup_{c \in (h(y'),a)} c = a,$$

and hence, since the opposite inequality is obvious, we obtain (1.20).

The proof of the implication $3^o \Rightarrow 7^o$ is similar, observing that for c and ψ_c as in 3° we have

$$\sup \psi_c(G) \leq \inf \psi_c(A_c) \leq \psi_c(y'). \tag{1.22}$$

$5^o \Rightarrow 6^o$. Since $y' \in A_a$, there exists $c \in (h(y'),a)$. Then

$$\sup_{\substack{\psi \in G^s \\ \psi(y') \in \psi(G)}} \inf_{\substack{y \in F \\ \psi(y) \in \psi(G)}} h(y) \leq h(y') < c < a, \tag{1.23}$$

and hence, if 5° holds, we obtain 6°.

ii) By i), it is enough to show that if G is F^*-connected, then $6^o \Rightarrow 7^o$. But, if $\psi \in F^*$ and $\psi(y') \notin \psi(G)$, then, since $\psi(G)$ is an interval in R, we have either $\sup \psi(G) \leq \psi(y')$ (whence $\psi \in G^s$) or $\inf \psi(G) \geq \psi(y')$, which is equivalent to $\sup (-\psi)(G) \leq (-\psi)(y')$ (whence $-\psi \in G^s$). Hence, by the obvious inequality $\geq$ in (1.19), if 6° holds, we obtain 7°.

iii) If G and the sets A_c with $c \in (h(y'),a)$ are F^*-connected, then, by [4], Theorem 1.3, we have $5^o \Rightarrow 3^o$ and by i), ii) above we have $3^o \Rightarrow 5^o \Leftrightarrow 6^o \Leftrightarrow 7^o$.

iv) If G and the sets S_c with $c \in [h(y'),a)$ are F^*-connected, then, by [4], Theorem 1.3, we have $5^o \Rightarrow 4^o$ and by ii) above we have $4^o \Rightarrow 3^o \Rightarrow 5^o \Leftrightarrow 6^o \Leftrightarrow 7^o$.

v) Replacing ψ_c by $-\psi_c$ and ψ by $-\psi$ we obtain the equivalences $n^o \Leftrightarrow n'$ $(n=1,\ldots,8)$. This completes the proof of Theorem 1.3.

From Theorems 1.2 and 1.3 there follows

THEOREM 1.4. Under the assumptions of Theorem 1.2, we have (1.18)-(1.20) and (1.18')-(1.20').

REMARK 1.3. Theorem 1.4 has been obtained, essentially, in [3], Corollary 2.3 and its proof (see also [3], Remark 2.7 (a)).

1.3 RESULTS OF WEAK DUALITY IN TERMS OF CLOSED STRIPS CONTAINING G

THEOREM 1.5. Let F be a locally convex space, G a subset of F and $h: F \to \bar{R}$ a functional with $A_a \neq \emptyset$, and let $y' \in A_a$. Consider the following statements:

1°. For each $c \in (h(y'), a)$ there exists $\psi_c \in G^s$ satisfying (1.4).

2°. For each $c \in [h(y'), a)$ there exists $\psi_c \in G^s$ satisfying (1.5).

3°. There holds

$$\inf h(G) = \sup_{\psi \in G^s} \inf_{\substack{y \in F \\ \psi(y) \in \overline{\psi(G)}}} h(y). \tag{1.24}$$

4°. There holds

$$\inf h(G) = \sup_{\substack{\psi \in G^s \\ \psi(y') \notin \overline{\psi(G)}}} \inf_{\substack{y \in F \\ \psi(y) \in \overline{\psi(G)}}} h(y). \tag{1.25}$$

5°. There holds

$$\inf h(G) = \sup_{\substack{\psi \in G^s \\ \sup \psi(G) < \psi(y')}} \inf_{\substack{y \in F \\ \psi(y) \in \overline{\psi(G)}}} h(y). \tag{1.26}$$

1'-5', obtained from 1°-5° similarly to the corresponding procedure of Theorem 1.3.

i) We have the implications $2^o \Rightarrow 1^o \Rightarrow 5^o \Rightarrow 4^o$ and the equivalence $3^o \Leftrightarrow 4^o$.

ii) If G is F^*-connected, then $3^o \Leftrightarrow 4^o \Rightarrow 5^o$.

iii) If G and the sets A_c with $c \in (h(y'),a)$ are F^*-connected, then $1^o \Leftrightarrow 3^o \Leftrightarrow 4^o \Leftrightarrow 5^o$.

iv) If G and the sets S_c with $c \in [h(y'),a)$ are F^*-connected, then $1^o \Leftrightarrow \ldots \Leftrightarrow 5^o$.

v) We have $n^o \Leftrightarrow n'$ $(n=1,\ldots,5)$.

PROOF. i) The implications $2^o \Rightarrow 1^o$ and $5^o \Rightarrow 4^o \Rightarrow 3^o$ are obvious (since $A_c \subset S_c$, $\{\psi \in G^s \mid \sup \psi(G) < \psi(y')\} \subset$ $\subset \{\psi \in G^s \mid \psi(y') \notin \overline{\psi(G)}\} \subset G^s$ and since the inequalities $\geq$ in (1.25) and (1.24) are obvious). The proofs of the implications $1^o \Rightarrow 5^o$ and $3^o \Rightarrow 4^o$ are the same as those of Theorem 1.3, implications $1^o \Rightarrow 8^o$ and $5^o \Rightarrow 6^o$, observing that $\sup \psi(G) = \sup \overline{\psi(G)}$ $(\psi \in F^*)$.

ii) If G is F^*-connected, the proof of $4^o \Rightarrow 5^o$ is similar to that of Theorem 1.3 ii), implication $6^o \Rightarrow 7^o$, observing that if $\psi \in F^*$ and $\psi(y') \notin \overline{\psi(G)}$, then either $\sup \psi(G) < \psi(y')$ or $\inf \psi(G) > \psi(y')$.

iii) and iv) follow from [4], Theorem 1.5 and ii) above. Finally, the proof of v) is similar to that of Theorem 1.3 v). This completes the proof of Theorem 1.5.

REMARK 1.4. From Theorems 1.2 and 1.5 there follows a result similar to Theorem 1.4. Such a result has been proved, essentially, in [3], Theorem 2.3 and its proof.

1.4 RESULTS OF WEAK DUALITY IN TERMS OF SUPPORT HYPERPLANES OF G

If we have (1.24), then, since $\sup \psi(G) \in \overline{\psi(G)}$, there holds also

$$\inf h(G) \leq \sup_{\psi \in G^s} \inf_{\substack{y \in F \\ \psi(y)=\sup \psi(G)}} h(y). \tag{1.27}$$

Thus, it is natural to ask whether the opposite inequality holds (similarly to the obvious inequalities $\geq$ in the preceding formulae of weak duality), and whether one can replace in Theorem 1.5 the closed strips $\bar{B}_{\psi,G} = \{y \in F \mid \psi(y) \in \overline{\psi(G)}\}$ (see [4]) by the support hyperplanes $H_{\psi,G}$ defined by (1.12) or by the support hyperplanes

$$H'_{\psi,G} = \{y \in F \mid \psi(y) = \inf \psi(G)\}. \tag{1.12'}$$

Since the set G^s is too large and since for $\psi \in G^s$ the hyperplanes (1.12), (1.12') are too "thin", the answer is negative. Indeed, even when G is a closed convex set and h is a finite continuous convex functional on a finite-dimensional space F (so (1.24) holds), the inequality in (1.27) may be strict, as shown by

EXAMPLE 1.1. Let $F = R^2$, the euclidean plane, and let

$$G = \{y \in F \mid \|y-(2,0)\| \leq 1\}, \tag{1.28}$$

$$h(y) = \|y\| = (|\eta_1|^2 + |\eta_2|^2)^{\frac{1}{2}} \qquad (y = (\eta_1,\eta_2) \in F), \tag{1.29}$$

so $a = \inf h(G) = 1$. For each c with $0 < c < a = 1$ let

$$\psi_c(y) = -\eta_1 \qquad (y = (\eta_1,\eta_2) \in F). \tag{1.30}$$

Then

$$\sup \psi_c(G) = -1 < \inf \psi_c(S_c) = \inf_{\substack{y \in F \\ \|y\| \leq c}} (-\eta_1) = -c,$$

so (1.5) holds. However, for $\psi_o = -\psi_c$ we have $\sup \psi_o(G) = 3$

(so $\psi_o \in G^s$) and

$$\inf h(G) = 1 < \inf_{\substack{y \in F \\ \psi_o(y)=\sup \psi_o(G)}} h(y) = \inf_{\substack{(\eta_1,\eta_2)\in F \\ \eta_1=3}} \|(\eta_1,\eta_2)\| = 3,$$

and hence the inequality in (1.27) is strict.

Nevertheless, we shall show now that for the subset $\{\psi \in G^s \mid \sup \psi(G) < \psi(y')\}$ of G^s, occurring in (1.26), the situation is different, under certain additional assumptions. To this end, let us first give the following generalization of [2], Lemma 2.1:

PROPOSITION 1.1. Let F be a locally convex space, G a subset of F with $G^s \neq \emptyset$, $h: F \to \bar{R}$ a functional with $A_a \neq \emptyset$, and $\psi \in G^s$, such that either i) $\psi(A_{a+\frac{1}{n}})$ is an interval $(n=1,2,\dots)$, or ii) $\psi(S_{a+\frac{1}{n}})$ is an interval $(n=1,2,\dots)$ or iii) $\bar{G} \cap A_a \neq \emptyset$ and $\psi(A_a)$ is an interval, or iv) $\bar{G} \cap S_a \neq \emptyset$ and $\psi(S_a)$ is an interval. If there exists $y' \in A_a$ such that

$$\sup \psi(G) \le \psi(y'), \tag{1.31}$$

then

$$a = \inf h(G) \ge \inf_{\substack{y \in F \\ \psi(y)=\sup \psi(G)}} h(y). \tag{1.32}$$

PROOF. i) Take $g_n \in G$ such that $h(g_n) < a + \frac{1}{n}$, so $\psi(g_n) \in \psi(A_{a+\frac{1}{n}})$. By $y' \in A_a \subset A_{a+\frac{1}{n}}$ we have $\psi(y') \in \psi(A_{a+\frac{1}{n}})$, whence, by $\psi(g_n) \le \sup \psi(G)$ and (1.31), and since $\psi(A_{a+\frac{1}{n}})$ is an interval, we obtain

$$\sup \psi(G) \in [\psi(g_n), \psi(y')] \subset \psi(A_{a+\frac{1}{n}}). \tag{1.33}$$

Thus, for each n there exists $y_n \in F$ with $h(y_n) < a + \frac{1}{n}$, such that $\psi(y_n) = \sup \psi(G)$, which proves (1.32). The proof of ii) is similar.

iv) Take $y_o \in \bar{G} \cap S_a$, so $\psi(y_o) \in \psi(\bar{G}) \cap \psi(S_a)$. By $y' \in A_a \subset S_a$ we have $\psi(y') \in \psi(S_a)$, whence, by $\psi(y_o) \leq \sup \psi(\bar{G}) = \sup \psi(G)$ and (1.31), and since $\psi(S_a)$ is an interval, we obtain

$$\sup \psi(G) \in [\psi(y_o), \psi(y')] \subset \psi(S_a). \tag{1.34}$$

Thus, there exists $y_1 \in F$ with $h(y_1) \leq a$, such that $\psi(y_1) = \sup \psi(G)$, which proves (1.32). The proof of iii) is similar, which completes the proof of Proposition 1.1.

We shall also need the following proposition and corollary, corresponding to [4], Proposition 1.1 and Corollary 1.1 respectively:

PROPOSITION 1.2. Let F be a locally convex space, G a subset of F, $h: F \to \bar{R}$ a functional, $c \in R$ with $A_c \neq \emptyset$, and $\psi_c \in F^*$.

i-i') If either (1.4) or (1.4') holds, then

$$\min \left\{ \inf_{\substack{y \in F \\ \psi_c(y) = \sup \psi_c(G)}} h(y), \inf_{\substack{y \in F \\ \psi_c(y) = \inf \psi_c(G)}} h(y) \right\} \geq c. \tag{1.35}$$

ii) Conversely, if $\psi_c(A_c)$ is an interval and there exists $y' \in A_c$ such that

$$\sup \psi_c(G) \leq \psi_c(y'), \tag{1.36}$$

and if we have

$$\inf_{\substack{y \in F \\ \psi_c(y) = \sup \psi_c(G)}} h(y) \geq c, \tag{1.37}$$

then (1.4) holds.

ii') If $\psi_c(A_c)$ is an interval and there exists $y' \in A_c$ such that

$$\psi_c(y') \leq \inf \psi_c(G), \tag{1.36'}$$

and if we have

$$\inf_{\substack{y \in F \\ \psi_c(y) = \inf \psi_c(G)}} h(y) \geq c, \tag{1.37'}$$

then (1.4') holds.

PROOF. Similarly to [4], proof of Proposition 1.1, it is immediate that we have (1.37) if and only if

$$\sup \psi_c(G) \neq \psi_c(y) \qquad (y \in A_c). \tag{1.38}$$

i-i') If (1.4) or (1.4') holds, then obviously, we have (1.38). Alternatively, i-i') follows also from $\sup \psi_c(G), \inf \psi_c(G) \in \overline{\psi_c(G)}$ and [4], Proposition 1.4 i).

ii) By $y' \in A_c$ and (1.36), (1.38), and since $\psi_c(A_c)$ is an interval, we obtain (1.4) (whence, in particular, $\sup \psi_c(G) < \psi_c(y')$).

Finally, ii') is equivalent to ii), considering $\psi'_c = -\psi_c$. This completes the proof of Proposition 1.2.

Even when F, G and h have "nice" properties, as in Example 1.1, one cannot omit the assumptions (1.36) and (1.36') in Proposition 1.2 ii) and ii') respectively, as shown by

EXAMPLE 1.2. Let F, G and h be as in Example 1.1 and for each c with $0 < c < 1 = a = \inf h(G)$ let

$$\psi_c(y) = \eta_2 \qquad (y = (\eta_1, \eta_2) \in F). \tag{1.39}$$

Then

$$\inf_{\substack{y \in F \\ \psi_c(y) = \sup \psi_c(G)}} h(y) = 1 = \inf_{\substack{y \in F \\ \psi_c(y) = \inf \psi_c(G)}} h(y),$$

$$\inf \psi_c(G) = -1 < \inf \psi_c(A_c) = -c < \sup \psi_c(A_c) = c < \sup \psi_c(G) = 1,$$

so we have (1.35), but neither (1.4), nor (1.4'). The same example motivates also the assumptions (1.36) and (1.36') in the following corollary of Proposition 1.2:

COROLLARY 1.1. Let F be a locally convex space, G a subset of F, $h: F \to \bar{R}$ a functional, $c \in R$ with $S_c \neq \emptyset$, and $\psi_c \in G^s$.

i-i') If either (1.5) or (1.5') holds, then we have (1.35).

ii) Conversely, if $\psi_c(S_c)$ is an interval and there exists $y' \in S_c$ satisfying (1.36) and if we have

$$\inf_{\substack{y \in F \\ \psi_c(y) = \sup \psi_c(G)}} h(y) > c, \tag{1.40}$$

then (1.5) holds.

ii') If $\psi_c(S_c)$ is an interval and there exists $y' \in S_c$ satisfying (1.36') and if we have

$$\inf_{\substack{y \in F \\ \psi_c(y) = \inf \psi_c(G)}} h(y) > c, \tag{1.40'}$$

then (1.5') holds.

PROOF. Parts i), i') follow from $A_c \subset S_c$ and Proposition 1.2 i) and i') respectively.

The proof of part ii) is similar to that of Proposition 1.2 ii), observing that if (1.40) holds, then $\sup \psi_c(G) \neq \psi_c(y) \quad (y \in S_c)$.

Finally, ii') is equivalent to ii), considering $\psi'_c = -\psi_c$. This completes the proof of Corollary 1.1.

Now we are ready to prove

THEOREM 1.6. Let F be a locally convex space, G a subset of F and $h: F \to \bar{R}$ a functional with $A_a \neq \emptyset$, such that either a) $A_{a+\frac{1}{n}}$ is F^*-connected $(n=1,2,\ldots)$, or b) $S_{a+\frac{1}{n}}$ is F^*-connected $(n=1,2,\ldots)$, or c) $\bar{G} \cap A_a \neq \emptyset$ and A_a is F^*-connected, or d) $\bar{G} \cap S_a \neq \emptyset$ and S_a is F^*-connected. Let $y' \in A_a$ and consider the following statements:

1°. For each $c \in (h(y'), a)$ there exists $\psi_c \in G^s$ satisfying (1.4).

2°. For each $c \in [h(y'),a)$ there exists $\psi_c \in G^s$ satisfying (1.5).

3°. There holds

$$\inf h(G) = \sup_{\substack{\psi \in G^s \\ \sup \psi(G) < \psi(y')}} \inf_{\substack{y \in F \\ \psi(y) = \sup \psi(G)}} h(y). \tag{1.41}$$

1'-3', obtained from 1°-3° similarly to the corresponding procedure of Theorem 1.3.

i) We have the implications $2^o \Rightarrow 1^o \Rightarrow 3^o$.

ii) If the sets A_c with $c \in (h(y'),a)$ are F^*-connected, then $1^o \Leftrightarrow 3^o$.

iii) If the sets S_c with $c \in [h(y'),a)$ are F^*-connected, then $1^o \Leftrightarrow 2^o \Leftrightarrow 3^o$.

iv) We have $n^o \Leftrightarrow n'$ $(n=1,2,3)$.

PROOF. i) The implication $2^o \Rightarrow 1^o$ is obvious (since $A_c \subset S_c$).

$1^o \Rightarrow 3^o$. By a), b), c) or d), $y' \in A_a$, 1°, Proposition 1.1 and $\sup \psi(G) \in \overline{\psi(G)}$ we have

$$\inf h(G) \geq \sup_{\substack{\psi \in G^s \\ \sup \psi(G) < \psi(y')}} \inf_{\substack{y \in F \\ \psi(y) = \sup \psi(G)}} h(y), \tag{1.42}$$

$$\inf_{\substack{y \in F \\ \psi(y) = \sup \psi(G)}} h(y) \geq \inf_{\substack{y \in F \\ \psi(y) \in \overline{\psi(G)}}} h(y) \qquad (\psi \in G^s), \tag{1.43}$$

whence we obtain 3° (by Theorem 1.5).

ii) Assume that the sets A_c with $c \in (h(y'),a)$ are F^*-connected and that 3° holds. Then, by (1.41), for any such $c < a$ there exists $\psi_c \in G^s$ satisfying (1.11) (whence (1.36)) and (1.37). Hence, since A_c is F^*-connected, by Proposition 1.2 ii) we obtain (1.4).

iii) If the sets S_c with $c \in [h(y'), a)$ are F^*-connected and if 3° holds, then, similarly to the above proof of ii), using now corollary 1.1 ii), we obtain (1.5).

Finally, the proof of iv) is similar to that of Theorem 1.3 v). This completes the proof of Theorem 1.6.

REMARK 1.5. Geometrically, formula (1.41) of Theorem 1.6 means that

$$\inf h(G) = \sup_{H \in \mathcal{H}_{G,y'}} \inf h(H), \tag{1.44}$$

where $\mathcal{H}_{G,y'}$ denotes the collection of all hyperplanes H in F which support G and have a translate separating strictly G from y'.

Finally, let us make some complementary observations to Proposition 1.1, collected in

REMARK 1.6. Under the assumptions of Proposition 1.1, but replacing (1.31) by the stronger condition

$$\sup \psi(G) \leq \inf \psi(A_a), \tag{1.45}$$

we have

$$\inf h(G) \geq \inf_{\substack{y \in F \\ \psi(y) = \sup \psi(G)}} h(y) = \inf_{\substack{y \in F \\ \psi(y) \in \overline{\psi(G)}}} h(y). \tag{1.46}$$

Moreover, replacing (1.45) by the stronger condition

$$\sup \psi(G) < \psi(y) \qquad (y \in A_a), \tag{1.47}$$

we have even

$$\inf h(G) = \inf_{\substack{y \in F \\ \psi(y) = \sup \psi(G)}} h(y) = \inf_{\substack{y \in F \\ \psi(y) \in \overline{\psi(G)}}} h(y). \tag{1.48}$$

Indeed, if (1.45) holds, then for any $y' \in A_a$ we have (1.31) and hence, by Proposition 1.1, we obtain (1.32). Furthermore, if the inequality $\geq$ in (1.43) is strict, then there exists $y_o \in F$

with $\psi(y_o) \in \overline{\psi(G)}$, such that

$$\inf_{\substack{y \in F \\ \psi(y)=\sup \psi(G)}} h(y) > h(y_o). \tag{1.49}$$

But, by (1.32) and (1.49), we have $y_o \in A_a$ and hence, by (1.45),

$$\sup \psi(G) \leq \inf \psi(A_a) \leq \psi(y_o). \tag{1.50}$$

On the other hand, by $\psi(y_o) \in \overline{\psi(G)}$ we have $\psi(y_o) \leq \sup \psi(G)$, whence, by (1.49), we obtain $\psi(y_o) < \sup \psi(G)$, in contradiction with (1.50). This proves (1.46). Finally, if (1.47) holds, then we have (1.45), whence also (1.46). On the other hand, by (1.47) and Proposition 1.2 i) (with $c=a$) we have the opposite inequality to (1.46) and hence the equality (1.48) (alternatively, (1.48) also follows from (1.32), (1.43), (1.47) and [4], Proposition 1.4 i) with $c=a$). Moreover, under some additional assumptions (see Proposition 1.2 ii) and [4], Proposition 1.4 ii)), one can also give results of converse type.

§2. RESULTS OF STRONG DUALITY

2.1 RESULTS OF STRONG DUALITY IN TERMS OF CLOSED HALF-SPACES CONTAINING G

THEOREM 2.1. Let F be a locally convex space, G a subset of F and $h: F \to \bar{R}$ a functional with $A_a \neq \emptyset$, and let $y' \in A_a$. The following statements are equivalent:

1°. There exists $\psi_o \in G^s$ satisfying

$$\sup \psi_o(G) < \psi_o(y) \qquad (y \in A_a). \tag{2.1}$$

2°. There holds

$$\inf h(G) = \max_{\psi \in G^s} \inf_{\substack{y \in F \\ \psi(y) \leq \sup \psi(G)}} h(y). \tag{2.2}$$

3°. There holds

$$\inf h(G) = \max_{\substack{\psi \in G^s \\ \sup \psi(G) < \psi(y')}} \inf_{\substack{y \in F \\ \psi(y) \leq \sup \psi(G)}} h(y). \tag{2.3}$$

1'-3', obtained from 1°-3° similarly to the corresponding procedure of Theorem 1.3.

PROOF. The equivalences $1^{\circ} \Leftrightarrow 2^{\circ} \Leftrightarrow 1' \Leftrightarrow 2'$ hold by [4], Theorem 2.1, and the proofs of the equivalences $2^{\circ} \Leftrightarrow 3^{\circ} \Leftrightarrow 3'$ are similar to those of Theorem 1.1, equivalences $3^{\circ} \Leftrightarrow 4^{\circ} \Leftrightarrow 4'$.

REMARK 2.1. By the inequalities $\geq$ in (2.2) and (2.3), one can also express 2° and 3° of Theorem 2.1 in the following equivalent forms, respectively:

4°. There exists $\psi_o \in G^s$ satisfying

$$\inf h(G) = \inf_{\substack{y \in F \\ \psi_o(y) \leq \sup \psi_o(G)}} h(y). \tag{2.4}$$

5°. There exists $\psi_o \in G^s$ satisfying (2.4) and

$$\sup \psi_o(G) < \psi_o(y'). \tag{2.5}$$

Similar remarks can be also made for 2', 3' and for the other results of §2.

From Theorem 2.1 and [4], Theorem 2.2, there follows

THEOREM 2.2. Let F be a locally convex space, G a convex subset of F and $h: F \to \bar{R}$ a functional with A_a non-empty, convex and open, and let $y' \in A_a$. Then we have (2.3).

2.2 RESULTS OF STRONG DUALITY IN TERMS OF STRIPS CONTAINING G

We recall that the core of a set G in a linear space F is the subset of G defined by

$$\text{core } G = \{y \in F \mid \forall\, y \in F,\ \exists\, \varepsilon > 0,\ \forall\, \lambda \in [-\varepsilon, \varepsilon],\ \lambda y + (1-\lambda) g \in G\}. \quad (2.6)$$

THEOREM 2.3. Let F be a locally convex space, G a subset of F and $h: F \to \bar{R}$ a functional with $A_a \neq \emptyset$, and let $y' \in A_a$. Consider the following statements:

1°. There exists $\psi_o \in G^s$ satisfying (2.1).

2°. There exists $\psi_o \in G^s$ satisfying either (2.1) or

$$\psi_o(g) < \inf \psi_o(A_a) \qquad (g \in G). \quad (2.7)$$

3°. There holds

$$\inf h(G) = \max_{\psi \in G^s} \inf_{\substack{y \in F \\ \psi(y) \in \psi(G)}} h(y). \quad (2.8)$$

4°. There holds

$$\inf h(G) = \max_{\substack{\psi \in G^s \\ \psi(y') \notin \psi(G)}} \inf_{\substack{y \in F \\ \psi(y) \in \psi(G)}} h(y). \quad (2.9)$$

5°. There holds

$$\inf h(G) = \max_{\substack{\psi \in G^s \\ \sup \psi(G) \leq \psi(y')}} \inf_{\substack{y \in F \\ \psi(y) \in \psi(G)}} h(y). \quad (2.10)$$

6°. There holds

$$\inf h(G) = \max_{\substack{\psi \in G^s \\ \sup \psi(G) < \psi(y')}} \inf_{\substack{y \in F \\ \psi(y) \in \psi(G)}} h(y). \quad (2.11)$$

1'-6', obtained from 1°-6° similarly to the corresponding procedure of Theorem 1.3.

i) We have the implications $1^o \Rightarrow 2^o \Rightarrow 5^o \Rightarrow 3^o$ and $1^o \Rightarrow 6^o \Rightarrow 5^o$ and the equivalence $3^o \Leftrightarrow 4^o$.

ii) If G is F^*-connected, then $3^o \Leftrightarrow 4^o \Leftrightarrow 5^o$.

iii) If G and A_a are F^*-connected, then $2^o \Leftrightarrow \ldots \Leftrightarrow 5^o$.

iv) If core $A_a \neq \emptyset$ and $y' \in$ core A_a, then $2^o \Rightarrow 6^o$.

v) We have $n^o \Leftrightarrow n'$ $(n=1,\ldots,6)$.

PROOF. i) The implications $1^o \Rightarrow 2^o$, $6^o \Rightarrow 5^o \Rightarrow 3^o$ and $4^o \Rightarrow 3^o$ are obvious (by the obvious inequalities $\geq$ in (1.19) and (1.17)).

$1^o \Rightarrow 6^o$. By 1° and [4], Proposition 1.2 i) (with c=a), we have

$$\inf h(G) = \inf_{\substack{y\in F \\ \psi_o(y)\in\psi_o(G)}} h(y). \tag{2.12}$$

Furthermore, by 1° and $y' \in A_a$ we have, in particular, (2.5), whence, by the obvious inequality $\geq$ in (1.20), we obtain (2.11).

The proof of the implication 2° - 5° is similar, observing that for ψ_o as in 2° we have

$$\sup \psi_o(G) \leq \inf \psi_o(A_a) \leq \psi_o(y'). \tag{2.13}$$

The proof of $3^o \Rightarrow 4^o$ is similar to that of Theorem 1.3, implication $5^o \Rightarrow 6^o$.

ii) By i), it is enough to show that if G is F^*-connected, then $4^o \Rightarrow 5^o$. The proof of this fact is similar to that of Theorem 1.3 ii).

iii) If G and A_a are F^*-connected, then by [4], Theorem 2.3, we have $2^o \Leftrightarrow 3^o$ and by ii) above we have $3^o \Leftrightarrow 4^o \Leftrightarrow 5^o$.

iv) If we have (2.1), then, by part i), implication $1^o \Rightarrow 6^o$, there holds 6°. Assume now that we have core $A_a \neq \emptyset$, $y' \in$ core A_a and that $\psi_o \in G^s$ satisfies (2.7). Then, by [4], Lemma 2.1, we obtain

$$\sup \psi_o(G) \leq \inf \psi_o(A_a) \leq \inf \psi_o(\text{core } A_a) < \psi_o(y'), \tag{2.14}$$

so (2.5) holds. But, by 2° and [4], Proposition 1.2 i) (with c=a) we have (2.12), whence by the obvious inequality $\geq$ in (1.20), we obtain (2.11).

Finally, replacing ψ_o by $-\psi_o$ and ψ by $-\psi$, we obtain v), which completes the proof of Theorem 2.3.

REMARK 2.2. Even when G is an open convex set and h is a convex functional on a finite-dimensional space F, with core $A_a \neq \emptyset$, one cannot replace in Theorem 2.3 iv) the assumption $y' \in$ core A_a by the weaker assumption $y' \in A_a$, as shown by the following simple example (of [4], Remark 2.4 a)): Let $F = R^2$ (the Euclidean plane), let

$$G = \{(\eta_1, \eta_2) \in F \mid \max(|\eta_1+1|, |\eta_2|) < 1\},$$

and let $h = \chi_M$, the indicator functional of the convex set

$$M = \{(\eta_1, \eta_2) \in F \mid \max(|\eta_1-1|, |\eta_2|) \leq 1\}.$$

Then $\inf h(G) = a = +\infty$ (since $G \cap M \neq \emptyset$) and for $\psi_o(y) = \eta_1$ $(y = (\eta_1, \eta_2) \in F)$ we have (2.7), but for $y' = (\eta_1', \eta_2') \in A_a$ such that $\eta_1' = 0$ there is no $\psi \in G^s$ satisfying $\sup \psi(G) < \psi(y')$, whence the right hand side of (2.11) is $-\infty$, so (2.11) does not hold.

From Theorem 2.3 and [4], Theorem 2.4, there follows

THEOREM 2.4. Let F be a locally convex space, G a convex subset of F with $G^s \neq \emptyset$ and $h: F \to \bar{R}$ a functional, with A_a non-empty and convex, such that either G or A_a is open, and let $y' \in A_a$. Then we have (2.10) and, in the case when Int $A_a \neq \emptyset$, $y' \in$ Int A_a, we have (2.11).

2.3 RESULTS OF STRONG DUALITY IN TERMS OF CLOSED STRIPS CONTAINING G

THEOREM 2.5. Let F be a locally convex space, G a subset of F and $h: F \to \bar{R}$ a functional with $A_a \neq \emptyset$, and let $y' \in A_a$. Consider the following statements:

1°. There exists $\psi_o \in G^s$ satisfying (2.1).

2°. There holds

$$\inf h(G) = \max_{\psi \in G^s} \inf_{\substack{y \in F \\ \psi(y) \in \overline{\psi(G)}}} h(y). \tag{2.15}$$

3°. There holds

$$\inf h(G) = \max_{\substack{\psi \in G^s \\ \psi(y') \notin \overline{\psi(G)}}} \inf_{\substack{y \in F \\ \psi(y) \in \overline{\psi(G)}}} h(y). \tag{2.16}$$

4°. There holds

$$\inf h(G) = \max_{\substack{\psi \in G^s \\ \sup \psi(G) < \psi(y')}} \inf_{\substack{y \in F \\ \psi(y) \in \overline{\psi(G)}}} h(y). \tag{2.17}$$

$1'$-$4'$, obtained from 1°-4° similarly to the corresponding procedure of Theorem 1.3.

i) We have the implications $1^o \Rightarrow 4^o \Rightarrow 3^o$ and the equivalence $2^o \Leftrightarrow 3^o$.

ii) If G is F^*-connected, then $2^o \Leftrightarrow 3^o \Leftrightarrow 4^o$.

iii) If G and A_a are F^*-connected, then $1^o \Leftrightarrow \ldots \Leftrightarrow 4^o$.

iv) We have $n^o \Leftrightarrow n'$ $(n=1,\ldots,4)$.

We omit the proof.

2.4 RESULTS OF STRONG DUALITY IN TERMS OF SUPPORT HYPERPLANES OF G

As shown by Example 1.2, we may have $\psi_c \in G^s \cap -G^s$ and

$$\inf h(G) = \inf_{\substack{y\in F \\ \psi_c(y)=\sup \psi_c(G)}} h(y) = \inf_{\substack{y\in F \\ \psi_c(y)=\inf \psi_c(G)}} h(y),$$

and yet strict inequality $<$ in (1.27) (see Example 1.1). However, we shall prove now

THEOREM 2.6. Let F be a locally convex space, G a subset of F and $h: F \rightarrow \bar{R}$ a functional with $A_a \neq \phi$, satisfying one of a), b), c) or d) of Theorem 1.6, and let $y' \in A_a$. Consider the following statements:

1°. There exists $\psi_o \in G^s$ satisfying (2.1).

2°. There holds

$$\inf h(G) = \max_{\substack{\psi\in G^s \\ \sup \psi(G)<\psi(y')}} \inf_{\substack{y\in F \\ \psi(y)=\sup \psi(G)}} h(y). \tag{2.18}$$

1'-2', obtained from 1°-2° similarly to the corresponding procedure of Theorem 1.3.

i) We have the implication $1^o \Rightarrow 2^o$.

ii) If A_a is F^*-connected, then $1^o \Leftrightarrow 2^o$.

iii) We have $n^o \Leftrightarrow n'$ $(n=1,2)$.

PROOF. Part i) follows from the second part of Remark 1.6.

ii) Assume that A_a is F^*-connected and that 2° holds. Then, by (2.18), there exists $\psi_o \in G^s$ satisfying (2.5) and

$$a = \inf h(G) = \inf_{\substack{y\in F \\ \psi_o(y)=\sup \psi_o(G)}} h(y). \tag{2.19}$$

Hence, since A_a is F^*-connected, by Proposition 1.2 ii) we obtain (2.1).

Finally, the proof of iii) is similar to that of Theorem 2.3 v). This completes the proof of Theorem 2.6.

From Theorem 2.6, using the separation theorem, combined with [4], Lemmas 2.2 i), 2.3 i) and [4], Theorem 2.6, there follows

THEOREM 2.7. Let F be a locally convex space, G a convex subset of F and $h: F \to \bar{R}$ a functional with $A_a \neq \emptyset$, and let $y' \in A_a$. If the sets $A_{a+\frac{1}{n}}$ $(n=1,2,\ldots)$ or $S_{a+\frac{1}{n}}$ $(n=1,2,\ldots)$ are convex and if either i) A_a is convex and open or ii) h is convex and G is open, with $a = \inf h(G) < +\infty$, then we have (2.18).

REMARK 2.3. Theorem 2.7 has been proved, essentially, in [2], Theorem 2.1 and Remark 2.2 (d), and their proofs.

REFERENCES

1. V. KLEE, Separation and suport properties of convex sets-a survey. In: Control theory and the calculus of variations, 235-303. Academic Press, New York, 1969.

2. I. SINGER, Generalizations of methods of best approximation to convex optimization in locally convex spaces. II: Hyperplane theorems. J. Math. Anal. Appl. 69 (1979), 571-584.

3. I. SINGER, Duality theorems for linear systems and convex systems. J. Math. Anal. Appl. 76 (1980), 339-368.

4. I. SINGER, Optimization by level set methods. I: Duality formulae. In: Optimisation: Théorie et algorithmes. Proc. Internat. Confer. held at Confolant, March, 1981. Lecture Notes in Pure and Appl. Math., Marcel Dekker, New York (to appear).

INCREST
Department of Mathematics
Bd. Păcii 220,
79622 Bucharest
and
Institute of Mathematics,
Str. Academiei 14
70109 Bucharest, Romania.

Functional Analysis, Holomorphy and Approximation Theory II, G.I. Zapata (ed.)

SPACES FORMED BY SPECIAL ATOMS II

Geraldo Soares de Souza

(Dedicated to Patricia, Lesley and Geraldo Jr., my children)

1. INTRODUCTION

We define the space $C_0^p = \{f\colon T \to \mathbb{R};\ f(t) = \sum_{\text{finite}} c_n b_n(t)\}$ for $\frac{1}{2} < p < \infty$ where T is the perimeter of the unit disk in the complex plane. C_0^p is endowed with the "norm" $\|f\|_{C_0^p} = \operatorname{Inf} \sum_{\text{finite}} |c_n|^p$ where the infimum is taken over all possible representations of f. Each b_n is a special p-atom, that is, a real-valued function b, defined on T, which is either $b(t) \equiv \frac{1}{2\pi}$ or $b(t) = -\frac{1}{|I|^{1/p}}\chi_R(t) + \frac{1}{|I|^{1/p}}\chi_L(t)$, where I is an interval on T, L is the left half of I and R is the right half. $|I|$ denotes the length of I and χ_E the characteristic function of E. Then we define the space C^p as being the completion of C_0^p under the "norm" $\|\ \|_{C_0^p}$. We may say $f \in C^p$ if there is a sequence (b_n) of special p-atoms and a sequence (c_n) of numbers such that $\sum_{n=1}^{\infty} |c_n|^p < \infty$ and $f(t) = \sum_{n=1}^{\infty} c_n b_n(t)$. The "norm" in C^p is $\|f\|_{C^p} = \operatorname{Inf} \sum_{n=1}^{\infty} |c_n|^p$, where the infimum is taken over all possible representations of f. In this paper we are interested in those p in the range $\frac{1}{2} < p < 1$. For $p = 1$ we may refer the interested reader to [5].

The spaces C^p need some further comments since they are defined in the abstract sense. Observe that C_0^p is a metric space under the metric $d(f,g) = \|f-g\|_{C_0^p}$ for $\frac{1}{2} < p < 1$, but not necessarily complete. Next consider the completion of C_0^p under the metric d. This completion is denoted by C^p, since $f \in C^p$ implies that f is an equivalence class of Cauchy sequences under the metric d, say $f = (f_n)_{n\geq 1}$ where $\|f_n - f_m\|_{C_0^p} \to 0$ as $n,m \to \infty$, then the "norm" of f in C^p is defined by $\|f\|_{C^p} = \lim_{n\to\infty} \|f_n\|_{C_0^p}$. The notation "norm" means $\|\cdot\|_{C^p}$ is not a genuine norm.

In this paper we study some properties of the spaces C^p and the computation of its dual spaces, which is the key result of this paper.

To make the presentation reasonably self-contained, we shall include a resumé of pertinent results and definitions.

2. PRELIMINARIES

DEFINITION 2.1 The Lipschitz space $\mathrm{Lip}\ \alpha$ is defined by $\mathrm{Lip}\ \alpha = \{g: T \to R,$ continuous, $g(x+h) - g(x) = O(h^\alpha)\}$ for $0 < \alpha < 1$. The $\mathrm{Lip}\ \alpha$ norm is given by

$$\|g\|_{\mathrm{Lip}\ \alpha} = \sup_{\substack{h>0 \\ x}} \left|\frac{g(x+h) - g(x)}{h^\alpha}\right|.$$

DEFINITION 2.2 The generalized Lipschitz space Λ_α is defined by $\Lambda_\alpha = \{G: T \to \mathbb{R},$ continuous, $G(x+h) + G(x-h) - 2G(x) = O(h^\alpha)\}$ for $0 < \alpha < 2$. The Λ_α norm is given by

$$\|G\|_{\Lambda_\alpha} = \sup_{\substack{h>0 \\ x}} \left|\frac{G(x+h) + G(x-h) - 2G(x)}{h^\alpha}\right|.$$

We would like to point out that Λ_α can be defined as $\Lambda_\alpha = \{G: T \to \mathbb{R},$ G absolutely continuous and $G' \in \mathrm{Lip}(\alpha-1)\}$ for $1 < \alpha < 2$, and for $0 < \alpha < 1$, Λ_α is the same as $\mathrm{Lip}\ \alpha$, for these two claims we state the following theorem.

THEOREM 2.3 (A. Zygmund). $g \in \mathrm{Lip}(\alpha-1)$ if and only if $G \in \Lambda_\alpha$, where $1 < \alpha < 2$ and $G(x) = \int_0^x g(t)dt$. Moreover there are two absolute constants N and M such that $N\|g\|_{\mathrm{Lip}(\alpha-1)} \leq \|G\|_{\Lambda_\alpha} \leq$

$\le M\|g\|_{Lip(\alpha-1)}$. If $0 < \alpha < 1$ then $g \in Lip(\alpha-1)$ if and only if $g \in \Lambda_\alpha$, moreover the norms are equivalent.

R.R. Coifman [1] observed that a distribution f is the real part of a boundary function $F \in H^p(\mathbb{D})$. ($F \in H^p(\mathbb{D})$ for $\frac{1}{2} < p \le 1$ if and only if $\|F\|_{H^p} = \sup_{r<1} \left(\int_T |F(re^{i\theta})|^p d\theta\right)^{1/p} < \infty$ where $\mathbb{D} = \{z \in C;\ |z| < 1\}$) if and only if there is a sequence (a_n), of p-atoms and a sequence (c_n), of numbers, such that $\sum_{n=1}^{\infty} |c_n|^p < \infty$ and $f(t) = \sum_{n=1}^{\infty} c_n a_n(t)$. (A real valued function defined on T is called a p-atom whenever a is supported on an interval $I \subset T$, $|a(t)| \le |I|^{-1/p}$ and $\int_I a(t)dt = 0$.) Moreover, letting $\lambda(f)$ equal the infimum of $\sum_{n=1}^{\infty} |c_n|^p$ over all such representations of f, there exists absolute constants M and N such that $M\|F\|^p_{H^p} \le \lambda(f) \le N\|F\|^p_{H^p}$, we shall denote the set of such f as ReH^p and $\|f\|_{ReH^p} = \lambda(f)$. We point out this result is true for $0 < p \le 1/2$, however we have to slightly modify the definition of p-atoms. We are not going to deal with p in this range. The interested reader is directed to [1].

Notice for $\frac{1}{2} < p < 1$, the boundary values of H^p must be taken in the sense of distributions, because, as the example $F(z) = \frac{1}{2\pi} \cdot \frac{1+z}{1-z}$ shows, one cannot uniquely recover F from the pointwise boundary value of its real part.

3. SOME PROPERTIES OF C^p

In this section we state and prove some properties of the space C^p.

LEMMA 3.1. C^p is an embedding in ReH^p for $\frac{1}{2} < p \le 1$, that is, the inclusion mapping is a bounded linear operator.

PROOF. Obvious from the definition of ReH^p and C^p, that $\|f\|_{ReH^p} \le \|f\|_{C^p}$.

LEMMA 3.2. If χ_I is the characteristic function of an interval I and $I \subset [0,2\pi]$, then $\chi_I \in C^p$, moreover $\|\chi_I\|_{C^p} \leq A_p|I|^p$ where A_p depends only on p.

PROOF. One can easily observe that it suffices to prove this theorem for $I = [0, \frac{2\pi}{2^N}]$ where N is a fixed non-negative integer. The idea is to expand χ_I in Haar-Fourier series on $[0,2\pi]$. In fact, we recall that the Haar system on $[0,2\pi]$ is defined by

$$\psi_{nk}(t) = \begin{cases} (\frac{2^n}{2\pi})^{1/2} & \text{on } [\frac{k-1}{2^n}2\pi, \frac{k-1/2}{2^n}2\pi) \\ -(\frac{2^n}{2\pi})^{1/2} & \text{on } [\frac{k-1/2}{2^n}2\pi, \frac{k}{2^n}2\pi] \\ 0 & \text{elsewhere.} \end{cases}$$

Consequently, the expansion of χ_I is $\chi_I(t) = \sum_{n=0}^{\infty}\sum_{k=1}^{2^n} a_{nk}\psi_{nk}(t)$, where $a_{nk} = \int_{I_{nk}} \chi_I(t)\psi_{nk}(t)dt$, $I_{nk} = [\frac{k-1}{2^n}2\pi, \frac{k}{2^n}2\pi]$.

If we split I_{nk} as in the definition of (ψ_{nk}) then the geometry of I_{nk} and $[0, \frac{2\pi}{2^N}]$ shows that $a_{n1} \neq 0$ for $0 \leq n < N$ and $a_{nk} = 0$ otherwise, thus the expansion of χ_I in Haar-Fourier series becomes

$$\chi_I(t) = \sum_{n=0}^{N-1} a_{n1}\psi_{n1}(t), \tag{3.3}$$

so by computing the coefficients a_{n1} we get $a_{n1} = (\frac{2^n}{2\pi})^{1/2}\cdot s$ where $s = \frac{2\pi}{2^N}$. Substituting these values into (3.3) we have

$$\chi_I(t) = \sum_{n=0}^{N-1} s(\frac{2^n}{2\pi})^{1/2}\psi_{n1}(t). \tag{3.4}$$

Now observe that $b_n(t) = (\frac{2^n}{2\pi})^{\frac{2-p}{2p}}\cdot\psi_{n1}(t)$ are special p-atoms for $n = 0,1,\ldots,N-1$, then we may write (3.4) in the form

$$\chi_I(t) = \sum_{n=0}^{N-1} s(\frac{2^n}{2\pi})^{\frac{p-1}{p}} b_n(t) \tag{3.5}$$

therefore $\chi_I \in C^p$; moreover by definition of "norm" in C^p we have from (3.5) that $\|\chi_I\|_{C^p} \le [\sum_{n=0}^{N-1} (\frac{2\pi}{2^n})^{1-p}] s^p$, then taking $A_p = \sum_{n=0}^{N-1} (\frac{2\pi}{2^n})^{1-p}$ we have $\|\chi_I\|_{C^p} \le A_p |I|^p$. Thus the theorem is proved for intervals of the form $[0, \frac{2\pi}{2^n}]$. Now if $I = [0,s]$ where s is an arbitrary number in $(0,2\pi]$, we can write the dyadic expansion of s and apply the above argument. Finally if I is any interval, say $I = (\alpha,\beta]$, $0 < \alpha < \beta \le 2\pi$, then $\chi_I = \chi_{[0,\beta]} - \chi_{[0,\alpha]}$, and so $\chi_I \in C^p$. On the other hand, observe that the operator $T_a f = f^a$ where $f^a(x) = f(x-a)$, maps C^p boundedly into C^p, in fact $\|T_a f\|_{C^p} \le \|f\|_{C^p}$, so if we take $f(t) = \chi_{(0,\beta-\alpha]}(t)$ then $f^\alpha(t) = \chi_{(\alpha,\beta]}(t)$ and therefore $\|\chi_{(\alpha,\beta]}\|_{C^p} \le \|\chi_{(0,\beta-\alpha]}\|_{C^p} \le A_p(\beta-\alpha)^p$. Thus the theorem is proved.

The next result is some sort of Hölder's inequality between C^p and $\mathrm{Lip}(\frac{1}{p} - 1)$ which will be very crucial in determining the linear functionals on C^p.

THEOREM 3.6. (Hölder's Type Inequality). If $f \in C^p$ and $g \in \mathrm{Lip}\,\alpha$ then $|\int_T f(t)g(t)dt| \le \|g\|_{\mathrm{Lip}\,\alpha} (\|f\|_{C^p})^{1/p}$ where $\alpha = \frac{1}{p} - 1$.

PROOF. Let $f(t) = \sum_{n=1}^{k} c_n b_n(t)$ where $b_n(t) = \frac{-1}{|I_n|^{1/p}} \chi_{R_n}(t) + \frac{1}{|I_n|^{1/p}} \chi_{L_n}(t)$, that is, f is finite linear combination of special p-atoms. Then

$$\int_T f(t)g(t)dt = \sum_{n=1}^{k} c_n \int_{I_n} b_n(t)g(t)dt =$$

$$= \sum_{n=1}^{k} c_n \int_{I_n} b_n(t)[g(t)-g(t_n)]dt$$

where $t_n \in I_n$, and thus using the fact that $g \in \mathrm{Lip}\,\alpha$ we get

$$\left|\int_T f(t)g(t)dt\right| \le \sum_{n=1}^{k} \frac{|c_n|}{|I_n|^{1/p}} \int_{I_n} |g(t)-g(t_n)|dt \le$$

$$\le \|g\|_{\text{Lip }\alpha} \sum_{n=1}^{k} \frac{|c_n|}{|I_n|^{1/p}} \int_{I_n} |t-t_n|^{\frac{1}{p}-1} dt.$$

Then $\left|\int_T f(t)g(t)dt\right| \le \|g\|_{\text{Lip }\alpha}(\sum_{n=1}^{k} |c_n|)$ so that for $\frac{1}{2} < p < 1$, it follows that $\left|\int_T f(t)g(t)dt\right| \le \|g\|_{\text{Lip }\alpha}(\sum_{n=1}^{k} |c_n|^p)^{1/p}$. Consequently we have $\left|\int_T f(t)g(t)dt\right| \le \|g\|_{\text{Lip }\alpha}(\|f\|_{C^p})^{1/p}$, therefore the theorem is proved for any f in C_0^p. Now by definition of C^p we have C_0^p is dense in C^p so the extension for any $f \in C^p$ is just routine, hence the theorem is proved.

The next results give us a different way to define a norm in the Lipschitz space Lip α for $\alpha = \frac{1}{p} - 1$.

COROLLARY 3.7. If $f \in C^p$ and $g \in \text{Lip }\alpha$ for $\alpha = \frac{1}{p} - 1$ then $A_p\|g\|_{\text{Lip }\alpha} \le \sup_{\|f\|_{C^p}\le 1} \left|\int_T f(t)g(t)dt\right| \le \|g\|_{\text{Lip }\alpha}$, where A_p depends only on p.

PROOF. Theorem 3.6 tells us that $\sup_{\|f\|_{C^p}\le 1} \left|\int_T f(t)g(t)dt\right| \le \|g\|_{\text{Lip }\alpha}$. On the other hand, if $f(t) = \frac{-1}{(2h)^{1/p}} \chi_{[\beta-h,\beta)}(t) + \frac{1}{(2h)^{1/p}} \chi_{(\beta,\beta+h]}(t)$ then $\int_T f(t)g(t)dt = \frac{G(\beta+h) + G(\beta-h) - 2G(\beta)}{(2h)^{1/p}}$ where $G(x) = \int_0^x g(t)dt$, so that $\sup_{\|f\|_{C^p}\le 1} \left|\int_T f(t)g(t)dt\right| \ge \frac{1}{2^{1/p}} \frac{|G(\beta+h) + G(B-h) - 2G(\beta)|}{h^{1/p}}$. Consequently $\sup_{\|f\|_{C^p}\le 1} \left|\int_T f(t)g(t)dt\right| \ge \frac{1}{2^{1/p}} \|G\|_{\Lambda_{\alpha+1}}$ and by Theorem 2.3 we get $\sup_{\|f\|_{C^p}\le 1} \left|\int_T f(t)g(t)dt\right| \ge A_p\|g\|_{\text{Lip }\alpha}$, and so combining these two inequalities involving Lip α-norm we get the desired result.

4. DUALITY

Consider the mapping $\psi_g: C^p \to \mathbb{R}$ defined by $\psi_g(f) = \int_T f(t)g(t)dt$, where g is a fixed function in $\mathrm{Lip}\,\alpha$ for $\alpha = \frac{1}{p} - 1$. One can easily see that ψ_g is a linear functional on C^p. Moreover, despite the fact that C^p is not a normed space it is obvious that by the usual argument $|\psi_g(f)| \leq \|g\|_{\mathrm{Lip}\,\alpha}(\|f\|_{C^p})^{1/p}$ is equivalent to the continuity of ψ_g, that is, ψ_g is a bounded linear functional on C^p. Consequently, we have that for each $g \in \mathrm{Lip}\,\alpha$, ψ_g is a bounded linear functional on C^p. At this point, a natural question is: Are these all the linear functionals on C^p? We anticipate that the answer is yes; in order to formulate the theorem which leads to this answer, we need some notation.

Throughout this paper X^* will denote the dual space of X, that is, the space of bounded linear functionals ψ on X with the norm

$$\|\psi\| = \sup_{\|f\|_X \leq 1} |\psi(f)|.$$

We recall our definition of bounded linear functional just means $\|\psi\| < \infty$.

THEOREM 4.1 (Duality theorem). If $\psi \in (C^p)^*$ then there is a unique $g \in \mathrm{Lip}\,\alpha$, $\alpha = \frac{1}{p} - 1$ and $\frac{1}{2} < p < 1$ such that $\psi = \psi_g$, that is, $\psi(f) = \int_T f(t)g(t)dt$ for all $f \in C^p$. Conversely if $\psi(f) = \int_T f(t)g(t)dt$ then $\psi \in (C^p)^*$. Moreover there is a constant A_p such that $A_p\|g\|_{\mathrm{Lip}\,\alpha} \leq \|\psi\| \leq \|g\|_{\mathrm{Lip}\,\alpha}$, where A_p depends only on p. Therefore the mapping $\varphi: \mathrm{Lip}\,\alpha \to (C^p)^*$ defined by $\varphi(g) = \psi_g$ is a Banach space isomorphism.

PROOF. If $\psi(f) = \int_T f(t)g(t)dt$, then we already have seen that Theorem 3.6 implies that ψ is a bounded linear functional, that is, $\psi \in (C^p)^*$, so it remains to prove the other direction. In fact, let $\psi \in (C^p)^*$ and define $G(s) = \psi(\chi_{[0,s]})$ for $s \in [0,2\pi]$. Observe that $G(s+h) - G(s) = \psi(\chi_{(s,s+h]})$ and thus Lemma 3.2 and the boundedness of ψ tells us that $|G(s+h) - G(s)| \le A_p h^p$ where A_p is an absolute constant, therefore G is continuous. On the other hand, using the definition of G we get

$$(4.2) \qquad \frac{1}{2^{1/p}} \cdot \frac{G(s+h) + G(s-h) - 2G(s)}{h^{1/p}} =$$

$$= \psi\left(\frac{1}{(2h)^{1/p}} \chi_{(s,s+h]} - \frac{1}{(2h)^{1/p}} \chi_{(s-h,s]}\right) .$$

Consequently, since $b(t) = \frac{1}{(2h)^{1/p}} \chi_{(s,s+h]}(t) - \frac{1}{(2h)^{1/p}} \chi_{[s-h,s]}(t)$ is a special p-atom we have $\|b\|_{C^p} \le 1$. Therefore using the boundedness of ψ in (4.2) we get $\left|\frac{G(s+h) + G(s-h) - 2G(s)}{h^{1/p}}\right| \le 2^{1/p}\|\psi\| < \infty$, so that $\|G\|_{\Lambda_\alpha} < \infty$, and therefore $G \in \Lambda_\alpha$ for $\alpha = 1/p$. So by the remarks made right after Definition 2.2 we have that G is absolutely continuous, therefore there exists a function g on T such that $G(s) = \int_0^s g(t)dt$. Thus by Theorem 2.3 we have $g \in \mathrm{Lip}(\frac{1}{p} - 1)$. On the other hand this implies that $\psi(\chi_{[0,s]}) = \int_T \chi_{[0,s]}(t)g(t)dt$ and thus if I is any interval in $[0,2\pi]$ we have that $\psi(\chi_I) = \int_T \chi_I(t)g(t)dt$. Consequently if $b(t) = \frac{-1}{|I|^{1/p}} \chi_R(t) + \frac{1}{|I|^{1/p}} \chi_L(t)$ is a special p-atom then $\psi(b) = \int_T b(t)g(t)dt$ and therefore for any $f \in C^p$ we have $\psi(t) = \int_T f(t)g(t)dt$. By Corollary 3.7 we have that there is a constant A_p depending only on p such that $A_p\|g\|_{\mathrm{Lip}\,\alpha} \le \|\psi\| \le \|g\|_{\mathrm{Lip}\,\alpha}$ for $\alpha = \frac{1}{p} - 1$ and thus the theorem is proved.

In [3] the author also introduced the spaces B^p for $\frac{1}{2} < p < \infty$ defined by $B^p = \{f\colon T \to R;\ f(t) = \sum_{n=1}^{\infty} c_n b_n(t);\ \sum_{n=1}^{\infty} |c_n| < \infty\}$ where b_n's are special p-atoms defined in the introduction. We endow B^p with the norm $\|f\|_{B^p} = \text{Inf} \sum_{n=1}^{\infty} |c_n|$ where the infimum is taken over all possible representations of f. One of the main results about B^p is the following.

THEOREM 4.3 (Duality Theorem). ψ is a bounded linear functional on B^p for $\frac{1}{2} < p < 1$ if and only if there is a unique $g \in \text{Lip}\,\alpha$, $\alpha = \frac{1}{p} - 1$ such that $\psi(f) = \int_T f(t)g(t)dt$ for all $f \in B^p$. Moreover the mapping φ: $\text{Lip}\,\alpha \to (B^p)^*$ defined $\varphi(g) = \psi$ is an isometric isomorphism.

The proof of this theorem follows basically the same line as C^p, however we point out that a basic difference between them is that C^p is a Fréchet space while B^p is a Banach space. Certainly the fact that B^p is a Banach space makes it easier to work with this space. For example by using the duality theorem and the Hahn-Banach theorem one can prove without much difficulty that B^p can be identified with the space of analytic functions f on the disk $\mathbb{D} = \{z \in C, |z| < 1\}$ satisfying

$$\int_0^1 \int_0^{2\pi} |f(re^{i\theta})|(1-r)^{\frac{1}{p}-2} d\theta dr < \infty .$$

These spaces were introduced by P.L. Duren, B.W. Romberg and A.L. Shields in [2]. With this identification one can say that B^p is a real characterization of such spaces. We point out that [2] shows that B^p is the smallest Banach space containing H^p; again we emphasize that we are working with $\frac{1}{2} < p < 1$. Now if we put together these results about B^p and C^p we have as a consequence the famous duality theorem for H^p in [2]. In fact, Corollary 3.1 tells us that $C^p \subset H^p$ continuously, on the other hand one easily can see from definition of H^p and B^p that $H^p \subset B^p$ continuously. Therefore we have $C^p \subset H^p \subset B^p$ which implies $\text{Lip}\,\alpha \subset (H^p)^* \subset \text{Lip}\,\alpha$ for $\alpha = \frac{1}{p} - 1$. Thus we get the duality of H^p for $\frac{1}{2} < p < 1$.

That is, $(H^p)^* = \mathrm{Lip}(\frac{1}{p} - 1)$.

One natural question to ask is: Do there exist constants M and N dependent only on p such that $M\|f\|_{C^p} \leq \|f\|_{H^p} \leq N\|f\|_{C^p}$ for $\frac{1}{2} < p < 1$? In other words are C^p and H^p the same space as Fréchet space? The answer to this question is negative, and will come out shortly in a joint paper with Gary Sampson [7].

5. INTERPOLATION THEOREM

In this section we present an elementary theorem on the interpolation of operators acting on C^p spaces. In order to state it we need some definitions.

Let f be a real valued measurable function on T. For $y > 0$ let $m(f,y) = m(|f|,y) = |\{x \in T, |f(x)| > y\}|$. $m(f,y)$ is called distribution function of f, $|\cdot|$ means the Lebesgue measure on T. $m(f,y)$ is non-negative, non-increasing and continuous from the right.

A sublinear operator T is a mapping from linear space of measurable functions defined on a measure space into measurable functions defined on another measure space satisfying

i) $|T(f+g)(x)| \leq |Tf(x)| + |Tg(x)|$ almost everywhere.

ii) $|T\alpha f| = |\alpha||Tf|$, whenever α is a scalar and f is in domain of T.

A sublinear operator $T: X \to Y$ is said to be bounded if $\|T\| = \sup\{\|Tf\|_Y : \|f\|_X \leq 1\} < \infty$.

A function f defined on T for which $\|f\|_p = (\int_T |f(t)|^p dt)^{1/p} < \infty$, is said to be in L_p. We shall say that a measurable function f belongs to $L(p,\infty)$, usually called weak L_p space, if there exists a positive number A such that $m(f,y) \leq (\frac{A}{y})^p$. In both definitions p lies in $(0,\infty)$. The least constant

A in the definition of weak L_p is considered as "norm" in $L(p,\infty)$.

We recall that one can easily show that $\|f\|_p^p =$ $= p\int_0^\infty \alpha^{p-1} m(f,\alpha)d\alpha$ and that L_p is continuously embedded in $L(p,\infty)$, that is, $m(f,y) \le (\frac{\|f\|_p}{y})^p$.

THEOREM 5.1. Let $\frac{1}{2} < p_0 < p \le 1 < p_1 < \infty$. Suppose T is a sub-linear operator mapping $T: C^{p_0} \to L(p_0,\infty)$ boundedly with norm M_0 and $T: L_{p_1} \to L(p_1,\infty)$ with norm M_1, that is

$$(5.2) \qquad m(Tf,\alpha) \le (\frac{M_0}{\alpha}\|f\|_{C^{p_0}})^{p_0} \quad \text{and} \quad m(Tf,\alpha) \le (\frac{M_1}{\alpha}\|f\|_{p_1})^{p_1}$$

respectively.

Then $T: C^p \to L_p$ with $\|Tf\|_p \le KM_0^tM_1^{1-t}(\|f\|_{C^p})^{1/p}$, where K is an absolute constant depending only on p_0, p_1 and p, $\frac{1}{p} = \frac{t}{p_0} + \frac{1-t}{p_1}$, $0 < t < 1$.

PROOF. Let f be a special p-atom, that is, $f(t) = \frac{-1}{|I|^{1/p}}\chi_R(t) + \frac{1}{|I|^{1/p}}\chi_L(t)$. Therefore

$$(5.3) \qquad (\|f\|_{C^{p_0}})^{p_0} \le |I|^{\frac{p-p_0}{p}} \quad \text{and} \quad (\|f\|_{p_1})^{p_1} \le |I|^{\frac{p-p_0}{p}} .$$

We now evaluate $\|Tf\|_p$ using the definition of L_p-"norm" in terms of the distribution function. We have,

$$\frac{1}{p}\|Tf\|_p^p = \int_0^\infty \alpha^{p-1}m(Tf,\alpha)d\alpha =$$

$$= \int_0^\sigma \alpha^{p-1}m(Tf,\alpha)d\alpha + \int_\sigma^\infty \alpha^{p-1}m(Tf,\alpha)d\alpha, \quad \sigma > 0.$$

Then (5.2) implies that

$$\frac{1}{p}\|Tf\|_p^p \le (M_0\|f\|_{C^{p_0}})^{p_0}\int_0^\sigma \alpha^{p-p_0-1}d\alpha + (M_1\|f\|_{p_1})^{p_1}\int_\sigma^\infty \alpha^{p-p_1-1}d\alpha.$$

Using (5.3) and the hypothesis on p's we obtain

$$\frac{1}{p}\|Tf\|_p^p \le \frac{M_0^{p_0}}{p-p_0}|I|^{\frac{p-p_0}{p}}\sigma^{p-p_0} + \frac{M_1^{p_1}}{p_1-p}|I|^{\frac{p-p_1}{p}}\sigma^{p-p_1}.$$

As σ is arbitrary we may take $\sigma = B|I|^{-\frac{1}{p}}$, where B is a constant to be determined. We get

$$(5.4)\qquad \frac{1}{p}\|Tf\|_p^p \le \frac{M_0^{p_0}}{p-p_0}B^{p-p_0} + \frac{M_1^{p_1}}{p_1-p}B^{p-p_1}.$$

Since $B > 0$ is arbitrary, replace B by the value that makes the expression minimal, namely take

$$B = \left(\frac{M_1^{p_1}}{M_0^{p_0}}\right)^{\frac{1}{p_1-p_0}}.$$

Substitutin in (5.4)

$$(5.5)\qquad \|Tf\|_p \le \left(\frac{p}{p-p_0}+\frac{p}{p_1-p}\right)^{1/p}\cdot M_o^{\frac{p_0}{p}\cdot\frac{p_1-p}{p_1-p_0}}\cdot M_1^{\frac{p_1}{p}\cdot\frac{p-p_0}{p_1-p_0}}.$$

Observe that $\frac{p_0}{p}\cdot\frac{p_1-p}{p_1-p_0}+\frac{p_1}{p}\cdot\frac{p-p_0}{p_1-p_0} = 1$. Taking $t = \frac{p_0}{p}\cdot\frac{p_1-p}{p_1-p_0}$ and $1-t = \frac{p_1}{p}\cdot\frac{p-p_0}{p_1-p_0}$, then $\frac{1}{p} = \frac{t}{p_0}+\frac{1-t}{p_1}$, $0 < t < 1$, and thus (5.5) becomes $\|Tf\|_p \le \left(\frac{p}{p-p_0}+\frac{p}{p_1-p}\right)^{1/p}\cdot M_0^t\cdot M_1^{1-t}$. Consequently, if $h(t) = \sum_{n=1}^{k} c_n b_n(t)$ is a finite linear combination of special p-atoms, that is, b_n's are equal p-atoms of type f, we have

$$\|Th\|_p^p \le \sum_{n=1}^{k}|c_n|^p\|Tb_n\|_p^p$$

and thus

$$(5.6)\qquad \|Th\|_p \le KM_0^tM_1^{1-t}(\|h\|_{C^p})^{1/p}\quad \text{where}\quad K = \left(\frac{p}{p-p_0}+\frac{p}{p_1-p}\right)^{1/p}.$$

That is, the theorem is proved for a dense subspace of C^p, namely C_0^p, so it can be extended to all C^p, preserving the inequality (5.6) in C^p. Then, if $f \in C^p$, (5.6) implies that

$\|Tf\|_p \leq KM_0^t M_1^{1-t}(\|f\|_{C^p})^{1/p}$. The proof is completed.

The interested reader can find similar types of interpolation theorem in [8] and [9].

REFERENCES

1. R.R. COIFMAN, A real variable characterization of H^p, Studia Math., 51 (1974) 269-274.

2. P.L. DUREN, B.W. ROMBERG and A.L. SHIELDS, Linear functionals on H^p with $0 < p-1$, J. Reine Angin Math., 238 (1969) 32-60.

3. GERALDO SOARES DE SOUZA, Spaced formed by special atoms, Ph.D. dissertation, SUNY at Albany, 1980.

4. ---------------------- and RICHARD O'NEIL, Spaces formed with special atoms, Proceedings Conference on Harmonic Analysis, 1980, Italy. Rendiconti del Circolo Matematico di Palermo, 139-144, Serie II, #1. 1981.

5. ----------------------, Spaces formed by special atoms I, to appear, Rocky Mountain Journal of Mathematics.

6. ---------------------- and GARY SAMPSON, A real characterization of the pre-dual of the Bloch functions, to appear, London Journal of Mathematics.

7. ---------------------- and -----------, An analytic characterization of C^p, In preparation.

8. ----------------------, Two theorems on interpolation of operators, Journal of Functional Analysis, 46, 149-157, (1982).

9. ----------------------, An interpolation of operators, to appear, Anais da Academia Brasileira de Ciências 54, #3 (1982).

10. ----------------------, The dyadic special atom space, Proceeding Conference on Harmonic Analysis, Minneapolis, April 1981, Lectures Notes in Mathematics, #908, Springer-Verlag, 1981.

11. A. ZYGMUND, Trigonometric Series, 2nd red. ed., Vols. I,II, Cambridge University Press, New York, 1959.

Department of Mathematics
Syracuse University
Syracuse, New York 13210

Functional Analysis, Holomorphy and Approximation Theory II, G.I. Zapata (ed.)

A HOLOMORPHIC CHARACTERIZATION OF C^*-ALGEBRAS

Harald Upmeier

§1. INTRODUCTION

The theory of operator algebras (i.e. C^*-algebras and von Neumann algebras on complex Hilbert spaces) is of increasing importance to many branches of mathematics, e.g. integration theory, operator theory, algebraic topology and in particular mathematical physics and quantum mechanics. Since C^*-algebras provide a natural framework for the foundations of quantum mechanics and quantum field theory it is an important problem to <u>characterize</u> the class of C^*-algebras by certain properties, for instance motivated by physical experiments. So far two characterizations of operator algebras in different categories have been obtained. The first is A. Connes' characterization of von Neumann algebras in terms of self-dual homogeneous Hilbert cones [8], the second is the work of Alfsen and Shultz [2,1] characterizing the <u>state spaces</u> of C^*-algebras using the geometry of compact convex sets and their affine function spaces. Although the methods of these papers are quite different, both approaches have a common feature: the characterization of C^*-algebras in the respective category can be divided into two steps, the first being the characterization of the larger class of <u>Jordan C^*-algebras</u> (JB^*-algebras) and the second being the characterization of (associative) C^*-algebras among all (non-associative) Jordan C^*-algebras. (For the case of self-dual homogeneous Hilbert cones this point of view has been adopted by Bellissard and Iochum [5]). The second step involves some concept of "orientation" of a Jordan operator algebra.

The geometric objects associated with an operator algebra A in the papers mentioned above are the Hilbert cone associated with A via Tomita-Takesaki theory (if A has a predual) and the state space of A, respectively, both endowed with a natural affine geometry. On the other hand, the open unit ball D of a C^*-algebra A has an interesting holomorphic structure: D is homogeneous with respect to holomorphic automorphisms and is therefore a bounded symmetric domain. The aim of this paper is to give a holomorphic characterization of C^*-algebras in terms of the holomorphic structure of the associated open unit ball. The corresponding result for Jordan C^*-algebras is the main theorem in [6]. Here we obtain a characterization of C^*-algebras among all Jordan C^*-algebras which uses the structure of the Lie algebra of all derivations. A similar idea is underlying in [8]; the concept of Hilbert cones however is only appropriate for operator algebras having a Banach predual.
In §2 we give a short introduction into the theory of Jordan C^*-algebras and their holomorphic characterization including the construction of the Jordan triple product for bounded symmetric domains in complex Banach spaces. In §3 some properties of the Lie algebra of all derivations of a Jordan C^*-algebra are studied which will be needed in the sequel. The concept of "orientation" of a Jordan C^*-algebra is introduced in §4, and §5 contains the proof of the main result, saying that orientable Jordan C^*-algebras are C^*-algebras.

§2. JORDAN C^*-ALGEBRAS AND THEIR HOLOMORPHIC CHARACTERIZATION

Jordan algebras made their first appearance in mathematical physics and quantum theory, starting from the following observation (P. Jordan 1932): Let H be a complex Hilbert space. Then the

Banach space $\mathcal{H}(H)$ of all bounded hermitian operators on H (which can be interpreted as (bounded) observables of a quantum mechanical system) is not closed under the associative operator product xy, but with respect to the anti-commutator product

$$(2.1) \qquad x \circ y := \frac{1}{2}(xy + yx),$$

$\mathcal{H}(H)$ becomes a non-associative algebra. As a consequence the anti-commutator product of two observables has a physical interpretation whereas, in general, the operator product does not. The product (2.1) satisfies the following identities:

(J1) $\quad x \circ y = y \circ x \qquad$ (Commutativity)

(J2) $\quad x^2 \circ (x \circ y) = x \circ (x^2 \circ y) \qquad$ (Jordan-Identity).

2.2 DEFINITION. Let A be a (not necessarily associative) algebra over the real or complex numbers with product denoted by $x \circ y$ for all $x,y \in A$. Then A is called a Jordan algebra if the identities (J1) and (J2) are satisfied.

Since the algebras appearing in quantum mechanics are in general infinite-dimensional, it is desirable to consider Banach Jordan algebras. A particularly important class of Banach Jordan algebras are the so-called Jordan C^*-algebras (i.e. JB^*-algebras and JB-algebras) which have been introduced and thoroughly studied by Alfsen, Shultz and Størmer [3].

2.3 DEFINITION. Let X be a real Jordan algebra with product $x \circ y$ and unit element e. Then X is called a JB-algebra if X is a Banach space with respect to a norm $x \mapsto |x|$ such that for all $x,y \in X$ the following properties hold:

(i) $\quad |x \circ y| \leq |x| \cdot |y|$,

(ii) $\quad |x^2 + y^2| \geq |x|^2$.

In this case the JB-norm $|\cdot|$ on X is uniquely determined by (i) and (ii). Note that JB-algebras are "formally real", i.e. $x_1^2 + \ldots + x_n^2 = 0$ implies $x_1 = \ldots = x_n = 0$ for all $x_1, \ldots, x_n \in X$. The complex analogue of JB-algebras are the so-called JB*-algebras:

2.4 DEFINITION. Let Z be a complex Jordan algebra with product $z \circ w$, unit element e and involution $z \mapsto z^*$. Then Z is called a JB*-<u>algebra</u> if Z is a Banach space with respect to a norm $z \mapsto |z|$ such that for all $u, v \in Z$ the following properties hold:

(i) $|u \circ v| \leq |u| \cdot |v|$

(ii) $|\{uu^*u\}| = |u|^3$,

where

$$\{uv^*w\} := u \circ (v^* \circ w) - v^* \circ (w \circ u) + w \circ (u \circ v^*) \tag{2.5}$$

denotes the <u>Jordan triple product</u> of $u, v, w \in Z$.

Wright and Youngson [29,30] have shown that the concepts introduced above are equivalent: Given a JB-algebra X, the complexification $Z := X^{\mathbb{C}} = X \oplus iX$ becomes a JB*-algebra with respect to a unique "JB*-norm"; conversely the self-adjoint part

$$X := \{x \in Z : x^* = x\}$$

of a JB*-algebra Z is a JB-algebra under the restricted norm. JB*-algebras and JB-algebras are often called "Jordan C*-algebras" because of the following example.

2.6 EXAMPLE. Let Z be a unital C*-<u>algebra</u>, i.e. Z is a complex associative Banach $*$-algebra with unit such that product, involution and norm are related by the condition

$$|zz^*| = |z|^2$$

for all $z \in Z$. Then Z becomes a JB*-algebra under the anticommutator product (2.1) and the JB*-norm coincides with the C*-norm. To see this, note that for associative $*$-algebras the Jordan triple product (2.5) reduces to

$$\{uv^*w\} = \frac{1}{2}(uv^*w + wv^*u).$$

In particular, $\{zz^*z\} = zz^*z$. This implies, via the spectral theorem for hermitian operators and the C*-condition:

$$|\{zz^*z\}|^2 = |zz^*z|^2 = |(zz^*)^3| = |zz^*|^3 = |z|^6 ,$$

hence (2.4.ii) is fulfilled.

In a fundamental paper [12] Jordan, von Neumann and Wigner classified all formally-real (simple) Jordan algebras of finite dimension. A natural extension of this classification is the following list of all JB-factors of type I [22,3]:

2.7 EXAMPLE.

(i) Let $\mathbb{K}$ denote the field $\mathbb{R}$ of real numbers, the field $\mathbb{C}$ of complex numbers or the skew-field $\mathbb{H}$ of quaternions, respectively. Let E be a Hilbert space over $\mathbb{K}$. Then the set $\mathcal{H}(E)$ of all bounded $\mathbb{K}$-linear self-adjoint operators on E is a JB-algebra under the product (2.1) and the operator norm.

(ii) Let Y be a real Hilbert space of dimension ≥ 2 with scalar product $(x|y)$. Then the direct sum $X := \mathbb{R} \oplus Y$ becomes a JB-algebra called <u>spin factor</u> under the product

$$(\alpha_1 \oplus y_1) \circ (\alpha_2 \oplus y_2) := (\alpha_1\alpha_2 + (y_1|y_2)) \oplus (\alpha_1 y_2 + \alpha_2 y_1).$$

The name "spin-factor" stems from the fact, that in quantum mechanics spin systems obeying the canonical anti-commutation relations are in close connection with Jordan algebra representations of a suitable spin factor as defined above.

(iii) The set $\mathcal{H}_3(\mathbb{O})$ of all self-adjoint 3×3 octonion matrices is a JB-algebra under the product (2.1) which cannot be embedded into an associative algebra and is therefore called the <u>exceptional</u> Jordan algebra. The octonion matrices of higher rank do not form a Jordan algebra, since $\mathbb{O}$ is not associative [7; Ch. VIII, Lemma 8.2].

Jordan algebras and JB-algebras in particular have found several applications to quantum mechanics, for instance using the affine geometry of the state space of a JB-algebra or the projective geometry associated with the exceptional Jordan algebra. However, even more promising seems to be the relationship between Jordan C*-algebras and <u>complex analysis</u>, more precisely the Jordan algebraic description of bounded symmetric domains in complex Banach spaces. For the sake of completeness we give a short survey about this relationship. Given a complex Banach space Z and a domain $D \subset Z$, a mapping $f: D \to Z$ is said to be holomorphic if locally around each point $a \in D$ there exists a convergent expansion

$$f(z) = \sum_{m=0}^{\infty} P_m(z-a)$$

into a series of m-homogeneous continuous polynomials $P_m: Z \to Z$. The polynomials P_m are uniquely determined by f and a, namely

$$P_m(z) = \frac{f^{(m)}(a)}{m!}(z,\ldots,z),$$

where $f^{(m)}(a)$ is the m-th derivative of f in $a \in D$, considered as a multilinear mapping [18; §5]. A bijective mapping which is holomorphic in both directions is called biholomorphic. The set of all biholomorphic automorphisms of a domain D forms a group, called the automorphism group of D and denoted by $\mathrm{Aut}(D)$. As a matter of notation, denote by $\mathcal{L}(E)$ the algebra of all bounded linear operators on a real or complex Banach space E.

Generalizing a well-known theorem of H. Cartan for domains in $\mathbb{C}^n$, the author [24] and independently J.P. Vigué [28] have shown that for a bounded domain D in a complex Banach space Z, the automorphism group $G = \mathrm{Aut}(D)$ carries in a natural way the structure of a real Banach Lie group. The essential idea behind this result is the construction of the Lie algebra of G as the set of all complete holomorphic vector fields on D. A holomorphic vector field ξ on a domain $D \subset Z$ can be written as

$$\xi = h(z)\frac{\partial}{\partial z},$$

where $h: D \to Z$ is a holomorphic mapping and the symbol $\frac{\partial}{\partial z}$ indicates that ξ is viewed as a holomorphic differential operator, associating to each holomorphic mapping $f: D \to Z$ the holomorphic mapping $\xi f\colon D \to Z$ defined by

$$(\xi f)(z) := f'(z)h(z),$$

where $f'(z) \in \mathcal{L}(Z)$ is the derivative of f in $z \in D$. Given another holomorphic vector field

$$\eta = k(z)\frac{\partial}{\partial z},$$

the commutator vector field

$$[\xi,\eta] := (k'(z)h(z) - h'(z)k(z))\frac{\partial}{\partial z}$$

has the characteristic property

$$[\xi,\eta]f = \xi(\eta f) - \eta(\xi f)$$

for all holomorphic mappings $f\colon D \to Z$. It follows that the complex vector space $\mathcal{H}(D)$ of all holomorphic vector fields on D becomes a Lie algebra under the commutator product. Regarded as an ordinary differential equation, each holomorphic vector field on D generates a <u>local flow</u> on D. A vector field is called complete if this local flow can be enlarged to a global flow on D. An equivalent formulation is given in

2.8 DEFINITION. A holomorphic vector field $\xi = h(z)\frac{\partial}{\partial z}$ on D is called <u>complete</u> if there exists an analytic real 1-parameter group $t \mapsto g_t \in \mathrm{Aut}(D)$ satisfying the differential equation

$$\frac{\partial g_t(z)}{\partial t} = h(g_t(z))$$

for all $z \in D$. In this case the transformations g_t are uniquely determined for all $t \in \mathbb{R}$ and are denoted by

$$g_t = \exp(t\xi),$$

since locally g_t is given by an exponential series. The set of all complete holomorphic vector fields on a domain D is denoted by $\mathrm{aut}(D)$. For domains in general, $\mathrm{aut}(D)$ is not closed under formation of sums and commutators. In case D is bounded, however, $\mathfrak{g} := \operatorname{aut}(D)$ turns out to be a real Lie subalgebra of $\mathfrak{I}(D)$ [24; Satz 2.6]. Moreover $\mathfrak{g}$ is a Banach Lie algebra and via the exponential mapping $\xi \mapsto \exp(1\cdot\xi)$, this Banach Lie algebra induces on $\operatorname{Aut}(D)$ the structure of a real Banach Lie group with Lie algebra $\mathfrak{g}$. As a consequence of Liouville's theorem, we have

$$\mathfrak{g} \cap i\mathfrak{g} = 0.$$

A crucial property of bounded domains is the following result.

2.9 CARTAN'S UNIQUENESS THEOREM. Let $D \subset Z$ be a bounded domain. Suppose $g \in \operatorname{Aut}(D)$ satisfies $g(a) = a$ and $g'(a) = \mathrm{id}_Z$ for some $a \in D$. Then $g = \mathrm{id}_D$.

PROOF. We may assume $a = o$. Assume, $g \neq \mathrm{id}_D$. Choose $k \geq 2$ minimal, such that around o there is an expansion

$$g(z) = z + P_k(z) + h(z),$$

where $P_k \neq o$ is a k-homogeneous continuous polynomial and h vanishes in o of order $>k$. By induction, the n-th iterate g^n of g has the expansion

$$g^n(z) = z + nP_k(z) + h'(z),$$

where h' has order $>k$. Since the transformations g^n leave the bounded domain D invariant, it follows from Cauchy's inequalities [18; §6, Prop. 3] that $\{nP_k : n \geq 0\}$ is a bounded set of k-homogeneous polynomials. This implies $P_k = o$, a contradiction.

Q.E.D.

By differentiation, Theorem 2.9 implies that each vector field $\xi = h(z)\frac{\partial}{\partial z} \in \operatorname{aut}(D)$ satisfying $h(a) = o$ and $h'(a) = o$ for some $a \in D$ must vanish identically.

The possibility of describing a bounded domain D algebraically (e.g. in terms of $\operatorname{Aut}(D)$ or $\operatorname{aut}(D)$) can only be expected if D admits sufficiently many holomorphic automorphisms. The class of bounded symmetric domains defined below fulfills this requirement in an ideal manner.

2.10 DEFINITION. A bounded domain D in a complex Banach space Z is called symmetric if for each point $a \in D$ there is a mapping $s_a \in \mathrm{Aut}(D)$ with the following properties:

$$s_a(a) = a, \qquad s_a'(a) = -\mathrm{id}_Z .$$

By Theorem 2.9 the automorphism s_a, called the symmetry in a, is uniquely determined. Moreover, $s_a^2 = \mathrm{id}_D$ and a is an isolated fixed point of s_a. It can be shown [28,14] that each bounded symmetric domain D is homogeneous, i.e. $\mathrm{Aut}(D)$ is transitive on D, and is biholomorphically equivalent to a bounded domain D satisfying $o \in D$ and $e^{it}D = D$ for all $t \in \mathbb{R}$ (domains of this kind are called circular). Conversely, every circular homogeneous domain D is symmetric, since the symmetry $z \mapsto -z$ at $o \in D$ can be transported to any other point in D. In particular, the unit disk

$$\{z \in \mathbb{C} : |z| < 1\}$$

is a bounded symmetric domain.

It turns out that there is a natural way of associating with each bounded symmetric domain a Jordan triple product generalizing the triple product (2.5). The following Lemmas are the crucial steps towards this algebraic construction.

2.11 LEMMA. Let D be a bounded circular domain. Then each vector field $\xi = h(z)\frac{\partial}{\partial z} \in \mathfrak{g} := \mathrm{aut}(D)$ is polynomial of degree ≤ 2, i.e. $h = h_o + h_1 + h_2$, where $h_k : Z \to Z$ are k-homogeneous continuous polynomials. Moreover, the vector fields

$$\xi_1 := h_1 \frac{\partial}{\partial z} \quad \text{and} \quad \xi_{-1} := \xi - \xi_1$$

belong to $\mathfrak{g}$ and there is a Cartan decomposition

(2.12) $$\mathfrak{g} = \mathfrak{k} \oplus \mathfrak{p} ,$$

where $\mathfrak{k} := \{\xi_1 : \xi \in \mathfrak{g}\}$ and $\mathfrak{p} := \{\xi_{-1} : \xi \in \mathfrak{g}\}$.

PROOF. Since D is circular, $\delta := iz \frac{\partial}{\partial z} \in \mathfrak{g}$. Therefore $\theta := \mathrm{ad}(\delta) \in \mathfrak{L}(\mathfrak{g})$, since $\mathfrak{g}$ is a Lie algebra. Let $\xi = \sum_{m=0}^{\infty} \xi_m$ be the expansion of $\xi \in \mathfrak{g}$ around o into m-homogeneous polynomial vector fields ξ_m. By Euler's relation

$$\theta\xi_m = [\delta, \xi_m] = i(m-1)\xi_m .$$

This implies for any polynomial $p \in \mathbb{R}[\theta]$,

$$p(\theta)\xi = \sum_{m=0}^{\infty} p(i(m-1))\xi_m \in \mathfrak{g} .$$

Choose $p(\theta) = \theta(\theta^2+1)$. Then $p(i(m-1)) = 0$ for $m \leq 2$, hence $p(\theta)\xi = 0$ as a consequence of Theorem 2.9. But $p(i(m-1)) \neq 0$ for all $m > 2$, which implies $\xi_m = 0$. Hence $\xi = \xi_0 + \xi_1 + \xi_2$, as asserted. Further, $\xi_{-1} = -\theta^2\xi \in \mathfrak{g}$ and hence $\xi_1 = \xi - \xi_{-1} \in \mathfrak{g}$. It is clear that (2.12) is a Cartan decomposition of $\mathfrak{g}$ relative to the involutive automorphism $\mathrm{Ad}(s)$, where $s(z) := -z$. Q.E.D.

2.13 LEMMA. Let D be a bounded symmetric circular domain. Then there is a unique mapping $Z \times Z \times Z \to Z$ denoted by $(u,a,v) \mapsto \{ua^*v\}$ having the following properties:

(i) $\{ua^*v\}$ is complex bilinear symmetric in the outer variables and conjugate linear in the inner variable $a \in Z$.

(ii) The subspace $\mathfrak{p} \subset \mathfrak{g}$ defined in 2.11 has the form

$$\mathfrak{p} = \{(u - \{zu^*z\}) \frac{\partial}{\partial z} : u \in Z\}.$$

Moreover, defining the operator $u \square v^*$ on Z by $(u \square v^*)z := \{uv^*z\}$, the following Jordan triple identity is valid:

(iii) $[u \square v^*, x \square y^*] = \{uv^*x\} \square y^* - x \square \{yu^*v\}^*$

for all $u,v,x,y \in Z$, where $[\lambda,\mu] := \lambda\mu - \mu\lambda$ denotes the commutator of linear operators.

PROOF. Since D is homogeneous, it follows from (2.12) that the evaluation mapping $\mathfrak{p} \to Z$, defined by $h(z)\frac{\partial}{\partial z} \mapsto h(o)$, is surjective. By definition, the vector fields in $\mathfrak{p}$ have the form $(u-q_u(z))\frac{\partial}{\partial z}$, where $q_u: Z \to Z$ is a 2-homogeneous polynomial, which is uniquely determined by u (since D is bounded) and depends conjugate linearly on u (since $\mathfrak{g} \cap i\mathfrak{g} = o$). Define

$$\{ua^*v\} := \frac{1}{2}(q_a(u+v) - q_a(u) - q_a(v)).$$

Then (i) and (ii) are satisfied. Property (iii) is a direct consequence of the fact that $\mathfrak{p}$ is a Lie triple system, i.e., $[\mathfrak{p}[\mathfrak{p},\mathfrak{p}]] \subset \mathfrak{p}$.

A Banach space Z with a composition $\{ua^*v\}$ satisfying properties (i) and (iii) is called a <u>Banach Jordan triple system</u>. Using Jordan triple systems, W. Kaup [14] has obtained an algebraic characterization of all symmetric Banach manifolds. In particular, the Banach Jordan triple system associated to each bounded symmetric domain D via 2.13 characterizes D uniquely. A particularly important class of Jordan triple systems are the JB^*-algebras with Jordan triple product (2.5). Therefore the problem arises which bounded symmetric domains correspond to JB^*-algebras. It turns out that the appropriate holomorphic condition relates to the notion of <u>tube domain</u>.

2.14 DEFINITION. Let X be a real Banach space with complexification $Z := X^{\mathbb{C}}$. Let $\Omega \subset X$ be an open convex cone. Then the domain

$$D_\Omega := \{z \in Z : \frac{z-z^*}{2i} \in \Omega\}$$

is called the <u>tube domain</u> with the base Ω. Here $z \mapsto z^*$ denotes the conjugation of Z with respect to X.

In case $X := \mathbb{R}$ and $\Omega := \{x \in \mathbb{R} : x > o\}$, the domain D_Ω is the familiar upper half-plane in $\mathbb{C}$. Therefore tube domains are often called "generalized half-planes". For any JB-algebra X, the set of squares $\{x^2 : x \in X\}$ is a convex cone having non-empty interior Ω. The associated tube domain D_Ω is called the <u>upper half-plane</u> of the JB*-algebra Z. The holomorphic characterization of JB*-algebras is given in the following theorem, the main result of [6], which generalizes the pioneering work by M. Koecher [17].

2.15 THEOREM. A bounded symmetric domain $D \subset Z$ belongs to a JB*-algebra if and only if D is biholomorphically equivalent to a tube domain. More precisely, the open unit ball $D := \{z \in Z : |z| < 1\}$ of a JB*-algebra is a bounded symmetric domain which is biholomorphically equivalent to the upper half-plane D_Ω of Z. Conversely, each bounded symmetric domain which is biholomorphically equivalent to a tube domain can be realized as the open unit ball of a JB*-algebra.

For the <u>proof</u> of Theorem 2.15 which uses the Gelfand-Naimark embedding theorem for JB-algebras [3] and a detailed analysis of the Lie algebra of all complete holomorphic vector fields on tube domains, the reader is referred to [16,6]. It should be noted that the biholomorphic equivalence between the open unit ball D of a JB*-algebra Z and its upper half-plane D_Ω is given by means of the so-called <u>Cayley transform</u> $\sigma: D \to D_\Omega$ which is completely defined in Jordan algebraic terms, namely

$$\sigma(z) = (z+ie) \circ (e+iz)^{-1} ,$$

where e is the unit element of Z und product and inverse are taken in the Jordan theoretic sense.

§3. DERIVATIONS OF JORDAN C*-ALGEBRAS

The Jordan algebraic characterization of bounded symmetric domains D and symmetric tube domains developed in §2 was based on a detailed analysis of the Lie algebra $\mathfrak{g} = \text{aut}(D)$ of all complete holomorphic vector fields on D. It is shown in §4 that for domains D equivalent to a tube domain there is a natural notion of "orientation" associated with $\mathfrak{g}$ characterizing the class of all C*-algebras. This concept is similar to the one introduced in [8], but has to be modified in order to apply to JB*-algebras in general which need not be Banach dual spaces. The appropriate modifications are motivated by some results on derivations of Jordan C*-algebras [25,26,27].

Let D be a bounded symmetric domain in a complex Banach space Z. Without loss of generality we may assume that D is circular. Let $\mathfrak{g} = \mathfrak{k} \oplus \mathfrak{p}$ be the Cartan decomposition (2.12) of the Lie algebra $\mathfrak{g} := \text{aut}(D)$. By definition of the Jordan triple product, the vector fields in $\mathfrak{p}$, which are sometimes called "infinitesimal transvections", have the form

$$(u - \{zu^*z\})\frac{\partial}{\partial z}$$

for $u \in Z$ and are therefore explicitly given in terms of the Jordan triple product associated with D. Our first aim is to give a Jordan triple characterization of the Lie algebra $\mathfrak{k}$:

3.1 LEMMA. A linear vector field $\lambda z \frac{\partial}{\partial z}$ belongs to $\mathfrak{k}$ if and only if $\lambda \in \mathcal{L}(Z)$ is a derivation of the Jordan triple system Z, i.e. for all $u,v,w \in Z$:

$$\lambda\{uv^*w\} = \{(\lambda u)v^*w\} + \{u(\lambda v)^*w\} + \{uv^*(\lambda w)\}. \tag{3.2}$$

PROOF. Suppose, $\lambda \in \mathcal{L}(Z)$ satisfies the identity (3.2). Then for all $t \in \mathbb{R}$, the invertible transformation $g_t := \exp(t\lambda) \in GL(Z)$ is a Jordan triple automorphism, i.e.

$$g_t\{uv^*w\} = \{(g_t u)(g_t v)^*(g_t(w)\}.$$

The domain D has several characterizations in terms of the Jordan triple product (cf. [14; §3]) which imply that $g_t \in \mathrm{Aut}(D)$ for all $t \in \mathbb{R}$. By definition of $\mathfrak{k}$, it follows that $\lambda z \frac{\partial}{\partial z} \in \mathfrak{k}$.

Conversely, suppose $\xi := \lambda z \frac{\partial}{\partial z} \in \mathfrak{k}$. Then $[\xi, \mathfrak{p}] \subset \mathfrak{p}$, since ξ is linear. This implies

$$\lambda\{zu^*z\} = 2\{(\lambda z)u^*z\} + \{z(\lambda u)^*z\}$$

for all $z, u \in Z$. By polarization, (3.2) follows. Q.E.D.

The Lie algebra of all derivations of a Jordan triple system Z is denoted by $\mathrm{aut}(Z)$. By Lemma 3.1,

$$\mathfrak{k} = \{\lambda z \frac{\partial}{\partial z} : \lambda \in \mathrm{aut}(Z)\}$$

has also a direct interpretation in terms of the Jordan triple system associated with D. Using the commutator notation for linear operators, the defining identity (3.2) of a derivation can be reformulated as follows:

$$[\lambda, u \,\square\, v^*] = (\lambda u) \,\square\, v^* + u \,\square\, (\lambda v)^*$$

for all $u, v \in Z$. Here $u \,\square\, v^* \in \mathcal{L}(Z)$ is defined as $(u \,\square\, v^*)z :=$ $:= \{uv^*z\}$. Using the fact that $[\mathfrak{p}, \mathfrak{p}] \subset \mathfrak{k}$ or checking the defining Jordan triple identity 2.13.iii one can easily show that

$$u \,\square\, v^* - v \,\square\, u^* \in \mathrm{aut}(Z)$$

whenever $u, v \in Z$. The linear subspace generated by these operators is an ideal of the Lie algebra $\mathrm{aut}(Z)$, called the ideal of all <u>inner derivations</u> of Z and denoted by $\mathrm{int}(Z)$. The inner derivations will play an essential role in the sequel.

Suppose now that $D \subset Z$ is a bounded symmetric domain of "tube type", i.e. having a realization as a tube domain. By

Theorem 2.15 we may assume that D is the open unit ball of a JB*-algebra structure on Z. Further, D is biholomorphically equivalent to the upper half plane of Z defined by

$$D_\Omega := \{z \in Z : \frac{z-z^*}{2i} \in \Omega\},$$

where $z \mapsto z^*$ denotes the involution of Z and Ω is the interior of the convex cone $\{x^2 : x \in X\}$ of all squares in the JB-algebra $X := \{x \in Z : x^* = x\}$ associated with Z. Using a complexified version of Cartan's uniqueness theorem for vector fields [15] it has been shown in [15,16] that the Lie algebra $\mathfrak{h} := \mathrm{aut}(D_\Omega)$ of all complete holomorphic vector fields on the (unbounded) tube domain D_Ω consists of polynomial vector fields of degree ≤ 2 and has a direct sum decomposition

$$(3.3) \qquad \mathfrak{h} = \mathfrak{h}_{-1} \oplus \mathfrak{h}_0 \oplus \mathfrak{h}_1 ,$$

where $\mathfrak{h}_j$ consists of all vector fields in $\mathfrak{h}$ which are homogeneous of degree $j+1$. Analogous to the case of bounded domains the components in (3.3) can be interpreted in terms of the Jordan algebras Z and X, namely

$$\mathfrak{h}_{-1} = \{u \frac{\partial}{\partial z} : u \in X\},$$

$$\mathfrak{h}_1 = \{\{zu^*z\} \frac{\partial}{\partial z} : u \in X\} \quad \text{and}$$

$$\mathfrak{h}_0 = \{\lambda z \frac{\partial}{\partial z} : \lambda \in \mathrm{aut}(\Omega)\},$$

where $\mathrm{aut}(\Omega)$ denotes the Lie algebra of all <u>infinitesimal transformations</u> ("derivations") λ of the cone Ω generating a 1-parameter group $g_t = \exp(t\lambda)$ of linear automorphisms of Ω. Our next aim is to show, that the Lie algebras $\mathrm{aut}(Z)$ and $\mathrm{aut}(\Omega)$ can be regarded as "dual" Lie algebras in the sense of symmetric space theory.

3.4 DEFINITION. Let X be a JB-algebra. A linear map $\delta : X \to X$ satisfying $\delta(x \circ y) = (\delta x) \circ y + x \circ (\delta y)$ for all $x, y \in X$ is called a derivation of X. Denote by aut(X) the Lie algebra of all derivations of X.

Modifying the proof for C^*-algebras given in [19; Lemma 4.1.3] it is easy to show that (everywhere defined) derivations of JB-algebras are norm-continuous.

3.5 DEFINITION. Let X be a JB-algebra. Denote by

$$M_x y := x \circ y$$

the multiplication operator on X induced by $x \in X$.

(Note that left and right multiplications cannot be distinguished since X is commutative).

In terms of the Jordan triple product (2.5) on X or on the complexified JB*-algebra $Z := X^{\mathbb{C}}$ the multiplication operators have the form

$$M_x = x \,\square\, e^*,$$

where e is the unit element of X. Similarly as for Jordan triple derivations, aut(X) consists of all operators $\delta \in \mathcal{L}(X)$ satisfying the commutator identity

$$[\delta, M_x] = M_{\delta x},$$

whenever $x \in X$.

3.6 LEMMA. Let X be a JB-algebra with open positive cone Ω and complexification Z. Put $M_X := \{M_x : x \in X\}$. Then there exist direct sum decompositions

$$\mathrm{aut}(Z) = \mathrm{aut}(X) \oplus iM_X$$

and

$$\mathrm{aut}(\Omega) = \mathrm{aut}(X) \oplus M_X .$$

(Note that aut(Z) consists of all Jordan triple derivations of the JB*-algebra Z satisfying (3.2)).

PROOF. Obviously, aut(X) is contained in aut(Z) and in aut(Ω), since each Jordan algebra automorphism of X is a Jordan triple automorphism of Z (by complexification) and leaves the cone Ω invariant. Moreover every $x \in X$ satisfies $x \square e^* = e \square x^*$ by (2.13.iii), whence

$$iM_X = ix \square e^* = \frac{1}{2}(ix \square e^* - e \square (ix)^*) \in \mathrm{aut}(Z).$$

Further, for any $u \in \Omega$ the fundamental formula for Jordan algebras [7; Ch. III, Satz 1.5] implies that $P_u := 2M_u^2 - M_{u^2} \in GL(Z)$ leaves Ω invariant. For every $x \in X$ we have $\exp(x) \in \Omega$ and

$$\exp(2M_x) = P_{\exp(x)}$$

by [7; Ch.XI, Satz 2.2]. It follows that $M_X \subset \mathrm{aut}(\Omega)$. To prove the converse inclusions, suppose $\lambda \in \mathrm{aut}(Z)$. By (3.2),

$$(\lambda e)^* = \{e(\lambda e)^*e\} = \lambda\{ee^*e\} - 2\{(\lambda e)e^*e\} = \lambda e - 2(\lambda e) = -\lambda e.$$

Hence $\lambda e = ix$ for some $x \in X$. Let $\delta := \lambda - iM_x$. Then $\delta \in \mathrm{aut}(Z)$ satisfies $\delta e = 0$. This implies $\delta \in \mathrm{aut}(X)$. Since $\lambda = \delta + iM_x$, we have obtained the desired decomposition for $\mathrm{aut}(Z)$ which is obviously direct. Similarly, for each $\lambda \in \mathrm{aut}(\Omega)$ we have $x := \lambda e \in X$. Therefore $\delta := \lambda - M_x \in \mathrm{aut}(\Omega)$ satisfies $\delta e = 0$. By [16; Prop. 5.4.vii] this implies $\delta \in \mathrm{aut}(X)$. Q.E.D.

An important consequence of Lemma 3.6 is the fact, that the Lie algebra $\mathrm{aut}(\Omega)$ associated with the open positive cone Ω of a JB-algebra X is not only a real Banach Lie algebra under some appropriate norm, but carries also an <u>involution</u> $\lambda \mapsto \lambda^*$ satisfying the properties

$$(\lambda^*)^* = \lambda$$

$$[\lambda,\mu]^* = [\mu^*,\lambda^*].$$

In fact, if $\lambda \in \mathrm{aut}(\Omega)$ is uniquely decomposed as $\lambda = \delta + M_x$ with $\delta \in \mathrm{aut}(X)$ and $x \in X$, define

$$\lambda^* := -\delta + M_x.$$

The geometric meaning of this involution on $\mathrm{aut}(\Omega)$ is clarified by the following examples.

3.7 EXAMPLE. Let E be a Hilbert space over $\mathbb{K} \in \{\mathbb{R},\mathbb{C},\mathbb{H}\}$. Let $X := \mathcal{H}(E)$ be the JB-algebra of all hermitian $\mathbb{K}$-linear (bounded) operators on E introduced in 2.7. Then Ω is the cone of all positive definite $\mathbb{K}$-linear operators on E. Denote by $\mathcal{L}(E)$ the real W*-algebra of all $\mathbb{K}$-linear operators on E. Since by definition the Jordan product on X is the anti-commutator product induced from $\mathcal{L}(E)$, it follows that

$$2M_a x = ax + xa = ax + xa^*$$

for $a = a^* \in X$. On the other hand, it is well known (cf. [25; Lemma 2.6]) that each derivation δ of X has the form

$$\delta x = ax - xa = ax + xa^*$$

for all $x \in X$, where $a = -a^* \in \mathcal{L}(E)$. Associate to each $a \in \mathcal{L}(E)$ the derivation $\lambda_a \in \mathrm{aut}(\Omega)$ defined by

$$\lambda_a x := ax + xa^*. \tag{3.8}$$

Then Lemma 3.6 implies that $a \mapsto \lambda_a$ yields a surjective homomorphism $\mathcal{L}(E) \to \mathrm{aut}(\Omega)$ of real involutive Lie algebras.

3.9 EXAMPLE. Suppose Z is a (unital) C*-algebra with self-adjoint part X and open positive cone Ω. Then it is well known (cf. [25; Cor. 2.12]) that each Jordan derivation $\delta \in \mathrm{aut}(X)$ is a derivation of Z in the usual (associative) sense. The complex Lie algebra $\mathrm{Der}(Z)$ of all derivations of Z has an involution $D \mapsto D^*$, defined by

$$D^*z := -(Dz^*)^*$$

for all $z \in Z$. Let $\mathrm{ad}(a)z := az - za$ be the "inner" derivation of Z associated with $a \in Z$. Then

$$\mathrm{aut}(X) = \{\delta \in \mathrm{Der}(Z) : \delta^* = -\delta\}$$

and $\mathrm{ad}(a)^* = \mathrm{ad}(a^*)$ for all $a \in Z$. It follows that

$$\delta + M_a \mapsto \delta + \frac{1}{2}\,\mathrm{ad}(a) \tag{3.10}$$

defines a homomorphism $\mathrm{aut}(\Omega) \to \mathrm{Der}(Z)$ of real involutive Lie algebras.

Now assume in addition, that Z is a (unital) C^*-algebra having only inner derivations, that is

$$\mathrm{Der}(Z) = \{\mathrm{ad}(a) : a \in Z\}.$$

For example all W^*-algebras and all simple unital C^*-algebras have this property [19; Th. 4.1.6 and Th. 4.1.11]. Then $a \mapsto \lambda_a$ defined as in (3.8) yields a surjective homomorphism $Z \to \mathrm{aut}(\Omega)$ of real involutive Lie algebras.

For C^*-algebras Z in general, a representation of $\lambda \in \mathrm{aut}(\Omega)$ in the form (3.8) is still possible, but the "implementing operator" a has to be chosen from the second dual space Z^{tt} (a W^*-algebra) since derivations of Z are in general not inner. The problem of characterizing those operators $a \in Z^{tt}$ occuring in (3.8) is quite difficult and this difficulty is responsible for the rather complicated notion of "orientation" introduced in §4.

In §4 orientations will be defined in terms of the involutive Lie algebra $\mathrm{aut}(\Omega)$ of the cone Ω belonging to a JB-algebra X. Of course, using Lemma 3.6 an equivalent condition could be imposed on the "dual" Lie algebra $\mathrm{aut}(Z)$. In view of the decompositions (2.12) of $\mathfrak{g} = \mathrm{aut}(D)$ and (3.3) of $\mathfrak{h} := \mathrm{aut}(D_\Omega)$ the notion of orientation has also a holomorphic interpretation in terms of the Lie algebras $\mathfrak{g}$ and $\mathfrak{h}$ of complete holomorphic vector fields on D and D_Ω, respectively.

To define the concept of orientation we must have a closer look at inner derivations. By Lemma 3.6 the Lie algebra $\mathrm{aut}(X)$ of all derivations of a JB-algebra X can be viewed as the "nontrivial" part of aut(Z) and aut(Ω). Moreover, the subspace of aut(X) spanned by all commutators

$$[M_x, M_y] = [x \square e^*, y \square e^*]$$

of multiplication operators for $x, y \in X$ forms an ideal in aut(X) which is denoted by int(X). The elements of int(X) are called <u>inner derivations</u> of the JB-algebra X. Similar as in Lemma 3.6 there exists a decomposition

$$\text{(3.11)} \qquad \mathrm{int}(Z) = \mathrm{int}(X) \oplus iM_X .$$

Obviously the ideal $\mathrm{int}(\Omega) := \mathrm{int}(X) \oplus M_X$ of aut(Ω) is invariant under the involution.

Let us consider some examples of inner derivations:

3.12 EXAMPLE.

(i) Let Z be a C^*-algebra with self-adjoint part X. Given $a, b \in X$, an elementary calculation shows

$$[M_a, M_b]x = \frac{1}{4}[[a,b],x].$$

It follows that the concept of Jordan inner derivation is more restrictive than the usual notion of inner derivation, since the "implementing operator" is required to be a finite sum of commutators in Z. For W^*-algebras, this condition is (up to central elements) always fulfilled, hence in this case both notions coincide (cf. the more general statement in Theorem 3.15).

(ii) The JB-algebras $\mathcal{H}(E)$ associated with a Hilbert space E over $\mathbb{K} \in \{\mathbb{R}, \mathbb{C}, \mathbb{H}\}$ and the exceptional Jordan algebra $\mathcal{H}_3(\mathbb{O})$ have only inner derivations, cf. Theorem 3.15.

(iii) Let $X = \mathbb{R} \oplus Y$ be the spin factor defined by some real Hilbert space Y. Then the Lie algebra aut(X) can be identified with the Lie algebra $o(Y)$ of all skew adjoint bounded operators on Y. The ideal int(X) corresponds to the set of all operators in $\mathcal{o}(Y)$ having <u>finite rank</u>.

Before stating the fundamental "approximation theorem" for JB-algebra derivations, let us give a short survey about the structure theory for JB-algebras as developed in [22,23,3,20]. Given a JB-algebra X, the second dual space X^{tt} of the real Banach space is again a JB-algebra with respect to the (commutative) Arens product [20,11]. JB-algebras which are Banach dual spaces are called JBW-<u>algebras</u> (in analogy to associative W*-algebras). Any JBW-algebra X has an (essentially unique) decomposition into three orthogonal weakly closed ideals

$$(3.13) \qquad X = X_{rev} \oplus X_{spin} \oplus X_{exc} ,$$

such that the "reversible" part X_{rev} can be realized as the self-adjoint part of a <u>real</u> W*-algebra (on a complex Hilbert space) and X_{spin} (resp. X_{exc}) is a JBW-algebra having only factor representations of spin type (resp. exceptional type).

By Example 3.12.iii. infinite dimensional spin factors have a lot of outer derivations. On the other hand, the remaining types of JBW-algebras behave rather nicely with respect to inner derivations. This is expressed in the following theorem [25; Th. 3.5]:

3.14 THEOREM. Let X be a JBW-algebra. Then $\mathrm{aut}(X) = \mathrm{int}(X)$ if and only if the dimension of all spin factor representations of X remains bounded. In particular, reversible and purely exceptional JBW-algebras have only inner derivations.

The proof uses the extension theorem for derivations of reversible JC-algebras [25; Th. 2.5], Sakai's theorem on the innerness of derivations of von Neumann algebras [19, Th. 4.1.6] and results about commutators in von Neumann algebras.

Given an arbitrary JB-algebra X and a derivation $\delta \in \mathrm{aut}(X)$, we may extend δ to a derivation δ^{tt} of the second dual X^{tt} in a canonical way. Since derivations vanish on the center of a JB-algebra, the direct sum decomposition (3.13) of the JBW-algebra X^{tt} is invariant under the extended derivation δ^{tt}. Hence we may apply Theorem 3.14 to this situation. Modulo some technical arguments one obtains the so-called approximation theorem for derivations on an arbitrary JB-algebra X [25; Th. 4.2]:

3.15 THEOREM. Let X be a JB-algebra. Then the derivation algebra $\operatorname{aut}(X)$ of X is the closure of the ideal $\operatorname{int}(X)$ of all inner derivations with respect to the strong operator topology on $\mathcal{L}(X)$, i.e. the topology of pointwise norm-convergence.

Simple examples show that in general $\operatorname{int}(X)$ is not dense in $\operatorname{aut}(X)$ with respect to the topology of uniform norm-convergence. By Lemma 3.6 and (3.11), the approximation theorem can also be formulated in terms of the Lie algebra $\operatorname{aut}(Z)$ of all Jordan triple derivations of the JB*-algebra $Z = X^{\mathbb{C}}$ associated with the JB-algebra X. Equivalently, in holomorphic terms, Theorem 3.15 says that each "infinitesimal rotation" $\xi \in \mathfrak{k}$ can be pointwise approximated by linear combinations of commutators of infinitesimal transvections in $\mathfrak{p}$.

After having clarified the main properties of derivation algebras of Jordan C*-algebras, we will define in §4 the notion of orientation of a JB-algebra X to be a complex structure on $\operatorname{aut}(\Omega)$ (modulo center) given by a closed operator which is densely defined with respect to the strong operator topology. In §5 it will be shown that a JB*-algebra has an orientation in this sense if and only if it is a C*-algebra.

§4. ORIENTATIONS OF JORDAN C^*-ALGEBRAS

In order to define "orientations" on a JB^*-algebra in terms of the Lie algebra $\mathrm{aut}(\Omega)$ we have to clarify one technical question, namely the structure of the center of $\mathrm{aut}(\Omega)$. Recall the notion of center for Jordan algebras [7; Ch. I, §5].

4.1 DEFINITION. Let X be a JB-algebra. Then the <u>center</u> of X is the set of all elements $x \in X$, such that $[M_x, M_y] = 0$ whenever $y \in X$.

4.2 LEMMA. Let X be a JB-algebra with open positive cone Ω. Let $\mathrm{aut}(\Omega) = \mathrm{aut}(X) \oplus M_X$ be the Cartan decomposition of $\mathrm{aut}(\Omega)$ (cf. Lemma 3.6). Then the center of $\mathrm{aut}(\Omega)$ consists of all multiplication operators M_x with $x \in \mathrm{center}(X)$.

PROOF. Suppose $\lambda = \delta + M_x \in \mathrm{center}(\mathrm{aut}(\Omega))$. For each $y \in X$, we have $M_y \in \mathrm{aut}(\Omega)$ and therefore $0 = [\lambda, M_y] = [\delta, M_y] + [M_x, M_y] = M_{\delta y} + [M_x, M_y]$. Evaluation at the unit element $e \in X$ gives $\delta y = 0$. Since y is arbitrary, $\delta = 0$ and $x \in \mathrm{center}(X)$. Conversely, suppose $x \in \mathrm{center}(X)$. Since $A := \mathrm{center}(X)$ is the self-adjoint part of an abelian C^*-algebra, it follows that $\delta(A) = 0$ for all $\delta \in \mathrm{aut}(X)$. Therefore M_x commutes with δ and also with all operators M_y for $y \in X$ by Definition 4.1.

Q.E.D.

Since $A := \mathrm{center}(X)$ is a closed subalgebra of X, $\underline{X} := X/A$ is a Banach space under the natural norm

$$|\underline{x}| := \inf\{|x| : x \in \underline{x}\}.$$

Similarly, $\mathrm{aut}(\Omega)$ is a real Banach Lie algebra and the quotient Banach Lie algebra $\underline{\mathrm{aut}}(\Omega) = \mathrm{aut}(X) \oplus \underline{X}$ by identifying $x \in X$ with the multiplication operator M_x. By Lemma 4.2, the centers of

$aut(\Omega)$ and $int(\Omega)$ coincide, hence

$$\underline{int}(\Omega) := int(\Omega)/center = int(X) \oplus \underline{X}\ .$$

The involution of $aut(\Omega)$ leaves the center invariant and therefore induces an involution of $\underline{aut}(\Omega)$ such that $\underline{int}(\Omega)$ is a $*$-invariant ideal. Modulo centers, the homomorphisms (3.10) and (3.8) turn out to be isomorphisms:

4.3 EXAMPLE. Let Z be a unital C^*-algebra having only inner derivations. Let Ω be the open positive cone of Z. Then the homomorphism (3.10) induces an isomorphism

$$\underline{aut}(\Omega) \to Der(Z). \tag{4.4}$$

Similarly, the homomorphism $Z \to aut(\Omega)$ defined by (3.8) yields an isomorphism

$$Z/center \to \underline{aut}(\Omega).$$

Modulo the isomorphism $ad\colon Z/center \to Der(Z)$ the above isomorphisms are inverse to each other.

We are now in a position to define "orientations" on Jordan C^*-algebras.

4.5 DEFINITION. Let X be a JB-algebra with open positive cone Ω. Denote by $\underline{aut}(\Omega)$ the involutive Lie algebra of all infinitesimal automorphisms of Ω modulo its center. An <u>orientation</u> of X (or of the JB^*-algebra $Z = X^{\mathbb{C}}$ associated with X) is given by the following:

(i) A (not necessarily closed) ideal $\mathfrak{a} \subset \underline{aut}(\Omega)$ containing $\underline{int}(\Omega)$ and invariant under the involution.

(ii) A complex structure $J\colon \mathfrak{a} \to \mathfrak{a}$ making $\mathfrak{a}$ a complex involutive Lie algebra such that $J\colon \mathfrak{a} \to \underline{aut}(\Omega)$ is a closed operator.

X is called <u>orientable</u>, if there exists an orientation on X.

4.6 REMARK.

(i) By the approximation Theorem 3.15, the ideal $\mathfrak{a}$ is dense in $\underline{\text{aut}}(\Omega)$ under pointwise convergence, so that J is "densely defined" with respect to this topology. However, $\mathfrak{a}$ cannot always be chosen to be uniformly dense in $\underline{\text{aut}}(\Omega)$.

(ii) The technical difficulty that the complex structure J may only be defined on a proper ideal in $\underline{\text{aut}}(\Omega)$ is not present if X is a Banach dual space, since in this case $\underline{\text{aut}}(\Omega) = \underline{\text{int}}(\Omega)$.

Our first aim is to give an equivalent notion of orientation which is technically easier to handle.

4.7 THEOREM. Let X be a JB-algebra. Then the orientations on X in the sense of 4.5 are in 1-1 correspondence with continuous $\mathbb{R}$-linear mappings

$$J: \underline{X} \to \text{aut}(X)$$

satisfying the property

$$[\delta_{ia}, \delta_{ib}] = [\underline{b}, \underline{a}] = \delta_{i\,\delta(a,b)} \tag{4.8}$$

for all $a,b \in X$. Here we define $\delta_{ia} := J_{\underline{a}} \in \text{aut}(X)$ and $\delta(a,b) := \delta_{ia}b \in X$.

PROOF. Let $J: \mathfrak{a} \to \mathfrak{a}$ be an orientation. Since $\underline{\text{int}}(\Omega) \subset \mathfrak{a}$ it follows that $\mathfrak{a} = \mathfrak{b} \oplus \underline{X}$, where $\mathfrak{b}$ is an ideal in $\text{aut}(X)$ containing $\text{int}(X)$. Since $\mathfrak{a}$ is a complex involutive Lie algebra, we have $J\lambda^* = -(J\lambda)^*$. This implies $J\underline{X} \subset \mathfrak{b} \subset \text{aut}(X)$. Denote the restriction $J|\underline{X}$ by J again. Since $J: \mathfrak{a} \to \underline{\text{aut}}(\Omega)$ is a closed operator and $\underline{X} \subset \mathfrak{a}$ is a Banach space, it follows that

$$J: \underline{X} \to \text{aut}(X)$$

is a closed $\mathbb{R}$-linear operator. By the closed graph theorem J is continuous on $\underline{X}$. Further, since $\mathfrak{a}$ is a complex Lie algebra under

J, we have for all $a,b \in X$:

$$[\delta_{ia},\delta_{ib}] := [J\underline{a},J\underline{b}] = -[\underline{a},\underline{b}] = [\underline{b},\underline{a}]$$

and

$$\delta_{i\,\delta(a,b)} := J\,\underline{\delta(a,b)} = J\,\delta_{ia}b = J\,[\delta_{ia},\underline{b}]$$
$$= J\,[J\underline{a},\underline{b}] = [J\underline{a},J\underline{b}] = [\delta_{ia},\delta_{ib}].$$

Hence the property (4.8) is satisfied.

Conversely, suppose that $J\colon \underline{X} \to aut(X)$ is a continuous $\mathbb{R}$-linear mapping having the property (4.8). Define

$$\mathfrak{a} := \{\delta_{ia} : a \in X\} \oplus \underline{X} \subset \underline{aut}(\Omega).$$

Given $a,b \in X$, it follows from (4.8) that $[M_a,M_b] + [\delta_{ia},\delta_{ib}] \in$ $\in center(aut(\Omega)) \cap aut(X) = 0$. Therefore

(4.9) $$[M_a,M_b] = -[\delta_{ia},\delta_{ib}] = -\delta_{i\,\delta(a,b)}\,.$$

It follows that $\delta_{ia} = 0$ implies $a \in center(X)$. Hence $J\colon \underline{X} \to aut(X)$ is injective. To show that $-\delta(a,b) = \delta(b,a)$, suppose $a \in X$. Then $\delta_{i\,\delta(a,a)} = [\delta_{ia},\delta_{ia}] = 0$. By the above observation, $\delta(a,a) \in center(X)$. Since δ_{ia} is a derivation of a commutative C^*-algebra containing a, apply [19; Lemma 4.1.2]. Hence $\delta(a,a) = 0$ and skew-symmetry follows by polarization. We next assert that for every $\delta \in aut(X)$ and $b \in X$:

(4.10) $$[\delta,\delta_{ib}] = \delta_{i(\delta b)}\,.$$

In fact, by the approximation Theorem 3.15, given any $x \in X$ there is a net $\delta_j \in int(X)$ with the following properties, the symbol "$\to$" denoting norm convergence in X:

$$\delta_j x \to \delta x, \quad \delta_j b \to \delta b \quad \text{and} \quad \delta_j(\delta_{ib}x) \to \delta(\delta_{ib}x).$$

This implies by norm-continuity of δ_{ib} (cf. §3):

$$[\delta_j,\delta_{ib}]x = \delta_j(\delta_{ib}x) - \delta_{ib}(\delta_j x) \to [\delta,\delta_{ib}]x.$$

On the other hand, since $\delta_j \in \operatorname{int}(X) \subset J\underline{X}$ by (4.9), it follows from (4.8) that

$$[\delta_j, \delta_{ib}]x = \delta_{i\delta_j b}x = -\delta_{ix}(\delta_j b) \to -\delta_{ix}(\delta b) = \delta_{i(\delta b)}x$$

using the skew-symmetry of δ. Comparing both limits gives (4.10). Since $\underline{X} \subset \mathfrak{a}$ by definition, (4.10) implies that $\mathfrak{a}$ is an ideal in $\underline{\operatorname{aut}}(\Omega)$ containing $\underline{\operatorname{int}}(\Omega)$ by (4.9). The injective mapping $J\colon \underline{X} \to \operatorname{aut}(X)$ has a well-defined extension $J\colon \mathfrak{a} \to \mathfrak{a}$ satisfying $J(\delta_{ia}) := -\underline{a} \in \underline{X}$ for every $a \in X$. Obviously, $-J^2 = \mathrm{id}_{\mathfrak{a}}$. An elementary verification shows that $J[\lambda,\mu] = [\lambda, J\mu]$ for all $\lambda,\mu \in \mathfrak{a}$, hence $(\mathfrak{a}, J)$ is a complex involutive Lie algebra. Finally, the continuity of J on $\underline{X}$ implies that $J\colon \mathfrak{a} \to \mathfrak{a}$ is a closed operator, since the decomposition $\underline{\operatorname{aut}}(\Omega) = \operatorname{aut}(X) \oplus \underline{X}$ is topologically direct. Q.E.D.

An immediate corollary of Theorem 4.7 is

4.11 COROLLARY. Every (unital) C^*-algebra Z has a canonical orientation.

PROOF. Let X be the self-adjoint part of Z. Then $A := \operatorname{center}(X)$ is the self-adjoint part of the center of Z. For $a \in X$ define

$$\delta_{ia}x := \frac{i}{2}[a,x] = \frac{i}{2}(ax-xa).$$

Obviously the derivation $\delta_{ia} \in \operatorname{aut}(X)$ depends only on the residue class $\underline{a} \in \underline{X}$ modulo A. Further, for all $x \in X$

$$|\delta_{ia}x| \le |\underline{a}|\cdot|x|$$

which implies $|\delta_{ia}| \le |\underline{a}|$ for the operator norm on X. It follows that the well-defined mapping

$$J\colon \underline{X} \to \operatorname{aut}(X)$$

sending $\underline{a}$ to $J\underline{a} := \delta_{ia}$ is continuous. By definition, property (4.8) is satisfied. Q.E.D.

For C^*-algebras there is a simple criterion for the continuity of the canonical orientation:

4.12 PROPOSITION. Let $J:\mathfrak{a} \to \mathfrak{a}$ be the canonical orientation of a C^*-algebra Z. Then the following statements are equivalent:

(i) J is continuous.

(ii) The ideal $\mathrm{Inn}(X)$ of all inner $*$-derivations on Z (in the associative sense) is closed in $\mathrm{aut}(X)$.

(iii) $\mathfrak{a}$ is closed in $\underline{\mathrm{aut}}(\Omega)$.

PROOF. (i) $\Rightarrow$ (ii). Suppose J is continuous. Then the mapping

$$J\colon \underline{X} \to \{\delta_{ia} : a \in X\} = \mathrm{Inn}(X)$$

is a homeomorphism. Since $\underline{X}$ is a Banach space, $\mathrm{Inn}(X)$ is complete, hence closed in $\mathrm{aut}(X)$.

(ii) $\Rightarrow$ (iii). Obvious from $\mathfrak{a} = \mathrm{Inn}(X) \oplus \underline{X}$.

(iii) $\Rightarrow$ (i). Since $\mathfrak{a}$ is supposed to be a Banach space and J is by definition a closed operator, J is continuous according to the closed graph theorem. Q.E.D.

4.13 REMARK. By Proposition 4.12 the canonical orientation J of a unital C^*-algebra is known to be continuous for several classes of C^*-algebras, namely for simple and, more generally, primitive C^*-algebras [21], for AW^*-algebras and n-homogeneous C^*-algebras [4]. If Z is a W^*-algebra, it is even true that

$$J\colon \underline{X} \to \mathrm{aut}(X)$$

is an isometry, if $\underline{X}$ is equipped with the quotient norm and $\mathrm{aut}(X)$ carries the operator norm with respect to X [13,11]. However, for C^*-algebras in general, J is not continuous. A recent example of an AF C^*-algebra with J discontinuous can be found in [4].

§5. ORIENTABLE JB*-ALGEBRAS ARE C*-ALGEBRAS

In this section we prove the converse of 4.11: Every JB*-algebra Z with an orientation is a C*-algebra. Moreover, the possible C*-algebra structures on Z are in 1-1 correspondence with the orientations of Z. The proof of this result is carried out in several steps. Denote by X the JB-algebra associated with a JB*-algebra Z with open positive cone Ω. According to Theorem 4.7, orientations are considered as continuous maps $J: \underline{X} \to \mathrm{aut}(X)$ satisfying property (4.8).

5.1 LEMMA.

(i) A spin factor X is orientable if and only if $\dim X = 4$.

(ii) $\mathcal{H}_3(\mathbb{O})$ is not orientable.

PROOF. (i) Suppose, J is an orientation of a spin factor X of Hilbert dimension n, $n \geq 3$. Then aut(X) is not abelian by (4.9). Since $\mathrm{aut}(X) = \mathbb{R}$ for $n = 3$, it follows that $n \geq 4$. By definition of the spin factor product, $[M_a, M_b]$ has rank ≤ 2 whenever $a, b \in X$. Since $\underline{\mathrm{int}}(\Omega) \subset \mathfrak{a}$, it follows from (4.9) that rank $[\delta_1, \delta_2] \leq 2$ for all $\delta_1, \delta_2 \in \mathrm{int}(X)$. This implies $n \leq 4$. On the other hand, for $n = 4$, $X = \mathcal{H}_2(\mathbb{C})$ is orientable by 4.11.

(ii) Let $X = \mathcal{H}_3(\mathbb{O})$. Then it is well known that $\dim(\mathrm{aut}(X)) = 52$ and $\dim \underline{X} = 26$. Any orientation J of X would induce an isomorphism $J: \underline{X} \to \mathrm{aut}(X)$, a contradiction. Q.E.D.

Let X be a JB-algebra and let $\pi: X \to \tilde{X}$ be a JB-representation into a JBW-algebra $\tilde{X}$ which is weak* dense. Denote by Ω and $\tilde{\Omega}$ the open positive cones of X and $\tilde{X}$, respectively. The second dual space X^{tt} of X is a JBW-algebra and π has a unique extension to a JBW-representation $\pi: X^{tt} \to \tilde{X}$. The kernel of π in X^{tt} is a weak* closed ideal, thus of the form

$$\ker(\pi) = P_{1-c}X^{tt} ,$$

where $c \in X^{tt}$ is a central projection and $P_x := 2M_x^2 - M_{x^2}$ denotes the quadratic representation. Now suppose $\lambda \in \mathrm{aut}(\Omega)$. By Lemma 3.6, the second transpose $\lambda^{tt} \in \mathcal{L}(X^{tt})$ commutes with P_c. Therefore there is a unique element $\pi_*(\lambda) \in \mathrm{aut}(\tilde{\Omega})$ such that the following diagram commutes

$$\begin{array}{ccc} \tilde{X} & \xleftarrow{\ \pi\ } & X^{tt} \\ {\scriptstyle \pi_*(\lambda)}\downarrow & & \downarrow{\scriptstyle \lambda^{tt}} \\ \tilde{X} & \xleftarrow[\ \pi\ ]{} & X^{tt} \end{array}$$

Obviously, $\pi_*: \mathrm{aut}(\Omega) \to \mathrm{aut}(\tilde{\Omega})$ is a homomorphism of involutive Lie algebras and for all $a \in X$,

$$\pi_* M_a = M_{\pi(a)} \quad (5.2)$$

Since π preserves centers, there is an induced homomorphism $\underline{\mathrm{aut}}(\Omega) \to \underline{\mathrm{aut}}(\tilde{\Omega})$, again denoted by π_*.

5.3 LEMMA. Let $\pi: X \to \tilde{X}$ be a representation of an orientable JB-algebra X onto a JBW-algebra $\tilde{X}$. Then $\tilde{X}$ is orientable.

PROOF. Let $J: \underline{X} \to \mathrm{aut}(X)$ be the orientation of X and put $\delta_{ia} := J\underline{a}$ for $a \in X$. Suppose $\pi(a) \in \mathrm{center}(\tilde{X})$. Given $b \in X$ the skew symmetry of $\delta(a,b) := \delta_{ia}b$ implies

$$[\delta_{ia}, M_b] = [M_a, \delta_{ib}]$$

and therefore

$$\begin{aligned}[] [\pi_*\delta_{ia}, M_{\pi b}] &= [\pi_*\delta_{ia}, \pi_* M_b] = \pi_*[\delta_{ia}, M_b] \\ &= \pi_*[M_a, \delta_{ib}] = [M_{\pi a}, \pi_*\delta_{ib}]. \end{aligned}$$

Since $\pi_*\delta_{ib} \in \mathrm{aut}(\tilde{X})$ vanishes at $\pi(a)$ and π is surjective, evaluation at the unit element of $\tilde{X}$ yields $\pi_*\delta_{ia} = 0$. Hence

$\pi(a) \mapsto \pi_*\delta_{ia}$ induces a well-defined mapping

$$\tilde{J}: \underline{X} \to \operatorname{aut}(\tilde{X}).$$

To see that $\tilde{J}$ is continuous, assume that J has norm N and observe that, for all $a, x \in X$

$$|\pi_*(\delta_{ia})\pi(x)| = |\pi\delta_{ia}(x)| \leq |\delta_{ia}(x)| \leq N|a||x|.$$

However, since the left hand side depends only on $\underline{\pi a}$ and $\pi(x)$, we have $|\pi_*(\delta_{ia})\pi(x)| \leq N\,|\underline{\pi a}|\,|\pi(x)|$ which proves that $\tilde{J}$ is continuous. (4.8) follows from the fact that $\pi(\delta_{ia}b) = (\pi_*\delta_{ia})(\pi b)$ and π_* is a homomorphism. Q.E.D.

5.4 COROLLARY. Each JB-algebra X with an orientation is a JC-algebra, i.e. a norm-closed unital subalgebra of $\mathcal{H}(H)$ for some complex Hilbert space H.

PROOF. Since $\mathcal{H}_3(\mathbb{O})$ is not orientable by Lemma 5.1, it follows from Lemma 5.3 that X has no factor representations onto $\mathcal{H}_3(\mathbb{O})$. By [3; Th. 9.5] the exceptional ideal of X vanishes, hence X is a JC-algebra. Q.E.D.

Recall that a JC-algebra $X \subset \mathcal{H}(H)$ is called reversible if $x_1 \ldots x_n + x_n \ldots x_1 \in X$ whenever $x_1, \ldots, x_n \in X$.

5.5. COROLLARY. Every orientable JC-algebra $X \subset \mathcal{H}(H)$ is reversible.

PROOF. Let $\pi: X \to \tilde{X}$ be a dense JB-representation into a spin factor $\tilde{X}$. Then $\pi(X)$ is a norm closed subspace of $\tilde{X}$ which is weak* dense. Since $\tilde{X}$ is a Hilbert space, $\pi(X) = \tilde{X}$. By Lemma 5.3, $\tilde{X}$ is orientable, hence $\dim \tilde{X} = 4$ by Lemma 5.1. By [1; Lemma 4.1, Lemma 4.4 and Lemma 4.5] it follows that the I_2-part of the weak closure of X is reversible. This implies that X is reversible [1; Th. 4.6]. Q.E.D.

5.6 LEMMA. Let δ be a derivation of a reversible JC-algebra $X \subset \mathcal{H}(H)$. Then

$$|\varphi(\delta(x)| \leq 2|\delta| \; \varphi(x^2)^{1/2}$$

for all $x \in X$ and every state φ of X.

PROOF. Let $A := C^*(X)$ be the C^*-algebra generated by X. By the extension theorem for reversible JC-algebras [25; Th. 2.5] there exists a $*$-derivation D on A satisfying $D|X = \delta$ and $|D| \leq 2|\delta|$. By [31; Th. 2] there exists a self-adjoint element a in the weak closure B of A such that

$$\delta x = i[a,x]$$

for all $x \in X$ and $2|a| = |D| \leq 2|\delta|$. Let ψ be any state of B extending φ (cf. [9; Lemma 2.10.1]). Then, by the Cauchy-Schwarz inequality,

$$|\varphi(\delta x)| = |\psi[a,x]| \leq |\psi(ax)| + |\psi(xa)| \leq 2\psi(a^2)^{1/2} \; \psi(x^2)^{1/2} \leq$$
$$2|a| \; \varphi(x^2)^{1/2} \leq 2|\delta| \; \varphi(x^2)^{1/2}.$$

Q.E.D.

Given a JB-algebra X with orientation J, the crucial step in constructing a C^*-structure on $X^{\mathbb{C}}$ is the extension of the orientation to the second dual space X^{tt}. In order to construct this extension it is essential that J is assumed to be a closed operator. We have already seen (Cor. 4.11) that the canonical orientation of a C^*-algebra has this property.

Suppose $\mu: X \times X \to X$ is a <u>continuous</u> bilinear mapping. By [3; p.23] μ has a unique bilinear extension

$$\bar{\mu}: X^{tt} \times X^{tt} \to X^{tt}$$

having the same norm and satisfying the following properties:

(5.7) $a \mapsto \bar{\mu}(a,b)$ is weak* continuous for $a,b \in X^{tt}$,

(5.8) $b \mapsto \bar{\mu}(a,b)$ is weak* continuous for $a \in X$, $b \in X^{tt}$.

If μ is the Jordan product in X, the extension $\bar{\mu}$ is called the <u>Arens product</u> in X^{tt} making X^{tt} into a JBW-algebra. We will apply the "Arens process" to the skew symmetric bilinear mapping δ associated with an orientation.

5.9 THEOREM. Let X be a JB-algebra with orientation J. Then the second dual space X^{tt} is orientable.

PROOF. For $a \in X$ define $\delta_{ia} := J\underline{a} \in \mathrm{aut}(X)$ and let $\delta: X \times X \to X$ be the continuous skew symmetric bilinear mapping given by $\delta(a,b) := \delta_{ia}(b)$. Let $\bar{\delta}: X^{tt} \times X^{tt} \to X^{tt}$ be the Arens extension of δ, written as $\bar{\delta}_{ia} b := \bar{\delta}(a,b)$ for $a,b \in X^{tt}$. If $a \in X$, $\bar{\delta}_{ia}$ is the second transpose of the derivation $\delta_{ia} \in \mathrm{aut}(X)$, hence $\bar{\delta}_{ia} \in \mathrm{aut}(X^{tt})$. Applying (5.7) to $\bar{\delta}$ and to the commutative Arens product $\bar{\mu}$ [3; Cor. 3.4] it follows that

$$\bar{\delta}_{ia} \in \mathrm{aut}(X^{tt}) \tag{5.10}$$

whenever $a \in X^{tt}$, since X is weak* dense in X^{tt}. Let us denote strong and weak convergence in X^{tt} by $\xrightarrow[\beta]{}$ and $\xrightarrow[\sigma]{}$, respectively. Suppose (a_j) and (b_j) are nets in X^{tt} with (a_j) norm bounded. Then

$$a_j \xrightarrow[\sigma]{} a, \quad b_j \xrightarrow[\beta]{} b \quad \text{implies} \quad \bar{\delta}(a_j, b_j) \xrightarrow[\sigma]{} \bar{\delta}(a,b). \tag{5.11}$$

To prove (5.11) let φ be any state of X. Let $\bar{\delta}_j := \bar{\delta}_{ia_j}$. Then

$$|\varphi(\bar{\delta}(a_j,b_j) - \bar{\delta}(a,b))| \le |\varphi(\bar{\delta}_j(b_j-b))| + |\varphi(\bar{\delta}(a_j-a,b))|.$$

The second summand converges to o by (5.7). Further, $\bar{\delta}_j \in \mathrm{aut}(X^{tt})$ by (5.10) and X^{tt} is reversible, since X is reversible by Cor. 5.5. Hence we may apply Lemma 5.6 to obtain

$$|\varphi(\bar{\delta}_j(b_j-b))| \le 2|\bar{\delta}_j|\ \varphi((b_j-b)^2)^{1/2}.$$

By definition of strong convergence, this term converges to o

since $\bar{\delta}$ is bounded and (a_j) is norm bounded. Hence (5.11) follows.

As a consequence of (5.11), $\bar{\delta}$ is skew symmetric since δ is skew symmetric and the unit ball of X is strongly dense in the unit ball of X^{tt} by [3; Prop. 3.9]. It follows that for all $x \in X^{tt}$

$$\bar{\delta}_{ia} x = -\bar{\delta}_{ix} a = o$$

whenever $a \in \mathrm{center}(X^{tt})$, since derivations vanish on the center. Therefore $a \mapsto \bar{\delta}_{ia}$ induces a well defined mapping $\bar{J}: X^{tt}/\mathrm{center} \to \to \mathrm{aut}(X^{tt})$ which is continuous since $\bar{\delta}$ is. Using [3; Prop.3.9], property (5.11) for $\bar{\delta}$ and the strong continuity of the Arens product $\bar{\mu}$ [3; Prop. 3.7] as well as skew symmetry of $\bar{\delta}$, it is straightforward to show that property (4.8) for δ carries over to $\bar{\delta}$. By Theorem 4.7, $\bar{J}$ is an orientation of X^{tt}. Q.E.D.

Theorem 5.9 is the essential tool for constructing the C*-structure on an orientable JB*-algebra Z. The rest of the proof uses some results of the structure theory for JB-algebra state spaces developed in [2,1]. Let X be a JB-algebra with multiplication operators M_x and quadratic representation

$$P_x := 2\, M_x^2 - M_{x^2}\,.$$

Then $P_x y = \{xy^*x\}$ for all $x,y \in X$, cf. (2.5). Let $p \in X$ be a projection. Then $X_p := P_p X$ is a JB-algebra with multiplication $(x,y) \mapsto \{xp^*y\}$ and unit element p. For $a \in X_p$ we have by [3; (2.35)]

$$P_{1-p} a = P_{1-p}\, P_p a = o.$$

By [3; Lemma 2.11],

$$(5.12) \qquad [M_a, M_p] = o = [M_a, P_p].$$

Hence the restriction $\rho_p(M_a) := M_a|X_p$ is well defined and coincides with the multiplication by a in X_p.

5.13 LEMMA. If X is orientable, then X_p is orientable for every projection $p \in X$.

PROOF. For $a \in X_p$ put $b := \delta_{ia}p$. By (4.8) and (5.12),

$$\delta_{ib} = [\delta_{ia}, \delta_{ip}] = -[M_p, M_b] = o.$$

Therefore, for every $x \in X$

$$[M_b, M_x] = -[\delta_{ix}, \delta_{ib}] = o.$$

This implies

$$\delta_{ip}a = -\delta_{ia}p = -b \in \text{center}(X).$$

As in the proof of Theorem 4.7, it follows that $b = \delta_{ip}a = o$, since δ_{ip} is a derivation. The derivation property of δ_{ia} implies

$$[\delta_{ia}, M_p] = o = [\delta_{ia}, P_p].$$

Hence $\rho_p(\delta_{ia}) := \delta_{ia}|X_p$ is a well defined derivation of X_p. In particular, for $a \in \text{center}(X_p)$ and $b \in X_p$,

$$\delta_{ia}b = -\delta_{ib}a = o.$$

Hence $\rho_p(\delta_{ia}) = o$. An elementary calculation shows that the mapping $\rho_p(M_a) \mapsto \rho_p(\delta_{ia})$ induces a well-defined orientation on X_p.

Q.E.D.

5.14 LEMMA. Let X be an orientable JB-algebra. Then X is of complex type (cf. [1; §3]).

PROOF. Let S be the state space of X and denote by F the face generated by two pure states φ and ψ. If φ and ψ are not equivalent in the sense of [1;§2], then F is the line segment generated by φ and ψ [1; Prop. 2.3]. If φ and ψ are equivalent, then by [1; Prop. 2.3] F is a Hilbert ball of dimension $k \geq 2$. By the proof of [1; Th. 3.11], there exists a projection

$p \in X^{tt}$ such that $V := X_p^{tt}$ is a spin factor with state space F. Since X is orientable, X^{tt} is orientable by Theorem 5.9, hence V is orientable by Lemma 5.13. Lemma 5.1 implies $k + 1 = \dim V = 4$. Hence $k = 3$. This means that S has the 3-ball property and by [1; Cor. 3.3], X is of complex type. Q.E.D.

We are now in a position to define the C^*-algebra structure associated with an orientation. Let Z be a JB^*-algebra with JB^*-norm $|\cdot|$ and involution $z \mapsto z^*$. Suppose the JB-algebra $X := \{x \in Z : x^* = x\}$ has an orientation J. Define $\delta_{ix} := J\underline{x}$ for $x \in X$ and denote by M_x the multiplication operator by x on X. For $x,y \in X$ define

$$(5.15) \qquad xy := M_x y + \frac{1}{i}\delta_{ix} y \in Z$$

and extend this product by bilinearity to a product on Z. We first list some elementary properties of (5.15).

5.16 LEMMA. For $x,y \in X$ the following properties hold:

(i) $\frac{xy + yx}{2} = M_x y$

(ii) $\frac{xy - yx}{2} = \frac{1}{i}\delta_{ix} y$

(iii) $(xy)^* = yx$.

PROOF. These properties follow from the fact, that $M_x y$ is symmetric and $\delta_{ix} y$ is skew symmetric in (x,y). Q.E.D.

It follows from 5.16.iii, that the complexification $Z := X^{\mathbb{C}}$ of a JB-algebra X with orientation becomes an involutive algebra under the (extended) product (5.15) and the JB^*-involution. By 5.16.i, the anti-commutator product in Z coincides with the original Jordan product. Let $\pi: X \to \tilde{X}$ be a weak* dense representation into a JB-factor of type I (called type I factor representation). Since $\tilde{X} = X_c^{tt}$ for some central projection $c \in X^{tt}$, it follows from Theorem 5.9 and Lemma 5.13, that each orientation on X induces a unique orientation of $\tilde{X}$ having the property

$$\pi_*(\delta_{ix}) = \tilde{\delta}_{i\pi(x)} \tag{5.17}$$

for all $x \in X$.

5.18 LEMMA. Let $\pi\colon X \to \tilde{X}$ be a type I factor representation of a JB-algebra X with orientation. Endow $\tilde{X}$ with the orientation induced by π. Then the complexification

$$\pi\colon Z \to \tilde{Z}$$

is a homomorphism of involutive algebras.

PROOF. Let $x,y \in X$. Then by (5.2) and (5.17),

$$\begin{aligned}\pi(xy) &= \pi(M_x y) + \frac{1}{i}\pi(\delta_{ix} y)\\ &= M_{\pi(x)}\pi(y) + \frac{1}{i}\tilde{\delta}_{i\pi(x)}\pi(y) = \pi(x)\pi(y).\end{aligned}$$

Q.E.D.

We can now prove the converse of 4.11:

5.19 THEOREM. Let X be a JB-algebra with orientation. Then the complexification $Z := X^{\mathbb{C}}$ becomes a C^*-algebra under the product (5.15), the JB^*-involution and the JB^*-norm. The underlying Jordan algebra of this C^*-algebra is the original JB^*-algebra structure on Z and the canonical orientation coincides with the given one.

PROOF. For each type I factor representation $\pi\colon X \to \tilde{X}$, there exists a complex Hilbert space H with $\tilde{X} = \mathcal{H}(H)$, since X is of complex type by Lemma 5.14. By (4.4) and [8; Prop. 4.12], $\tilde{X}$ has only two orientations under both of which $\tilde{Z} := \tilde{X}^{\mathbb{C}}$ becomes a C^*-algebra. Consider the associator

$$[abc] := (ab)c - a(bc)$$

in Z relative to the product (5.15). By Lemma 5.18,

$$\pi[abc] = [\pi a,\pi b,\pi c] = o,$$

hence Z is associative, since the type I factor representations form a faithful family on X [3; Cor. 5.7]. Similarly, for any $z \in Z$ we have by Lemma 5.18,

$$|\pi(z^*z)| = |(\pi z)^*(\pi z)| = |\pi z|^2,$$

hence

$$|z^*z| = \sup_\pi |\pi(z^*z)| = (\sup_\pi |\pi z|)^2 = |z|^2.$$

The inequality $|zw| \leq |z| \cdot |w|$ for all $z, w \in Z$ is shown in a similar way. Hence Z is a C^*-algebra. The remaining statements follow from Lemma 5.16. Q.E.D.

It follows from Cor. 4.11 and Theorem 5.19 that a JB^*-algebra Z is orientable if and only if it is a C^*-algebra; moreover the possible C^*-algebra structures on Z are in 1-1 correspondence with the orientations on Z. Together with Theorem 2.15, we obtain the desired holomorphic characterization of C^*-algebras:

5.20 MAIN THEOREM. Let D be a bounded domain in a complex Banach space Z. Then D is biholomorphically equivalent to the open unit ball of a (unital) C^*-algebra if and only if the following conditions are satisfied:

(i) D is a bounded symmetric domain which is biholomorphically equivalent to a tube domain.

(ii) The JB^*-algebra structure on Z associated with D is orientable.

Moreover the possible C^*-algebra structures on Z are in 1-1 correspondence to the orientations on Z.

REFERENCES

1. ALFSEN, E.M., HANCHE-OLSEN, H., SHULTZ, F.W., State spaces of C^*-algebras. Acta Math. 144 (1980), 267-305.

2. ALFSEN, E.M., SHULTZ, F.W., State spaces of Jordan algebras. Acta math. 140 (1978), 155-190.

3. ALFSEN, E.M., SHULTZ, F.W., STØRMER, E., A Gelfand-Neumark theorem for Jordan algebras. Advances Math. 28 (1978), 11-56.

4. ARCHBOLD, R.J., On the norm of an inner derivation of a C^*-algebra. Math. Proc. Camb. Phil. Soc. 84 (1978), 273-291.

5. BELLISSARD, J., IOCHUM, B., Homogeneous self-dual cones versus Jordan algebras. The theory revisited. Ann. Inst. Fourier, Grenoble, 28 (1978), 27-67.

6. BRAUN, R., KAUP, W., UPMEIER, H., A holomorphic characterization of Jordan C^*-algebras. Math. Z. 161 (1978), 277-290.

7. BRAUN, H., KOECHER, M., Jordan-Algebren. Grundl. der math. Wiss. 128. Berlin-Heidelberg-New York: Springer 1966.

8. CONNES, A., Caractérisation des espaces vectoriels ordonnés sous-jacents aux algèbres de von Neumann. Ann. Inst. Fourier, Grenoble, 24 (1974), 121-155.

9. DIXMIER, J., C^*-algebras. Amsterdam-New York-Oxford: North Holland 1977.

10. EFFROS, E.G., STØRMER, E., Positive projections and Jordan structure in operator algebras. Math. Scand. 45 (1979), 127-138.

11. HANCHE-OLSEN, H., A note on the bidual of a JB-algebra. To appear in Math. Z.

12. JORDAN, P., von NEUMANN, J., WIGNER, E., On an algebraic generalization of the quantum mechanical formalism. Ann. of Math. 35 (1934), 29-64.

13. KADISON, R.V., LANCE, E.C., RINGROSE, J.R., Derivations and automorphisms of operator algebras II. J. Funct. Anal. 1 (1967), 204-221.

14. KAUP, W., Algebraic characterization of symmetric complex Banach manifolds. Math. Ann. 228 (1977), 39-64.

15. KAUP, W., UPMEIER, H., An infinitesimal version of Cartan's uniqueness theorem. Manuscripta math. 22 (1977), 381-401.

16. KAUP, W., UPMEIER, H., Jordan algebras and symmetric Siegel domains in Banach spaces. Math. Z. 157 (1977), 179-200.

17. KOECHER, M., An elementary approach to bounded symmetric domains. Houston: Rice University 1969.

18. NACHBIN, L., Topology on Spaces of Holomorphic Mappings. Erg. d. Math. 47. Berlin-Heidelberg-New York: Springer 1969.

19. SAKAI, S., C^*-algebras and W^*-algebras. Erg. d. Math. 60. Berlin-Heidelberg-New York: Springer 1971.

20. SHULTZ, F.W., On normed Jordan algebras which are Banach dual spaces. J. Funct. Anal. 31 (1979), 360-376.

21. STAMPFLI, J., The norm of a derivation. Pac. J. Math. 33 (1970), 737-748.

22. STØRMER, E., Jordan algebras of type I. Acta Math. 115 (1966), 165-184.

23. STØRMER, E., Irreducible Jordan algebras of self-adjoint operators. Trans. Amer. Math. Soc. 130 (1968), 153-166.

24. UPMEIER, H., Über die Automorphismengruppen von Banach-Mannigfaltigkeiten mit invarianter Metrik. Math. Ann. 223 (1976), 279-288.

25. UPMEIER, H., Derivations of Jordan C^*-algebras. Math. Scand. 46 (1980), 251-264.

26. UPMEIER, H., Derivation algebras of JB-algebras. Manuscripta math. 30 (1979), 199-214.

27. UPMEIER, H., Automorphism groups of Jordan C^*-algebras. Math. Z. 176 (1981), 21-34.

28. VIGUÉ, J.-P., Le groupe des automorphismes analytiques d'un domaine borné d'un espace de Banach complexe. Application aux domaines bornés symétriques. Ann. scient. Éc. Norm. Sup. 4e série 9 (1976), 203-282.

29. WRIGHT, J.D.M., Jordan C^*-algebras. Mich. Math. J. 24 (1977), 291-302.

30. YOUNGSON, M.A., A Vidav Theorem for Banach Jordan Algebras. Math. Proc. Camb. Phil. Soc. 84 (1978), 263-272.

31. ZSIDÓ, L., The norm of a derivation in a W^*-algebra. Proc. Amer. Math. Soc. 38 (1973), 147-150.

DEPARTMENT OF MATHEMATICS
UNIVERSITY OF PENNSYLVANIA
PHILADELPHIA, PA 19104, USA

Functional Analysis, Holomorphy and Approximation Theory II, G.I. Zapata (ed.)

A PROPERTY OF FRÉCHET SPACES

Manuel Valdivia

SUMMARY

Let F be a dense subspace of a Fréchet space E. If F is not barreled there is an infinite dimensional closed subspace G of E such that $F \cap G = \{0\}$. Some consequences of this result are given.

The linear spaces we shall use are defined over the field K of the real or complex numbers. Given a dual pair $\langle L,M\rangle$, $\sigma(L,M)$ and $\mu(L,M)$ are the weak and the Mackey topology on L, respectively. If A is a $\sigma(L,M)$-bounded absolutely convex subset of L we denote by L_A the linear hull of A endowed with the normed topology derived from the gauge of A. The word "space" is used to denote a Hausdorff locally convex topological space. If E is a space, E' is its topological dual. If $x \in E$ and if $u \in E'$ we shall write sometimes $\langle u,x\rangle$ instead of $u(x)$. A space E is a Baire-like space if given an increasing sequence (A_n) of closed absolutely convex sets of E such that $\bigcup_{n=1}^{\infty} A_n$ is absorbing in E there is a natural number n_o such that A_{n_o} is a neighbourhood of the origin. N denotes the set of the natural numbers.

I shall need for the proof of Theorem 1 the following result:

a) <u>If</u> E <u>is an infinite dimensional metrizable separable space, then</u> E <u>has the property of quasi-complementation</u>.

This result was obtained by Mackey in [2] for normed space but the proof given there remains valid for metrizable spaces.

THEOREM 1. Let F be a dense subspace of a Fréchet space E. If F is not barrelled there is a subspace G of E such that the following conditions are satisfied:

1. G is closed;
2. the dimension of G is infinite;
3. $G \cap F = \{0\}$.

PROOF. Let T be a barrel in F which is not a neighbourhood of the origin and let T^o be its polar set in E'. If (U_n) is a fundamental system of absolutely convex neighbourhood of the origin in F we take

$$2^2 x_1 \in U_1, \quad x_1 \notin T, \quad u_1 \in T^o \text{ with } |u_1(x_1)| > 1.$$

Supposing that we have obtained

$$x_j \in \frac{1}{2^{2j}} U_j, \quad x_j \notin T, \quad u_j \in T^o,$$

such that

$$|u_j(x_j)| > 1, \quad u_j(x_i) = 0, \quad i \neq j, \quad i,j = 1,2,\ldots,n,$$

let H and L be the orthogonal subspaces of $\{x_1,x_2,\ldots,x_n\}$ and $\{u_1,u_2,\ldots,u_n\}$ in E' and F respectively. Since $\{u_1,u_2,\ldots,u_n\}$ generate a linear space X which is a topological complement of $H \cap E'_{T^o}$ in E'_{T^o} there is a $\lambda > 1$ such that

$$\lambda(X + T^o \cap H) \supset T^o. \tag{1}$$

On the other hand, L is a finite codimensional closed subspace of F and therefore $T \cap L$ is not a neighbourhood of the origin in L and thus there is

$$y_{n+1} \in \frac{1}{\lambda 2^{2(n+1)}} L \cap U_{n+1}, \quad y_{n+1} \notin T.$$

By using (1) we can find $u_{n+1} \in T^o \cap H$ and $\lambda_j \in K$, $j = 1,2,\ldots,n$, so that

$$|\lambda(\sum_{j=1}^{n} \lambda_j u_j + u_{n+1})(y_{n+1})| > 1.$$

By setting $\lambda y_{n+1} = x_{n+1}$ it follows that

$$x_{n+1} \in \frac{1}{2^{2(n+1)}} U_{n+1}, \quad x_{n+1} \notin T, \quad u_{n+1} \in T^{o},$$

$$u_j(x_{n+1}) = u_{n+1}(x_j) = 0, \quad j=1,2,\dots,n,$$

$$|u_{n+1}(x_{n+1})| = |\lambda(\sum_{j=1}^{n} \lambda_j u_j + u_{n+1})(y_{n+1})| > 1.$$

Then, sequences (x_p) and (u_p) in F and E′ can be select such that

$$x_p \in \frac{1}{2^{2p}} U_p, \quad x_p \notin T, \quad u_p \in T^{o},$$

$$|u_p(x_p)| > 1, \quad u_p(x_q) = 0, \quad p \neq q, \quad p,q = 1,2,\dots .$$

For every positive integer n we find a sequence

$$1(n), 2(n), \dots, k(n), \dots$$

of even positive integer numbers which are pairwise differente such that if h, k, m and n are positive integers with $m \neq n$ then $k(n) \neq h(m)$. Let V be the linear hull of

$$\{u_{2n-1} + \frac{1}{k(n)} u_{k(n)} : n,k = 1,2,\dots\}$$

and let $\bar{V}$ be its closure in $E'[\sigma(E',F)]$. The closure of V in $E'_{T^{o}}$ contains

$$\{u_1, u_3, \dots, u_{2n-1}, \dots\}$$

and therefore

$$u_{2n-1} \in \bar{V}, \quad n = 1,2,\dots .$$

Let A be the closed absolutely convex hull in E of

$$\{2^{2}x_1, 2^{4}x_2, \dots, 2^{2n}x_n, \dots\}.$$

Since the sequence $(2^{2n}x_n)$ converges to the origin in E the set

A is compact. Let P be the linear hull

$$\{u_{2n-1} : n = 1,2,\ldots\}.$$

We take a non-zero element of P

$$u = \sum_{j=1}^{n} \alpha_j u_{2j-1}, \quad \alpha_j \in K, \quad j = 1,2,\ldots,n.$$

Then there is a positive integer p, $1 \le p \le n$, so that $\alpha_p \neq 0$. Given a positive number ε with $2\varepsilon < |\alpha_p|$, suppose the existence of $v \in V$ such that

$$\sup\{|(v-u)(x)| : x \in A\} < \varepsilon.$$

We can write

$$v = \sum_{j=1}^{m} \sum_{k\in P_j} \beta_{j,k} \left(u_{2j-1} + \frac{1}{k} u_k\right)$$

with $m \ge n$, P_j a finite subset of even positive integer numbers and $\beta_{j,k} \in K$, $k \in P_j$, $j = 1,2,\ldots,m$. The coefficient of u_{2p-1} in $u-v$ coincides with

$$\alpha_p - \sum_{k\in P_p} \beta_{p,k}$$

and therefore

$$\varepsilon > \sup\{|(u-v)(x)| : x \in A\} \ge |(u-v)(x_{2p-1})| \ge |\alpha_p - \sum_{k\in P_p} \beta_{p,k}|,$$

from where we get

$$|\alpha_p| \le \varepsilon + |\sum_{k\in P_p} \beta_{p,k}|. \tag{2}$$

In $u-v$, the coefficient of u_k, $k \in P_p$, is $\frac{1}{k}\beta_{p,k}$. If we take an element ε_k of K of modulus one such that

$$\varepsilon_k \beta_{p,k} = |\beta_{p,k}|, \quad k \in P_p$$

we have that, since $P_i \cap P_j = \emptyset$, $i\neq j$, $i,j=1,2,\ldots,m$, and

$$\sum_{k\in P_p} \varepsilon_k 2^k x_k \in A,$$

$$\varepsilon > \sup\{|(u-v)(x)| : x \in A\} \ge |(u-v)(\sum_{k\in P_p} \varepsilon_k 2^k x_k)| \ge$$

$$\geq \sum_{k \in \mathcal{P}_p} \frac{1}{k} |\beta_{p,k}| \cdot 2^k \geq \sum_{k \in \mathcal{P}_p} |\beta_{p,k}|$$

and therefore $|\alpha_p| < 2\varepsilon$, according to (2), which is a contradiction. We can conclude that $P \cap M = \{0\}$ if M is the closure of V in $E'[\mu(E',E)]$. Then M is $\sigma(E',F)$-dense in $\bar{V}$, closed in $\bar{V}$ for $\sigma(E',E)$ and of infinite codimension in that space. Let R be the orthogonal subspace of M in E and let S be the orthogonal subspace of M in F. If $\bar{S}$ denotes the closure of S in E it follows that $\bar{S}$ is of infinite codimension in R. Let Y be an algebraic complement of $\bar{S}$ in R. We take linearly independent vectors

$$\{z_1, z_2, \ldots, z_n, \ldots\} \tag{3}$$

in Y. If Z denotes the closed linear hull of (3) in E then Z is separable and $\bar{S} \cap Z$ of infinite codimension in Z. According to a) we obtain a quasicomplement subspace G of $\bar{S} \cap Z$ in Z. Then G is closed in E, it has infinite dimension and $G \cap F = G \cap (\bar{S} \cap Z) = \{0\}$. q.e.d.

In order to prove the following lemma let $\mathfrak{G}$ be a family of subsets of a space E which are bounded closed absolutely convex and satisfying the following conditions:

a. If Q is a finite part of E there is an $A \in \mathfrak{G}$ such that $A \supset Q$.

b. If $A_1, A_2 \in \mathfrak{G}$ there is $A_3 \in \mathfrak{G}$ with $A_3 \supset A_1 \cup A_2$.

c. If $A \in \mathfrak{G}$ and if $\lambda > 0$ then $\lambda A \in \mathfrak{G}$.

We denote by $E'[\tau]$ the space E' endowed with the topology of the uniform convergence on the elements of $\mathfrak{G}$.

LEMMA. Let F be a subspace of E satisfying the following conditions:

1. $F \cap A$ is closed, $A \in \mathcal{G}$

2. $E_A \cap F$ is of finite codimension in E_A, $A \in \mathcal{G}$.

Then, F is closed if $E'[\tau]$ is complete.

PROOF. Let x be a vector in E which is not in F. Let $\{x_i : i \in I\}$ a family of vectors in E such that $\{x, x_i : i \in I\}$ is a Hamel basis of an algebraic complement of F in E. Let $\mathfrak{F}$ be the family of all finite subsets of I. If $f \in \mathfrak{F}$, $p \in N$ and $A \in \mathcal{G}$ we find a continuous linear form $u(f,p,A)$ on E such that it takes the value 1 on x and

$$|\langle u(f,p,A), z\rangle| < 1, \quad \forall\, z \in A \cap F + \{\sum_{j \in f} \alpha_j x_j : |\alpha_j| \le p\}.$$

We give an order relation $\le$ in $(\mathfrak{F}, N, \mathcal{G})$: if $f_1, f_2 \in \mathfrak{F}$; $p_1, p_2 \in N$ and $A_1, A_2 \in \mathcal{G}$ then $(f_1,p_1,A_1) \le (f_2,p_2,A_2)$ if and only if $f_1 \subset f_2$, $p_1 \le p_2$ and $A_1 \subset A_2$.

We consider the net

$$\{u(f,p,A) : (f,p,A) \in (\mathfrak{F}, N, \mathcal{G}), \le\}. \tag{4}$$

We take $\varepsilon > 0$ and $B \in \mathcal{G}$. We find a positive integer m such that $\frac{2}{m} < \varepsilon$. Since $E_B \cap F$ is of finite codimension in E_B there is a set $D \in \mathcal{G}$, a part $g \in \mathfrak{F}$ and an integer $q \in N$ such that

$$D \cap F + \{\beta x + \sum_{j \in g} \beta_j x_j : |\beta| \le q, |\beta_j| \le q, j \in g\}$$

contains B. If $z \in B$ we have that

$$z = y + \gamma x + \sum_{j \in g} \gamma_j x_j, \quad y \in D \cap F, |\gamma| \le q, |\gamma_j| \le q, j \in g.$$

If

$$(g, qm, mD) \le (f_s, p_s, A_s), \qquad s = 1,2,$$

it follows that

$$|\langle u(f_1,p_1,A_1) - u(f_2,p_2,A_2), mz - m\gamma x\rangle|$$

$$= |\langle u(f_1,p_1,A_2), mz\rangle - \langle u(f_2,p_2,A_2), mz\rangle| \le 2,$$

from where we get

$$|\langle u(f_1,p_1,A_1) - u(f_2,p_2,A_2),z\rangle| \le \frac{2}{m} < \varepsilon ,$$

implying that (4) is Cauchy in $E'[\tau]$. Let u be the limit of (4) in $E'[\tau]$. Since

$$\langle u(f,p,A),x\rangle = 1, \quad (f,p,A) \in (\mathfrak{F},N,\mathfrak{a}),$$

it follows that $u(x) = 1$. On the other hand, if $z \in F$, given $\varepsilon > 0$ there is a positive integer r and $A \in \mathfrak{a}$ such that $\frac{1}{r} < \varepsilon$, $z \in A$. Then if $B \supset rA$, $f \in \mathfrak{F}$ and $p \in N$ it follows that

$$|\langle u(f,p,B),z\rangle| \le \frac{1}{r} < \varepsilon ,$$

and thus $u(z) = 0$. Accordingly x does not belong to the closure of F in E. q.e.d.

THEOREM 2. Let E be a barrelled space such that $E'[\mu(E',E)]$ is complete. If F is a subspace of E with codimension in E less than $2^{\aleph_0}$, then F is barrelled.

PROOF. We suppose first that F is closed. Let A be a weakly compact absolutely convex subset of E. E_A is a Banach space such that $E_{A\cap F}$ is of codimension less than $2^{\aleph_0}$ in E_A. Every infinite dimensional Banach space has dimension larger or equal that $2^{\aleph_0}$, [1], and therefore $E_A/E_{A\cap F}$ is finite dimensional and thus $E_{A\cap F}$ is of finite codimension in E_A. If $\varphi: E \to E/F$ is the canonical surjection, $\varphi(A)$ generates a finite dimensional subspace in E/F and thus if $F^{\perp}$ is the subspace of $E'[\mu(E',E)]$ orthogonal to F the topology induced by $\mu(E',E)$ on $F^{\perp}$ is the weak topology and therefore $F^{\perp}$ is topologically isomorphic to a product of lines and thus it has a topological complement L in $E'[\mu(E',E)]$. Then F has a topological complement in E and therefore F is barreled. $F'[\mu(F',F)]$ is isomorphic to L and therefore $F'[\mu(F',F)]$ is complete.

We suppose now that F is dense in E. Let T be a barrel in F, let $\bar{T}$ be its closure in E and let H be the linear hull of $\bar{T}$. Let B be a weakly compact absolutely convex subset of E and let M be the closure of $E_{B\cap H}$ in E_B, which is a Banach space. If $E_{B\cap H}$ were not barreled there is an infinite dimensional subspace G of M such that $E_{B\cap H} \cap G = \{0\}$, according to Theorem 1. But this contradicts that $E_{B\cap H}$ had codimension less than $2^{\aleph_0}$ in M. We have now that $E_{B\cap H}$ is barreled. Then $\bar{T}$ absorbs $B \cap H$ and thus $B \cap H$ is closed in E. We apply the lemma to conclude that H is closed and therefore $H = E$. $\bar{T}$ is then a barrel in E and, since E is barrelled, a neighbourhood of the origin and $T = \bar{T} \cap F$ is a neighbourhood of the origin in F.

The general case reduces to the former ones.

q.e.d.

COROLLARY 1.2. Let E be an ultrabornological space. If F is a subspace of E of codimension less than $2^{\aleph_0}$ then F is barrelled.

If G is a barrelled space every countable codimensional subspace of G is barrelled, [3]. Using this fact we proved in [4] that every countable codimensional subspace of an ultrabornological space is bornological. Using Corollary 1.2 a slight modification of the proof given in [4] provides with the following

THEOREM . Let E be an ultrabornological space. If F is a subspace of E of codimension less than $2^{\aleph_0}$ then F is bornological.

THEOREM 4. Let E be a Baire-like space such that $E'[\mu(E',E)]$ is complete. If F is a subspace of E of codimension less than

$2^{\aleph_o}$ then F is Baire-like.

PROOF. If F is closed in E then F has a topological complement in E and thus F is Baire-like. If F is dense in E let (T_n) be an incrasing sequence of closed absolutely convex sets of F such that $\bigcup_{n=1}^{\infty} T_n$ is absorbing in F. Let $\bar{T}_n$ be the closure of T_n in E, $n = 1,2,...,$ and H the linear hull of $\bigcup_{n=1}^{\infty} \bar{T}_n$. If B denotes a weakly compact absolutely convex subset of E, $E_{B\cap H}$ is barrelled (see the proof of Theorem 2) and therefore a Baire-like space, [3], and thus there is a positive integer n_o such that $T_{n_o} \cap B \cap H$ is a neighbourhood of the origin in $E_{B\cap H}$. Then $\bar{T}_{n_o}$ absorbs $B \cap H$ and $B \cap H$ is closed in E. The conclusion follows as in Theorem 2.

q.e.d.

COROLLARY 1.4. Let E be an ultrabornological space and let F be a subspace of E of codimension less than $2^{\aleph_o}$. If E is a Baire-like space then F is a Baire-like space.

REFERENCES

1. LÖWIG, H., Über die Dimension linearer Räume. Studia Math. 5, 18-23 (1934).

2. MACKEY, G., Note on a Theorem of Murray. Bull. Amer. Math. Soc. 52, 322-325 (1946).

3. VALDIVIA, M., Absolutely convex sets in barrelled spaces. Ann. Inst. Fourier, Grenoble, 21, 2, 3-13 (1971).

4. VALDIVIA, M., On final topologies. J. Reine angew. Math. 251, 193-199 (1971).

Facultad de Matemáticas
Dr. Moliner 50
Burjasot - Valencia - Spain